T0315207

QUANTUM DYNAMICS
OF SIMPLE SYSTEMS

QUANTUM DYNAMICS OF SIMPLE SYSTEMS

The Forty Fourth Scottish Universities
Summer School in Physics,
Stirling, August 1994.

A NATO Advanced Study Institute.

Edited by

G-L Oppo – University of Strathclyde

S M Barnett – University of Strathclyde

E Riis – University of Strathclyde

M Wilkinson – University of Strathclyde

Series Editor

P Osborne – University of Edinburgh

CRC Press
Taylor & Francis Group
Boca Raton London New York

CRC Press is an imprint of the
Taylor & Francis Group, an **informa** business

CRC Press
Taylor & Francis Group
6000 Broken Sound Parkway NW, Suite 300
Boca Raton, FL 33487-2742

First issued in hardback 2017

© 1996 by The Scottish Universities Summer School in Physics
CRC Press is an imprint of Taylor & Francis Group, an informa business

No claim to original U.S. Government works

ISBN-13: 978-0-7503-0490-0 (pbk)
ISBN-13: 978-1-1384-6469-8 (hbk)

Visit the Taylor & Francis Web site at
http://www.taylorandfrancis.com

and the CRC Press Web site
http://www.crcpress.com

British Library Cataloguing-in-Publication Data:

A catalogue record for this book is available
from the British Library

Library of Congress Cataloging-in-Publication Data are available.

SUSSP Proceedings

/continued

SUSSP Proceedings (continued)

Lecturers

Boris F Altshuler	MIT, Boston, USA
Michael Berry	University of Bristol, UK
Eugene B Bogomolny	Institut de Physique Nucleaire, Orsay, France
Michel Brune	École Normale Supérieure, Paris, France
Alan J Duncan	Stirling University, Scotland
Ralph DeVoe	IBM, San Jose, California, USA
Shmuel Fishman	Israel Institute of Technology, Haifa, Israel
Roy J Glauber	Harvard University, Cambridge, USA
Serge Haroche	École Normale Supérieure, Paris, France
Ed A Hinds	Yale University, Yale, USA
H Jeff Kimble	California Institute of Technology, Pasadena, USA
Peter L Knight	Imperial College, London, UK
David A B Miller	AT&T Laboratories, Holmdel, USA
Stig Stenholm	University of Helsinki, Finland
Jook T M Walraven	University of Amsterdam, The Netherlands

Executive Committee

Prof W J Firth	University of Strathclyde	*Director*
Dr S M Barnett	University of Strathclyde	*Secretary*
Dr M Wilkinson	University of Strathclyde	*Treasurer*
Dr M Hooper	University of Strathclyde	*Deputy Treasurer*
Dr G-L Oppo	University of Strathclyde	*Chief-Editor*
Dr E Riis	University of Strathclyde	*Local Organiser*
Prof A I Ferguson	University of Strathclyde	*Social Convener*

Preface

It is well known that boys never grow up. They are constantly amused by toys both during their childhood and adulthood, the only difference being the level of sophistication and manoeuvrability of the toys. The older the player, the fancier the toy. So it goes in quantum physics. The present level of experimental sophistication allows one to explore domains unimaginable just ten years ago and to test the most fundamental laws of quantum mechanics. This led to a renewed interest in devising new tests, experiments, and devices where one can observe the interaction and localisation of just a few atoms and/or a few photons. These have been used to reveal new nonclassical effects, to question the limit of the principle of correspondence and to force quantum behaviour in semiconductors. In other words, an entire new class of *Quantum Toys*. The main objective of this school was to provide an overview of the present range of Quantum Toys and to teach and instruct newcomers about their use and exotic behaviours. During the two weeks of the 44th Scottish Universities Summer School in Physics (SUSSP44) we observed E. Bogomolny riding a quantum surfboard on surfaces of negative curvature, M. Berry extricating quantum chaology in a forest of Riemann zeroes, M. Brune and S. Haroche communicating via quantum teleportation and avoiding the demolition of quantum measurements. R. DeVoe and R. Glauber instructed us on how to trap and juggle one or two atoms, A. Duncan measured simultaneously photons here and there, S. Fishman avoided classical diffusion by quantum localisation and E. Hinds played with molecular watches of weird symmetries. H. Kimble kept squeezing the rubbery light, while P. Knight quantum jumped and decohered; D. Miller explored landscapes of quantum wells, S. Stenholm steered away from level crossing while, finally, J. Walraven tried to pack atoms tight enough to fit into a Bose piece of luggage. In short, we had fun, a lot of instructive quantum fun and we believe that the students felt at home in this sort of *Quantum Kindergarten.*

This book contains background material as well as discussions about the future developments of Quantum Toys. Far from being complete, it covers specific subjects of quantum dynamics in a competent and detailed way, with a specific emphasis on simple systems where few atoms or electrons are involved. We regret the absence of contributions from M. Berry and R. Glauber. We completely sympathise, however, with the reasons they provided for graciously declining our invitation to submit material for this book. The same, unfortunately, cannot be said about other authors who promised their notes. By the way, does anyone know where B. Altshuler is hiding now?

Ninety-five students from 20 countries as far as Australia, Israel, New Zealand, USA and Finland took part in the SUSSP44 school. The school was supported by NATO, the Scottish Universities and the University of Strathclyde. We also acknowledge support for travel and fees from the US National Science Foundation and for social events from

Scottish Enterprise, Coherent Ltd, Elliot Scientific Ltd, Newport Ltd and Spectra-Physics Ltd. We thank the staff of the University of Stirling for efficiency and kindness.

Special thanks to Mr. Seumas MacNeill who enlightened us about the history and sounds of the Scottish Bagpipes during a special evening lecture. Sincere thanks for help at the school to A. Chefles, N. Lütkenhaus, J. Jeffers, R. Kay, G. MacFarlane, P. MacKay, F. Shankland, A. Sinclair, M. Snadden, P. Walker and in particular to G. Yeoman who also converted many manuscripts to the Latex format.

Finally, a word of special appreciation to Kathleen and Allister Ferguson for the entertaining afternoon at their mansion and the organisation of the traditional Scottish Conference dinner.

Gian-Luca Oppo (Editor) and William J Firth (Director)
Glasgow, 1995

Contents

Introduction to models on constant negative curvature surfaces

Eugene Bogomolny

Division de Physique Théorique et Institut de Physique Nucléaire
Université de Paris

1 Introduction

Standard courses of classical and quantum mechanics often start with a lot of examples of systems which admit exact solutions. Harmonic oscillators, particles in Coulomb fields and other simple models are building stones of everyones knowledge. The spectacular development of the inverse scattering method gives a long list of integrable models with beautiful mathematics behind. However, standard textbooks practically never mention that complete integrability is a very rare property and almost all problems (even simple looking ones) cannot be solved exactly. They are nonintegrable not because we are not clever enough to find exact solutions but because their behaviour is so complicated that it cannot be described in terms of known (*i.e.* simple) functions. In the usual language 'very complicated' is synonymous to 'erratic', 'random', 'unpredictable', 'chaotic' and quite often one refers to nonintegrable systems as to chaotic systems without a precise definition of these words.

In classical mechanics the existence of chaotic trajectories has been known from the time of Poincaré and the theory of classical chaos is now one of the main parts of modern theory of dynamical systems. Much less is known about how classical chaotic motion manifests itself on quantum properties. The answer to this question is the main problem of the new and quickly growing domain of mathematical physics which had conditionally named 'Quantum Chaos'. This name is attributed to investigations of deterministic quantum systems whose classical limit has certain chaotic properties.

There are many different aspects of this problem. These lectures will deal with specific models of quantum chaos, namely, free motion on constant negative curvature surfaces generated by discrete groups. In many senses these models are the best mathematical models of classical and quantum chaotic motion and play, in the theory of classical and quantum chaos a role similar to the harmonic oscillator in the theory of integrable models.

The rich mathematical structure of these models has attracted the attention of mathematicians for centuries but it is quite difficult to read the existing (enormous) literature on this subject without a preliminary knowledge. The purpose of these lectures is to give a simple and self-contained introduction to these problems. The main attention is given to the explanation of ideas and technical details of the derivation of the famous Selberg trace formula which connects 'quantum' eigenvalues of the Laplace–Beltrami operator with classical periodic orbits. This formula is one of the most important in quantum chaos and it serves as a solid mathematical basis (and, sometime, confirmation) for the application of semiclassical Gutzwiller trace formulae.

The plan of the lectures is the following. Section 2 is devoted to the discussion of the connection between classical mechanics and geodesic motion on certain surfaces. The detailed derivation of the Maupertius' principle and the Jacobi equation is presented. Though this material is classical it is difficult to extract from standard text–books.

In Section 3 the simplest not-so-trivial model of free motion on surface of a plane torus with zero curvature is considered. Though it is a soluble model for which eigenfunctions and eigenvalues can be written explicitly it serves as a prototype for more complicated cases. Construction of this model, introduction of group of isometries, the invariant Laplace operator, automorphic boundary conditions, derivation of the trace formula—all steps are done in such a manner that they could be generalised to models on constant negative curvature. This is performed in detail in Section 4.

The lectures are more devoted to the explanation of ideas than to mathematical rigour and in many cases calculations are performed in the simplest manner but not necessarily the most effective ones.

2   Geometry and mechanics

The modern starting point of the classical mechanics is the least action principle. For all paths $x(t)$ one defines the action on this path as a functional

$$S(x(t)) = \int_{t_1}^{t_2} \mathcal{L}(\dot{x}(t), x(t)) dt \tag{1}$$

where $\mathcal{L}(\dot{x}(t), x(t))$ is called the Lagrangian and is a function of $x(t)$ and its time derivative $\dot{x}(t)$. The standard example is

$$\mathcal{L}(\dot{x}(t), x(t)) = \frac{1}{2}\dot{x}^2 + V(x). \tag{2}$$

(Here $x(t)$ is n -dimensional vector $x_j(t)$, $j = 1, \dots, n$.)

Let us consider the values of $S(x(t))$ on all paths $x(t)$ for which

$$x(t_1) = x_1, \quad x(t_2) = x_2. \tag{3}$$

For each such path, one can calculate the value of the action $S(x(t))$. The classical trajectory is a path for which the action functional has an extremum. If $S(x(t))$ is a function of one variable, 'classical' points are points of minima and maxima of S.

To find classical equations for classical trajectory $x_{cl}(t)$ we have to compare the values of action on all nearby paths

$$x(t) = x_{cl}(t) + \delta x(t) \tag{4}$$

where in order to obey conditions (3)

$$\delta x(t_1) = \delta x(t_2) = 0. \tag{5}$$

Expanding $\mathcal{L}(\dot{x}, x)$ up to linear terms one obtains

$$\mathcal{L}(\dot{x}, x) = \mathcal{L}(\dot{x}_{cl}, x_{cl}) + \frac{\partial \mathcal{L}}{\partial x}\delta x + \frac{\partial \mathcal{L}}{\partial \dot{x}}\delta \dot{x},$$

$$S(x(t)) = S(x_{cl}) + \int_{t_1}^{t_2} \frac{\partial \mathcal{L}}{\partial x}\delta x\, dt + \int_{t_1}^{t_2} \frac{\partial \mathcal{L}}{\partial \dot{x}}\delta \dot{x}\, dt.$$

But $\delta \dot{x} = d\delta x/dt$ and by integrating the last term by part one obtains

$$S(x_{cl} + \delta x) = S(x_{cl}) + \int_{t_1}^{t_2} \left(\frac{\partial \mathcal{L}}{\partial x} - \frac{d}{dt}\frac{\partial \mathcal{L}}{\partial \dot{x}} \right) \delta x\, dt + \frac{\partial \mathcal{L}}{\partial \dot{x}}\delta x \Big|_{t_1}^{t_2}. \tag{6}$$

The extrema of the action are given by the vanishing of the linear terms and classical equations of motion (called Lagrange's equations) read

$$\frac{d}{dt}\frac{\partial \mathcal{L}}{\partial \dot{x}_i} - \frac{\partial \mathcal{L}}{\partial x_i} = 0. \tag{7}$$

If the Lagrangian has the form (2) one obtains the well known Newton's equations

$$\frac{d^2 x_i}{dt^2} = -\frac{\partial V(x)}{\partial x_i}. \tag{8}$$

However, many other forms of Lagrangians are possible. All of them, through Lagrange's equations, will lead to some kind of classical motion. Let us consider two simple cases.

Example 1

$$\mathcal{L} = \sqrt{1 + \dot{x}^2}. \tag{9}$$

Then

$$\frac{\partial \mathcal{L}}{\partial \dot{x}} = \frac{\dot{x}}{\sqrt{1 + \dot{x}^2}},$$

and classical equation has the form

$$\frac{d}{dt}\frac{\dot{x}}{\sqrt{1 + \dot{x}^2}} = 0,$$

from which it follows that $\ddot{x} = 0$ and therefore

$$x(t) = at + b \tag{10}$$

It means that the classical trajectories are just straight lines connecting two given points. The values of parameters a and b can be computed from the initial and final values of x (see Equation (3)), i.e. $at_1 + b = x_1$, and $at_2 + b = x_2$, which gives

$$a = \frac{x_2 - x_1}{t_2 - t_1}. \tag{11}$$

It is also possible to compute the value of the action on the classical trajectory which connects two given points:

$$S_{cl} = \int_{t_1}^{t_2} \sqrt{1 + \dot{x}^2} = \sqrt{1 + a^2}(t_2 - t_1) = \sqrt{(x_2 - x_1)^2 + (t_2 - t_1)^2}. \tag{12}$$

Example 2

$$\mathcal{L} = \frac{1}{x}\sqrt{1 + \dot{x}^2}. \tag{13}$$

Here

$$\frac{\partial \mathcal{L}}{\partial \dot{x}} = \frac{\dot{x}}{x\sqrt{1 + \dot{x}^2}}, \qquad \frac{\partial \mathcal{L}}{\partial x} = -\frac{1}{x^2}\sqrt{1 + \dot{x}^2}.$$

In this case the classical equation (7) can be easily transformed to the form

$$\frac{d^2(x^2)}{dt^2} = -2$$

and 'classical' trajectories are semi-circles perpendicular to the abscissa axis

$$x^2 + (t - a)^2 = R^2, \tag{14}$$

where constants a and R have to be calculated from the boundary values (3)

$$x_1^2 + (t_1 - a)^2 = R^2, \qquad x_2^2 + (t_2 - a)^2 = R^2.$$

Later we shall need the explicit value of the classical action for this model. One has

$$x(t) = \sqrt{R^2 - (t - a)^2},$$

$$\mathcal{L} = \frac{1}{x}\sqrt{1 + \dot{x}^2} = \frac{R}{R^2 - (t - a)^2}.$$

Then

$$S_{cl} = \int_{t_1}^{t_2} \frac{R\,dt}{R^2 - (t - a)^2} = \frac{1}{2}\log\frac{R + t - a}{R - t + a}\Big|_{t_1}^{t_2}.$$

It is convenient to compute $\cosh(S_{cl})$

$$\cosh(S_{cl}) = \frac{1}{2}\left\{\frac{(R + t_2 - a)(R - t_1 + a)}{(R - t_2 + a)(R + t_1 - a)}\right\}^{1/2} + \frac{1}{2}\left\{\frac{(R - t_2 + a)(R + t_1 - a)}{(R + t_2 - a)(R - t_1 + a)}\right\}^{1/2}.$$

After simple transformations one obtains:

$$\cosh(S_{cl}) = 1 + \frac{(x_2 - x_1)^2 + (t_2 - t_1)^2}{2x_2 x_1}. \tag{15}$$

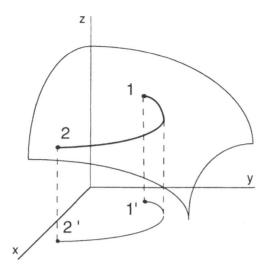

Figure 1. *Motion on a surface.*

The Maupertius Metric

Note that the previous two examples have a special form of Lagrangian, namely,

$$\mathcal{L}(\dot{x}, x; t) = \rho(x, t) \left\{ 1 + \left(\frac{dx}{dt} \right)^2 \right\}^{1/2}, \tag{16}$$

which has a nice geometrical interpretation. Let us consider a two-dimensional surface $z = z(x, y)$ (see Figure 1) The distance between two point on this surface can be computed by the standard formula

$$ds^2 = dx^2 + dy^2 + dz^2,$$

where dz should be calculated along the surface

$$dz = \frac{\partial z}{\partial x} dx + \frac{\partial z}{\partial y} dy.$$

It gives

$$ds^2 = E dx^2 + 2F dx dy + G dy^2, \tag{17}$$

where the first differential parameters

$$E = \left(\frac{\partial z}{\partial x} \right)^2 + 1, \qquad F = \frac{\partial z}{\partial x} \frac{\partial z}{\partial y}, \qquad G = \left(\frac{\partial z}{\partial y} \right)^2 + 1$$

are known functions of x and y. If we consider $x = x(t)$, $y = y(t)$ with $t_1 < t < t_2$ defining a line in the (x, y) plane, the integral

$$S_{12} = \int_{t_1}^{t_2} \left\{ E \left(\frac{dx}{dt} \right)^2 + 2F \frac{dx}{dt} \frac{dy}{dt} + G \left(\frac{dy}{dt} \right)^2 \right\}^{1/2} dt, \tag{18}$$

gives the length of the corresponding line over the surface.

Note that this length is invariant under an arbitrary re-parametrisation $t \rightarrow t(\tau)$ which means that the length is the internal property of the line on the surface and does not depend on the coordinates used for its description. In particular, using $t = x$ the length of the line on the surface can be written as

$$S_{12} = \int_{x_1}^{x_2} \left\{ E + 2F \frac{dy}{dx} + G \left(\frac{dy}{dx} \right)^2 \right\}^{1/2} dx, \tag{19}$$

If $F = 0$ and $G = E = \rho^2$ Equation (19) coincides with (16).

The geodesics of a surface are defined as lines on which the length between two points has an extremum (in the simplest case a minimum) with respect to all other curves on this surface. Our first example given by Equation (9) corresponds to a geodesics of a 2 dimensional Euclidean plane with the usual metric

$$ds^2 = dx^2 + dy^2, \tag{20}$$

and the second one (Equation (13)) states that on the surface with the metric

$$ds^2 = \frac{1}{y^2}(dx^2 + dy^2), \tag{21}$$

the geodesics are semi-circles with centers on the x axis. We shall see later that it corresponds to a free motion on a constant negative curvature surface.

Of course, defining the action as in Equation (18) gives us a classical (free) motion on a given surface. But what is the connection between the standard action written as a difference of kinetic and potential energy (2) and the free motion on a surface?

Let us consider a 2-dimensional system with a potential $V(x, y)$. The Newtonian equations of motion read

$$\frac{d^2 x}{dt^2} = -\frac{\partial V(x, y)}{\partial x}, \qquad \frac{d^2 y}{dt^2} = -\frac{\partial V(x, y)}{\partial y}. \tag{22}$$

The conservation of energy gives:

$$\frac{1}{2}(\dot{x}^2 + \dot{y}^2) + V(x, y) = E. \tag{23}$$

The equations of motion define the trajectory $x(t)$, $y(t)$. However, one can also consider y as function of x

$$y = y(x(t)). \tag{24}$$

It gives

$$\frac{dy}{dt} = y'\frac{dx}{dt}, \qquad \frac{d^2y}{dt^2} = y''\left(\frac{dx}{dt}\right)^2 + y'\frac{d^2x}{dt^2},$$

where d/dx is denoted by a prime. From the integral of energy

$$\left(\frac{dx}{dt}\right)^2 = \frac{2(E - V(x,y))}{1 + (y')^2}, \tag{25}$$

and Equations (22) one obtains

$$y''\frac{2(E - V(x,y))}{1 + (y')^2} - y'\frac{\partial V}{\partial x} + \frac{\partial V}{\partial y} = 0. \tag{26}$$

Note that this equation defines only $y(x)$ *i.e.* it gives only the shape of the classical trajectory. The time dependence $x(t)$ has to be computed from Equation (25).

We stress that Equation (26) is the equation for geodesics on a surface with the following metric

$$ds^2 = 2(E - V(x,y))(dx^2 + dy^2), \tag{27}$$

which is called Maupertius' metric. It can be checked as follows. First choose x as a parameter on this surface so that

$$\mathcal{L}(y', y; x) = \sqrt{2(E - V(x,y))(1 + (y')^2)}. \tag{28}$$

Then

$$\frac{\partial \mathcal{L}}{\partial y} = -\frac{\sqrt{1 + (y')^2}}{\sqrt{2(E - V)}}\frac{\partial V}{\partial y}, \qquad \frac{\partial \mathcal{L}}{\partial y'} = \frac{\sqrt{2(E - V)}}{\sqrt{1 + (y')^2}}y'.$$

The classical equation (7) now gives

$$\frac{d}{dx}\left(\frac{\sqrt{2(E - V)}}{\sqrt{1 + (y')^2}}y'\right) = -\frac{\sqrt{1 + (y')^2}}{\sqrt{2(E - V)}}\frac{\partial V}{\partial y}.$$

After simple transformations one obtains Equation (26).

Therefore, the classical equations of motion can be rewritten in the geometrical form as the equations for the free motion on a surface with the metric (27) and the classical trajectories are extrema of the 'length'

$$S(x(\tau), y(\tau)) = \int_{\tau_1}^{\tau_2} \rho(x,y)\left\{\left(\frac{dx}{d\tau}\right)^2 + \left(\frac{dy}{d\tau}\right)^2\right\}^{1/2} d\tau, \tag{29}$$

where $\rho(x,y) = \sqrt{2(E - V(x,y))}$.

Much less known is the fact that the linearisation of the classical equations of motion in a vicinity of a given trajectory can also be written in geometrical form. To see this let us consider Equations (22) and assume that we know a certain classical trajectory $x_0(t), y_0(t)$. Substituting

$$x(t) = x_0(t) + \zeta(t), \qquad y(t) = y_0(t) + \eta(t) \tag{30}$$

into Equations (22) and conserving only linear terms one obtains:

$$\frac{d^2\zeta}{dt^2} = -\frac{\partial^2 V}{\partial x^2}\zeta - \frac{\partial^2 V}{\partial x \partial y}\eta, \qquad \frac{d^2\eta}{dt^2} = -\frac{\partial^2 V}{\partial y^2}\eta - \frac{\partial^2 V}{\partial x \partial y}\zeta, \tag{31}$$

where the second derivatives of the potential have to be computed on the trajectory.

Let us introduce the tangent and normal vectors to our trajectory (see Figure 2

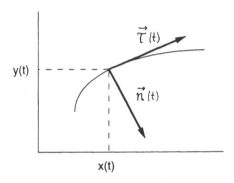

Figure 2. *Normal and tangent vectors to a classical trajectory*

$$\boldsymbol{\tau} = \frac{1}{\sqrt{\dot{x}_0^2 + \dot{y}_0^2}}(\dot{x}_0, \dot{y}_0), \qquad \mathbf{n} = \frac{1}{\sqrt{\dot{x}_0^2 + \dot{y}_0^2}}(\dot{y}_0, -\dot{x}_0). \tag{32}$$

From now on, we shall omit the index 0 implicitly assuming that $(x(t), y(t))$ denotes our reference trajectory.

The vectors $\boldsymbol{\tau}$ and $\mathbf{n}$ obey the relations

$$\boldsymbol{\tau} \cdot \boldsymbol{\tau} = 1, \quad \mathbf{n} \cdot \mathbf{n} = 1, \quad \boldsymbol{\tau} \cdot \mathbf{n} = 0, \quad \dot{\boldsymbol{\tau}} = k\mathbf{n}, \quad \dot{\mathbf{n}} = -k\boldsymbol{\tau}, \tag{33}$$

where k is related to the curvature of the trajectory:

$$k = \frac{\ddot{x}\dot{y} - \ddot{y}\dot{x}}{\dot{x}^2 + \dot{y}^2}.$$

Now one can project the deviations from the trajectory on two vectors $\boldsymbol{\tau}$ and $\mathbf{n}$

$$\delta\mathbf{x} = f\boldsymbol{\tau} + g\mathbf{n}, \tag{34}$$

or

$$\zeta = \frac{1}{\sqrt{\dot{x}^2 + \dot{y}^2}}(f\dot{x} + g\dot{y}), \qquad \eta = \frac{1}{\sqrt{\dot{x}^2 + \dot{y}^2}}(f\dot{y} - g\dot{x}). \tag{35}$$

It is easy to show that

$$\begin{aligned}
\delta\dot{\mathbf{x}} &= (\dot{f} - kg)\boldsymbol{\tau} + (\dot{g} + kf)\mathbf{n}, \\
\delta\ddot{\mathbf{x}} &= A\boldsymbol{\tau} + B\mathbf{n},
\end{aligned} \tag{36}$$

where

$$\begin{aligned}
A &= \ddot{f} - g\dot{k} - 2k\dot{g} - fk^2, \\
B &= \ddot{g} + f\dot{k} + 2k\dot{f} - gk^2.
\end{aligned}$$

But $A = \boldsymbol{\tau} \cdot \delta\ddot{\mathbf{x}}$, $B = \mathbf{n} \cdot \delta\ddot{\mathbf{x}}$, and from Equations (31) one obtains

$$A = \frac{-1}{\dot{x}^2+\dot{y}^2}\left[f\left(\dot{x}^2\frac{\partial^2 V}{\partial x^2}+2\dot{x}\dot{y}\frac{\partial^2 V}{\partial x\partial y}+\dot{y}^2\frac{\partial^2 V}{\partial y^2}\right) +g\left((\dot{y}^2-\dot{x}^2)\frac{\partial^2 V}{\partial x\partial y}+\dot{x}\dot{y}\left(\frac{\partial^2 V}{\partial x^2}-\frac{\partial^2 V}{\partial y^2}\right)\right)\right],$$

$$B = \frac{-1}{\dot{x}^2+\dot{y}^2}\left[g\left(\dot{y}^2\frac{\partial^2 V}{\partial x^2}-2\dot{x}\dot{y}\frac{\partial^2 V}{\partial x\partial y}+\dot{x}^2\frac{\partial^2 V}{\partial y^2}\right) +f\left((\dot{y}^2-\dot{x}^2)\frac{\partial^2 V}{\partial x\partial y}+\dot{x}\dot{y}\left(\frac{\partial^2 V}{\partial x^2}-\frac{\partial^2 V}{\partial y^2}\right)\right)\right].$$

We introduce the following notations:

$$u = \dot{x}\frac{\partial V}{\partial x} + \dot{y}\frac{\partial V}{\partial y}, \qquad v = \dot{x}\frac{\partial V}{\partial y} - \dot{y}\frac{\partial V}{\partial x}, \qquad \rho = \dot{x}^2 + \dot{y}^2. \tag{37}$$

Then

$$u^2 + v^2 = (\dot{x}^2 + \dot{y}^2)\left(\left(\frac{\partial V}{\partial x}\right)^2 + \left(\frac{\partial V}{\partial y}\right)^2\right),$$

$$\dot{\rho} = -2u, \qquad k = \frac{v}{\rho}, \qquad \dot{k} = \frac{\dot{v}}{\rho} + \frac{2uv}{\rho^2},$$

$$\dot{u} = \dot{x}^2\frac{\partial^2 V}{\partial x^2} + 2\dot{x}\dot{y}\frac{\partial^2 V}{\partial x\partial y} + \dot{y}^2\frac{\partial^2 V}{\partial y^2} - \frac{u^2 + v^2}{\rho},$$

$$\dot{v} = (\dot{x}^2 - \dot{y}^2)\frac{\partial^2 V}{\partial x\partial y} + \dot{x}\dot{y}\left(\frac{\partial^2 V}{\partial y^2} - \frac{\partial^2 V}{\partial x^2}\right).$$

Combining these relations one obtains two equations for functions f and g:

$$\ddot{f} + \frac{f}{\rho}\left(\dot{u} + \frac{u^2}{\rho}\right) - \dot{g}\frac{2v}{\rho} - g\frac{2}{\rho}\left(\dot{v} + \frac{u}{\rho}\right) = 0, \tag{38}$$

$$\ddot{g} + g\left(-\frac{u^2 + 2v^2}{\rho^2} - \frac{\dot{u}}{\rho} + \frac{\partial^2 V}{\partial x^2} + \frac{\partial^2 V}{\partial y^2}\right) + \frac{2v}{\rho}\left(\dot{f} + \frac{fu}{\rho}\right) = 0. \tag{39}$$

Now consider the energy

$$E = \frac{1}{2}(\dot{x}^2 + \dot{y}^2) + V(x, y).$$

In the linear approximation given by Equation (30) the energy changes by the amount

$$\delta E = \dot{x}\zeta + \dot{y}\dot{\eta} + \zeta\frac{\partial V}{\partial x} + \eta\frac{\partial V}{\partial y}.$$

Expressing ζ, η through f and g one obtains

$$\delta E = \sqrt{\rho}\dot{f} - \frac{2gv}{\sqrt{\rho}} + \frac{fu}{\sqrt{\rho}}. \tag{40}$$

One can easily check that the equation $\delta\dot{E} = 0$, which is the consequence of the conservation of energy, coincides with Equation (38) and only Equation (39) is non-trivial.

Now choose the deviations from the main trajectory such that the energy is conserved. Then the functions f and g are related by the condition $\delta E = 0$ or

$$\dot{f} = \frac{2gv}{\rho} - \frac{fu}{\rho}. \tag{41}$$

Substituting it into Equation (39) one obtains an equation for the function g:

$$\ddot{g} + g\left(\frac{2v^2 - u^2}{\rho^2} + \frac{\partial^2 V}{\partial x^2} + \frac{\partial^2 V}{\partial y^2}\right) = 0. \tag{42}$$

Let us introduce a new variable s equal to the length of the trajectory (see Equation (27))

$$\frac{ds}{dt} = \dot{x}^2 + \dot{y}^2 \equiv \rho, \tag{43}$$

and instead of g we consider a function y defined by the following equation

$$g = \frac{y}{\sqrt{\rho}}.$$

Then

$$\dot{g} = \sqrt{\rho}\frac{dy}{ds} + \frac{uy}{\rho^{3/2}},$$

and

$$\ddot{g} = \rho^{3/2}\frac{d^2y}{ds^2} + y\left(\frac{\dot{u}}{\rho^{3/2}} + \frac{u^2}{\rho^{5/2}}\right).$$

Finally, one finds that the equation for $y(s)$ can be written in the following form

$$\frac{d^2y}{ds^2} + Ky = 0, \tag{44}$$

where

$$K = \frac{1}{\rho^2}\left[\frac{2}{\rho}\left(\left(\frac{\partial V}{\partial x}\right)^2 + \left(\frac{\partial V}{\partial y}\right)^2\right) + \left(\frac{\partial^2 V}{\partial x^2} + \frac{\partial^2 V}{\partial y^2}\right)\right].$$

It is easy to see that this function can be rewritten in the following form

$$K = -\frac{1}{2\rho}\Delta \log \rho, \tag{45}$$

where as usual $\rho = 2(E - V(x,y))$.

This function has a clear geometrical meaning. It is the Gaussian curvature of a surface whose linear element is given by Equation (27). One of its important property is its invariance under re-parametrisation of the surface. Let us consider this topic in more detail. Our surface has been defined by the distance between two nearby points:

$$ds^2 = \rho(x,y)(dx^2 + dy^2). \tag{46}$$

But coordinates x and y are arbitrary and one can perform an arbitrary transformation $x = x(u,v)$, $y = y(u,v)$. Under this substitution the form of ds^2 will change. (In particular one can obtain cross terms $du\,dv$.) Let us restrict ourselves to transformations for which ds^2 preserves the shape (46) (called the conformal metric.) Introducing the complex coordinate $z = x + iy$, Equation (46) can be rewritten as

$$ds^2 = \rho(z, \bar{z})dz\,d\bar{z}. \tag{47}$$

It is evident that transformations which conserve this form of ds^2 are the following:

$$z \to f(z),$$

with an arbitrary analytical function $f(z)$. Then

$$ds^2 = \rho(z, \bar{z})\left|\frac{df(z)}{dz}\right|^2 dz\,d\bar{z},$$

which defines the transformation of $\rho(x,y)$. The Gaussian curvature K in the complex coordinates takes the form

$$K = -\frac{1}{8\rho}\frac{\partial^2}{\partial z \partial \bar{z}}\log \rho,$$

and as $\Delta F(z) = 0$ for any analytical function $F(z)$, K does not change under such transformation. This means that the Gaussian curvature of a surface depends only on the internal geometry and does not depend on different embedding of this surface in external spaces.

Everything discussed so far shows that the classical equation of motion can be written in the form of equations for the geodesics of Maupertius' metric of a certain surface. Even the equations for the linearised motion have a clear geometrical interpretation. However, the explicit construction of this surface is a difficult problem. A far simpler task is to take a surface and try to find the geodesic motion on it. First of all the latter is a problem of classical mechanics and from its solution one can also hope to gain a good understanding of other classical problems as well. Secondly, it is important to note that geodesic motion on 'good' surfaces possesses a rich mathematical structure which is absent in generic systems. For quantum chaos it is important in many cases to find easily and naturally the corresponding classical problem.

Therefore from now on we shall consider certain reasonable surfaces and try to investigate classical and quantum problems on them.

3   Zero curvature surfaces

The plane

The simplest of all possible models is the usual Euclidean 2-dimensional plane (x, y). Its metric element is given by usual Euclidean distance

$$ds^2 = dx^2 + dy^2. \tag{48}$$

The geodesics of this metric are straight lines and the distance between two points equals

$$d(2, 1) = \sqrt{(x_2 - x_1)^2 + (y_2 - y_1)^2}. \tag{49}$$

We have seen above how these trivial facts follow from the definitions of geodesics.

The important property of the plane (which is crucial for standard geometry) is the existence of a 3-parameter group of isometric transformations. This means the following. The distance between two points depends on the coordinates of these points. Is there a transformation of each point such that the distance between the transformed points equals the distance between the initial ones? It is well known that for the plane these isometries are translations

$$x \rightarrow x + a$$
$$y \rightarrow y + b$$

or rotations

$$x \rightarrow x \cos \alpha - y \sin \alpha$$
$$y \rightarrow x \sin \alpha + y \cos \alpha$$

or any combinations of them.

If one introduces the complex coordinates $z = x + iy$, then the distance between two points can be calculated from

$$d^2(z_2, z_1) = |z_2 - z_1|^2$$

and isometries can be written as transformations

$$z \rightarrow e^{i\alpha} z + q, \tag{50}$$

where q is complex and α is real. The existence of such isometries is a manifestation of the uniformity of the usual Euclidean geometry and, for example, allows for the introduction of the notion of equal figures.

In order to investigate quantum problems on this plane we need to introduce the notion of the invariant Laplace operator which is defined as a differential operator which commutes with the isometries. Let Δ be an invariant operator. This means that if $g(x, y) = \Delta f(x, y)$ is the result of the action of Δ on an arbitrary function $f(x, y)$, then for an invariant operator $g(x', y') = \Delta f(x', y')$, where x', y' are isometries of our space.

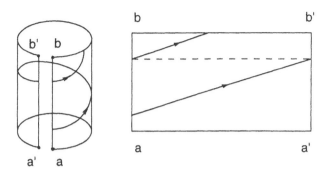

Figure 3. *A cylinder and its plane model.*

It is well known that for the plane, the invariant operator coincides with the standard Laplace operator

$$\Delta = \frac{\partial^2}{\partial x^2} + \frac{\partial^2}{\partial y^2}. \tag{51}$$

This fact can easily be verified by writing Δ in the complex coordinates

$$\Delta = \frac{1}{4}\frac{\partial^2}{\partial z \partial \bar{z}}.$$

The problem now is: what can we do with the plane? The simplest way is to consider the motion on the whole plane. This seems to be not very interesting because all classical trajectories are straight lines. Another possibility is to try to construct different surfaces from the plane.

The cylinder and the torus

One can build a cylinder from the plane (see Figure 3). by a simple gluing operation. The geodesics of the plane (straight lines) now become the geodesics of the cylinder. If one cuts the cylinder along a vertical line, one obtains a section of the plane but with opposite points identified. Classical trajectories are now segments of straight lines between two vertical lines. When the trajectory crosses one of vertical lines it appears from the opposite side as indicated in Figure 3. The simplest method to perform it in an automatic way is to say that all points whose x coordinates differ by L are identified (here L is the circumference of the cylinder). Therefore if one considers functions which are invariant under the transformation

$$x \rightarrow x + L,$$
$$y \rightarrow y$$

one can say that they are defined on the cylinder.

Our surface is however still of infinite volume. To construct a finite surface it is convenient to build a torus (see Figure 4). Now if one cuts the torus over two lines as

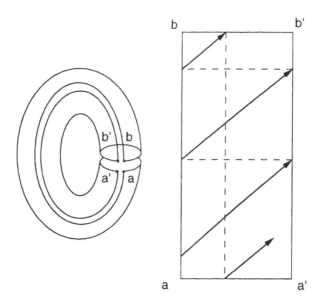

Figure 4. *A torus and its plane model.*

indicated on this figure, one obtains a rectangular part of the plane where all opposite sides are identified. What is nice about this construction is that we obtain a surface whose geodesics are simply the images of straight lines which are cut in a special manner. In this case one knows the local description of geodesics and the non-trivial dynamics comes from global properties of the surface.

How can this process of cutting and gluing be described mathematically? Let us reconsider Figure 4. It is evident that geodesics of the torus can be obtained from geodesics of the whole plane by identifying all points which differ by the following transformations:

$$
\begin{aligned}
x &\rightarrow x + A \\
y &\rightarrow y + B
\end{aligned}
\tag{52}
$$

where A and B are real. One can say that the plane with such identification is a torus.

For later developments, it is important to note that these transformations belongs to isometries (see Equation (50)) and therefore do not change the distance between points. One can also consider these two transformations as generators of a group and compute the whole group of such transformations. The most general transformation is now

$$
\begin{aligned}
x &\rightarrow x + nA, \\
y &\rightarrow y + mB,
\end{aligned}
\tag{53}
$$

where n and m are integers. They evidently form a (commutative) group.

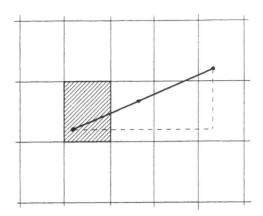

Figure 5. *Tessellation of the plane by images of a rectangle.*

Note that images of a point under these transformations cannot get too close to the initial point because n and m are integers. This is the defining property of the so-called discrete groups, the only groups we shall consider below. Under the action of this group the original rectangle will tessellate the whole plane (see Figure 5), *i.e.* the images of the initial rectangle under transformations (53) cover the whole plane without omissions and double coverings.

Mathematically this is equivalent to the statement that our rectangle is a fundamental domain of the discrete group (53). It means that two points inside it cannot be connected via a group transformation while for any point outside the domain one can find a point inside it which is a pre-image of that point.

The geodesics on the whole plane are just straight lines. Inside the fundamental domain (the rectangle) they are segments cut and glued by the action of the group. It is also interesting to know the periodic orbits of this motion. By definition periodic orbits are classical trajectories which close continuously after a certain time. In the whole plane this is equivalent to the statement that periodic orbits are segments of straight lines which go through a point (x, y) and through its image under one of the group transformations (53) with certain values (n, m). Therefore, if x_i and y_i are the coordinates of the initial point on a periodic orbit then the values of its final point, x_f and y_f, can be computed from the formula

$$x_f = x_i + nA$$
$$y_f = y_i + mB.$$

The geometrical length of this periodic orbit can be evaluated from the usual formula (49)

$$l_p = \sqrt{(An)^2 + (Bm)^2}. \tag{54}$$

Therefore, one knows the length of periodic orbits if one is given the transformation under which this periodic orbit is invariant.

Quantum mechanics on the torus

Up to now we have considered the classical motion on a torus defined by the factorisation of the Euclidean plane by the action of a discrete group of isometries. We now want to investigate the corresponding quantum problems that are the main purpose of these lectures.

It has already been mentioned that the natural quantum operator is the invariant Laplace operator and a typical 'quantum' problem is to find its eigenfunctions and eigenvalues:

$$-\Delta \Psi_n(x, y) = E_n \Psi_n(x, y). \tag{55}$$

In order to have a discrete spectrum one has to impose special boundary conditions on $\Psi_n(x, y)$. Following the previous discussion, it is natural to the following periodicity conditions

$$\begin{aligned} \Psi(x + A, y) &= \Psi(x, y), \\ \Psi(x, y + B) &= \Psi(x, y), \end{aligned} \tag{56}$$

which simply explain the fact that $\Psi(x, y)$ are defined on the torus. In other words, we shall consider eigenfunctions of the Laplace operator which are invariant under the action of our discrete group. Note that this problem is consistent because the Laplace operator is an invariant operator.

We now have a well defined quantum problem. Our particular example can be easily solved but we use it as a prototype for much more complicated models. The explicit solution of Equation (55) with boundary conditions (56) can be found in the following form

$$\Psi_n(x, y) = \exp(ik_1 x) \exp(ik_2 y), \tag{57}$$

where

$$E = k_1^2 + k_2^2,$$

and

$$Ak_1 = 2\pi N_1, \qquad Bk_2 = 2\pi N_2.$$

Here N_1 and N_2 are arbitrary integers (positive, negative or zero). Thus the eigenvalues of the Laplace operator on the torus with the periodic boundary conditions have the form

$$E_{N_1, N_2} = \left(\frac{2\pi}{A}\right)^2 N_1^2 + \left(\frac{2\pi}{B}\right)^2 N_2^2. \tag{58}$$

Everything is simple. However, we shall proceed as if we do not know the answer, the advantage being that the method could be generalised for models on constant negative curvature. We use two approaches: the Green function method and the Poisson summation method.

Green functions on the torus

Our starting point (as in general cases) are not the eigenfunctions but the Green function of the Laplace operator $G_E(x, y; x'y')$. The latter is defined as follows

$$(\Delta + E)G_E(x, y; x'y') = \delta(x - x')\delta(y - y'), \tag{59}$$

and

$$G_E(x + nA, y + mB; x', y') = G_E(x, y; x'y'). \tag{60}$$

There is a general formula gives the Green function through exact eigenfunctions of our quantum problem:

$$G_E(\mathbf{x}; \mathbf{x}') = \sum_n \frac{\bar{\Psi}_n(\mathbf{x}')\Psi_n(\mathbf{x})}{E - E_n + i\epsilon}, \tag{61}$$

where the summation is taken over all eigenvalues of the original problem. This formula works if our eigenfunctions form a complete orthogonal set of functions (with respect to a certain measure). This means that

$$\int \bar{\Psi}_n(\mathbf{x})\Psi_m(\mathbf{x})d\mathbf{x} = \delta_{mn}, \tag{62}$$

and

$$\sum_n \bar{\Psi}_n(\mathbf{x}')\Psi_n(\mathbf{x}) = \delta(\mathbf{x}' - \mathbf{x}). \tag{63}$$

These conditions are always satisfied for 'reasonable' problems and we shall not try to prove them in general. In our case they trivially follow from the theory of Fourier expansions.

In the denominator of Equation (61) there is a term in $i\epsilon$. Here ϵ is an arbitrary small positive number. It is added in order to stress that there are many Green functions which differ by the rule used for treating the pole terms in Equation (61). In our case we defines the Green function at real values of energy as the limits obtained with energies with positive imaginary part.

If one knows the exact Green function one can compute all quantum quantities. Let us consider the density of eigenvalues defined as follows:

$$d(E) = \sum_n \delta(E - E_n), \tag{64}$$

where the summation is taken over all energy levels of a system. To express this through the Green function it is convenient to use the identity:

$$\text{Im}\left(\frac{1}{x + i\epsilon}\right) = -\frac{\epsilon}{x^2 + \epsilon^2} \rightarrow -\pi\delta(x) \quad \text{as} \quad \epsilon \rightarrow 0$$

which can be easily proved by noting that for an arbitrary function $f(x)$

$$\lim_{\epsilon \rightarrow 0} \int_{-\infty}^{+\infty} \frac{\epsilon f(x)}{x^2 + \epsilon^2}dx = \lim_{\epsilon \rightarrow 0} \int_{-\infty}^{+\infty} \frac{f(\epsilon u)}{u^2 + 1}du = f(0)\int_{-\infty}^{+\infty} \frac{du}{u^2 + 1} = \pi f(0).$$

Then from Equation (61)

$$-\frac{1}{\pi}\mathrm{Im}(G_E(\mathbf{x},\mathbf{x})) = \sum_n |\Psi_n(\mathbf{x})|^2 \delta(E - E_n),$$

and using Equations (62) one concludes that

$$d(E) = -\frac{1}{\pi}\int \mathrm{Im}(G_E(\mathbf{x},\mathbf{x}))d\mathbf{x}. \qquad (65)$$

This formula is quite general and it is the starting point of all trace formulae.

The main difficulty of such an approach is the fact that one does not know the exact Green function of a general quantum problem. What is easy to compute is the Green function for the whole plane which corresponds to a free motion on the plane without any periodic boundary conditions. The first step of actual computation of the free Green function $G_E^0(\mathbf{x},\mathbf{x}')$ consists of noting that it should depends only on the distance between points $\mathbf{x}$ and $\mathbf{x}'$. It is a manifestation of the uniformity of the Euclidean plane. Denoting $r = \{(x - x')^2 + (y - y')^2\}^{1/2}$ one tries to find the Green function in the form

$$G_E^0(\mathbf{x},\mathbf{x}') = G(r). \qquad (66)$$

Then $\partial G(r)/\partial x_j = G'(r)z_j/r$ and $\Delta G = G'' + G'/r$ where the prime denotes the derivative with respect to r and $z_j = x_j - x_j'$. Note that ΔG depends only on r. This is a consequence of the invariance of the Laplacian under the isometries. One concludes that

$$G'' + \frac{1}{r}G' + EG = \delta(\mathbf{x}). \qquad (67)$$

Therefore, for non-zero r, $G(r)$ obeys the equation

$$G'' + \frac{1}{r}G' + k^2 G = 0, \qquad (68)$$

where we have introduced the momentum k by $E = k^2$.

This is the standard form of the equation for the Bessel function of zero order (see page 951 of [1]). Therefore its most general solution can be written as

$$G(r) = a H_0^{(1)}(kr) + b H_0^{(2)}(kr), \qquad (69)$$

where a and b are constant and $H_0^{(1,2)}$ are Hankel's functions with the following asymptotics behaviour as $x \to \infty$ (see page 962 of [1])

$$H_0^{(1,2)}(x) \to \sqrt{\frac{2}{\pi x}} e^{\pm i(x - \pi/4)}. \qquad (70)$$

It is now that the $i\epsilon$ prescription in Equation (61) becomes important. It is easy to see that one can consider the limit $E \to i\infty$ in the expression for the Green function without crossing poles in the denominator. However, from Equation (61), it follows that in this limit $G_E(\mathbf{x},\mathbf{x}') \to 0$. (Note that in general $G_E(\mathbf{x},\mathbf{x}') \to 0$ when $E \to i\epsilon\infty$).

The limit $E \to i\infty$ corresponds to the limit $\mathrm{Im}(k) \to \infty$. Therefore, the term with $H_0^{(2)}(x)$ disappears and

$$G_E^0(\mathbf{x}, \mathbf{x}') = a H_0^{(1)}(kr).$$

In order to find the constant a, one has to investigate the behaviour of the Green function at small r. As $x \to 0$ (see page 960 of [1])

$$H_0^{(1)} \to i\frac{2}{\pi} \log x.$$

However, it is well known that

$$\Delta \log r = 2\pi \delta(\mathbf{x}).$$

There are many ways of proving this equality. One of the simplest is the following. Let us consider a certain regularisation of $\log r$, for example $\log(r + \epsilon)$ with $\epsilon \to 0$. Then

$$\Delta \log(r + \epsilon) = \frac{d}{rdr}\left(r\frac{d}{dr}\log(r + \epsilon)\right) = \frac{\epsilon}{r(r + \epsilon)^2}.$$

This function, however, tends to zero as $\epsilon \to 0$ and $r \neq 0$ and

$$\int \frac{\epsilon}{r(r + \epsilon)^2} f(x) d^2 x = \int f(x)\frac{\epsilon}{(r + \epsilon)^2} dr d\phi = 2\pi f(0).$$

The last equality can easily be obtained after the substitution $r \to \epsilon u$.

As the Green function should obey Equation (59) the coefficient in front of $\log r$ has to be equal to $1/(2\pi)$. It fixes $a = 1/(4i)$ and the final expression for the free Green function has the form

$$G_E^0(\mathbf{x}, \mathbf{x}') = \frac{1}{4i} H_0^{(1)}(k|\mathbf{x} - \mathbf{x}'|). \tag{71}$$

This expression of course obeys Equation (59) but it is not invariant under the action of our group (60). In order to fulfill these conditions the most convenient way is to use the method of images. This means the following. Our free Green function gives the field in the point $\mathbf{x}$ when the source is placed in point $\mathbf{x}'$. However in our torus model, all points of the plane whose coordinates are connected by group transformations

$$T_{nm}(x, y) = (x + nA, y + mB)$$

should be considered as one point. Therefore, the exact Green function which obeys periodic boundary conditions (60) can be written in the following form

$$G_E(\mathbf{x}, \mathbf{x}') = \sum_{n,m} G_E^0(\mathbf{x}, T_{nm}\mathbf{x}'), \tag{72}$$

where the summation in taken over all group transformations. Explicitly

$$G_E(\mathbf{x}, \mathbf{x}') = \frac{1}{4i} \sum_{n,m=-\infty}^{+\infty} H_0^{(1)}\left(k\sqrt{(x - x' - nA)^2 + (y - y' - mB)^2}\right). \tag{73}$$

Why is this the exact Green function? First, it obeys the Laplace equation as each term in the sum obeys it. Second, when $x \to x'$ only in one term the argument of Hankel's function tends to zero. This is a consequence of the discrete character of our group. Third, x changing to $x + A$ (or y to $y + B$) corresponds only to a reordering of terms in the sum and, consequently, the sum is not changed.

Having obtained the expression for the exact Green function, one can compute the density of eigenvalues by using Equation (65)

$$
\begin{aligned}
d(E) &= -\frac{1}{\pi} \int_D \text{Im}(G_E(\mathbf{x}, \mathbf{x})) d^2x, \\
&= -\frac{1}{\pi} \int_D \text{Im} \left(\frac{1}{4i} \sum_{n,m=-\infty}^{+\infty} H_0^{(1)} \left(k\sqrt{(nA)^2 + (mB)^2} \right) \right) d^2x,
\end{aligned}
\tag{74}
$$

where the integral is taken over the initial rectangle, *i.e.* the fundamental domain of our group. However,

$$
\text{Im} \left(\frac{1}{4i} H_0^{(1)}(x) \right) = -\frac{1}{4} J_0(x),
$$

where $J_0(x)$ is the usual Bessel function of the first kind. Since the integrand in Equation (74) does not depend on x the integration over d^2x gives the area of the fundamental domain $\mu(D) = AB$ and one obtains:

$$
d(E) = \frac{\mu(D)}{4\pi} + \frac{\mu(D)}{4\pi} \sum_{n,m=-\infty}^{+\infty}{}' J_0 \left(k\sqrt{(nA)^2 + (mB)^2} \right).
\tag{75}
$$

The prime here means that the term $m = n = 0$ is omitted. However, we have seen before that

$$
l_p = \sqrt{(nA)^2 + (mB)^2}
\tag{76}
$$

is the length of periodic orbits which remains invariant under the group transformation $x \to x + nA$, $y \to y + mB$. Therefore we obtain the following result; for our torus model the density of (quantum) eigenvalues (*i.e.* the energy levels) can be written as a sum of two contributions

$$
d(E) = \bar{d}(E) + d_{\text{osc}}(E),
\tag{77}
$$

where $\bar{d}(E) = \mu(D)/(4\pi)$ is a smooth term (called the Weyl term) and $d_{\text{osc}}(E)$ can be written as a sum over all classical periodic orbits of our system

$$
d_{\text{osc}}(E) = \frac{\mu(D)}{4\pi} \sum_{\text{pp}} J_0(kl_p),
\tag{78}
$$

where the summation is taken over all periodic orbits.

This type of formulae which express the quantum density of states for a certain model as a sum over classical periodic orbits is called the trace formulae. They allow (in principle) the computation of quantum quantities through the information taken from classical mechanics.

The method of Poisson summation

In our simple example one knows the quantum spectrum exactly and it is not necessary to perform all these calculations leading to the trace formula (75). One can obtain the same result by a different method. By definition

$$d(E) = \sum_{N_1,N_2=-\infty}^{+\infty} \delta(E - E_{N_1,N_2}),$$
(79)

where

$$E_{N_1,N_2} = \left(\frac{2\pi}{A}\right)^2 N_1^2 + \left(\frac{2\pi}{B}\right)^2 N_2^2.$$
(80)

To rewrite the sum over all N_1 and N_2, it is convenient to use the Poisson summation formula which states

$$\sum_{n=-\infty}^{+\infty} f(n) = \sum_{m=-\infty}^{+\infty} \int e^{2\pi i m n} f(n) dn.$$
(81)

This formula can be demonstrated heuristically as follows. Consider the function

$$g(x) = \sum_{n=-\infty}^{+\infty} \delta(x - n).$$

which is a periodic function of x with period 1; consequently, it can be expanded in a Fourier series

$$g(x) = \sum_m e^{2\pi i m x} C_m.$$

The coefficients C_m can be calculated by the usual formulae

$$C_m = \int_0^1 g(x) e^{-2\pi i m x} dx.$$

However, in the interval $[0,1)$ the function $g(x)$ has only δ-function singularities at $x = 0$ and $C_m = 1$. This leads to

$$\sum_{n=-\infty}^{+\infty} \delta(x - n) = \sum_{m=-\infty}^{+\infty} e^{2\pi i m x}.$$
(82)

Integrating it with an arbitrary function $f(x)$, one obtains Equation (81). Therefore, the density of states can be rewritten as follows

$$d(E) = \sum_{M_1,M_2=-\infty}^{+\infty} \int_{-\infty}^{+\infty} \exp\left(2\pi i(M_1 N_1 + M_2 N_2)\right) \delta\left(E - \frac{4\pi^2 N_1^2}{A^2} - \frac{4\pi^2 N_2^2}{B^2}\right) dN_1 dN_2.$$

Instead of N_1, N_2, let us introduce the coordinates r and θ via

$$N_1 = \frac{A}{2\pi} r \cos\theta, \qquad N_2 = \frac{B}{2\pi} r \sin\theta.$$

The Jacobian of this transformation

$$\begin{vmatrix} \partial N_1/\partial r & \partial N_1/\partial \theta \\ \partial N_2/\partial r & \partial N_2/\partial \theta \end{vmatrix} = \frac{\mu(D)}{(2\pi)^2} r,$$

where $\mu(D) = AB$ is the area of our rectangle and

$$\begin{aligned} d(E) &= \frac{\mu(D)}{4\pi^2} \sum_{M_1,M_2=-\infty}^{+\infty} \int \exp(ir(M_1 A \cos\theta + M_2 B \sin\theta))\delta(k^2 - r^2)r\,dr\,d\theta \\ &= \frac{\mu(D)}{8\pi^2} \sum_{M_1,M_2-\infty}^{+\infty} \int_0^{2\pi} \exp\left(i\sin\nu\sqrt{(M_1 A)^2 + (M_2 B)^2}\right) d\nu, \end{aligned}$$

where we have transformed

$$M_1 A \cos\theta + M_2 B \sin\theta = \sin(\theta + \varphi)\sqrt{(M_1 A)^2 + (M_2 B)^2}$$

and changed the integration over θ to a new variable $\nu = \theta + \varphi$. However (see page 952 of [1]),

$$J_0(z) = \frac{1}{2\pi} \int_0^{2\pi} \exp(iz\sin\nu)d\nu,$$

and finally one obtains the same formula for the density of states as before (see Equation (75)).

When $kl_{\mathrm{p}} \gg 1$ one can use the asymptotic expression of Bessel's function which gives

$$d_{\mathrm{osc}}(E) = \frac{\mu(D)}{4\pi}\sqrt{\frac{2}{\pi k}}\sum_{\mathrm{pp}}\frac{1}{\sqrt{l_{\mathrm{p}}}}\cos\left(kl_{\mathrm{p}} - \frac{\pi}{4}\right), \qquad (83)$$

and $l_{\mathrm{p}} = \sqrt{A^2 n^2 + B^2 m^2}$.

The first remark is that this sum does not converge absolutely (even the sum $\sum_{\mathrm{pp}} 1/l_{\mathrm{p}}$ diverges). A simple method to treat such problem is to multiply both sides of the trace formula by a certain function $h(E)$ and integrate them over E. In this way one obtains

$$\sum_n h(E_n) = \frac{\mu(D)}{4\pi}\int_0^\infty h(E)dE + \frac{\mu(D)}{4\pi}\sum_{\mathrm{pp}}\int_0^\infty h(E)J_0(kl_{\mathrm{p}})dE. \qquad (84)$$

If the Fourier harmonics of $h(E)$ decay quickly, then all sums are finite and there are no problems with convergence. This is the form in which the mathematicians like to write trace formulae.

What have we learned from these calculations? We have presented two different methods of computing the trace formula for the free motion on a torus. The first one was based on the method of images and on the explicit expression for the free Green function. In the next sections we shall see how this can be generalised to models on constant negative curvature surfaces. All steps will be exactly the same but the calculations will be more difficult. The second method was based on the Poisson summation formula and can be applied only to integrable systems where one can write the exact (or approximate) expression for the energy levels. It is by this method that Berry and Tabor [2] obtained the semiclassical trace formula for integrable systems.

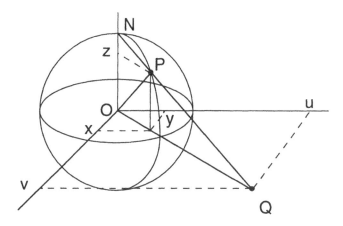

Figure 6. *Stereographic coordinates of a sphere.*

4   Constant negative curvature surfaces

In the previous Section we have discussed the classical and quantum problems connected with a compact surface of zero curvature, *i.e.* a torus. In his section we are concerned with models on surfaces of constant negative curvature but it is useful to first look at the geometry of a compact surface with positive curvature.

Geometry of the sphere

The surface of constant positive curvature is known to everyone. It is, of course, a sphere which in the usual three-dimensional space can be described by the equation

$$x^2 + y^2 + z^2 = R^2. \tag{85}$$

To find an internal description of this sphere it is convenient to consider the stereographic projection of it on a plane as indicated in Figure 6. To an arbitrary point of the sphere P one associates a point Q on the plane (x, y) which is the intersection of a straight line which goes from the north pole of the sphere to P.

Let us denote the coordinates of the point Q in the plane by u and v. Then, elementary geometrical considerations gives that the coordinates of the point P can expressed through the coordinates of Q by the following formulae:

$$x = \frac{2uR^2}{R^2 + u^2 + v^2}, \qquad y = \frac{2vR^2}{R^2 + u^2 + v^2}, \qquad z = R\frac{R^2 - u^2 - v^2}{R^2 + u^2 + v^2}. \tag{86}$$

The simple way to derive these relations is to note that angles $\widehat{NOP} = \widehat{OPM} = 2\widehat{NQO}$. The next step is to express the distance between 2 points on the sphere through u and v. One obtains

$$dx = \frac{2R^2}{R^2 + u^2 + v^2}\left(du - \frac{2u}{R^2 + u^2 + v^2}(udu + vdv)\right),$$

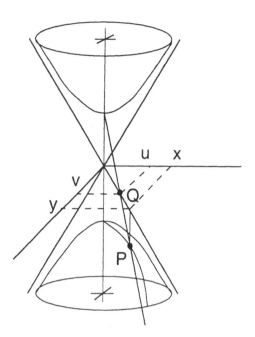

Figure 7. *Stereographic coordinates of a hyperboloid.*

$$dy = \frac{2R^2}{R^2 + u^2 + v^2}\left(dv - \frac{2v}{R^2 + u^2 + v^2}(udu + vdv)\right),$$

$$dz = \frac{-4R^3}{(R^2 + u^2 + v^2)^2}(udu + vdv).$$

Substituting these expressions into $ds^2 = dx^2 + dy^2 + dz^2$, the line element on the sphere gives, after simplification,

$$ds^2 = \frac{4R^4}{(R^2 + u^2 + v^2)^2}(du^2 + dv^2). \tag{87}$$

It is useful to check that computing the Gaussian curvature by Equation (45) one obtains R^{-2} as it should be.

The geometry of the hyperboloid

The hyperboloid defined by the equation (see Figure 7)

$$z^2 - x^2 - y^2 = R^2. \tag{88}$$

is a two sheeted surface intersecting the z axis at points $\pm R$. Let us consider only the $z < 0$ part of the hyperboloid and introduce the stereographic projection as indicated in Figure 7. Geometrical considerations similar to the above show that the coordinates

$(x, y\ z)$ of a point P on the hyperboloid are connected with the coordinates (u, v) of the point Q by the following formulae:

$$x = \frac{2uR^2}{R^2 - u^2 - v^2}, \qquad y = \frac{2vR^2}{R^2 - u^2 - v^2}, \qquad z = -R\frac{R^2 + u^2 + v^2}{R^2 - u^2 - v^2}. \quad (89)$$

Note that coordinates u, v are inside the circle of radius R. However Equations (89) can be obtained from Equations (87) just by changing the sign of R^2 and the sign of z. Therefore one does not need to perform the calculation of the linear element on the hyperboloid. It is sufficient to change the sign of R^2 in Equation (87) and one concludes that the linear element of our surface is

$$ds^2 = \frac{4R^4}{(R^2 - u^2 - v^2)^2}(du^2 + dv^2). \quad (90)$$

while its Gaussian curvature is $-R^{-2}$. The formal change of sign of R^2 when going from the sphere to the hyperboloid is the reason why the latter is often called the pseudosphere.

Without loss of generality we shall consider the case $R = 1$ when the linear element of a pseudosphere takes the form

$$ds^2 = \frac{4}{(1 - u^2 - v^2)^2}(du^2 + dv^2). \quad (91)$$

This model of a constant negative curvature surface is called the Poincaré disc. For our purposes it will be more convenient to use another model (called the Poincaré half-plane) which can be obtained from (91) by the following change of coordinates $(z = x + iy)$

$$u + iv = -i\frac{z - 1}{z + 1}.$$

In the new coordinates, which are not to be confused with the x, y coordinates of the embedding space, ds^2 can be expressed as

$$ds^2 = \frac{1}{y^2}(dx^2 + dy^2), \quad (92)$$

and the interior of the unit circle is transformed into the upper half-plane $y > 0$.

Note that we have already considered this model in Section 2. We have seen that geodesics of this metric are the circular arcs or vertical half-lines orthogonal to the x axis and the distance between two points (see Equation (15)) can be computed by the formula

$$\cosh(d(2, 1)) = 1 + \frac{(x_2 - x_1)^2 + (y_2 - y_1)^2}{2y_2 y_1}. \quad (93)$$

Now we shall try to ask the same questions as in the case of the Euclidean plane.

Isometries

The first and the most important question is the nature of isometries. Let us show that they are the fractional transformations of the form

$$z \to z' = \frac{az + b}{cz + d},\tag{94}$$

where a, b, c, and d are real numbers. One has

$$
\begin{aligned}
z'_1 - z'_2 &= \frac{az_1 + b}{cz_1 + d} - \frac{az_2 + b}{cz_2 + d} = (ad - bc)\frac{z_1 - z_2}{(cz_1 + d)(cz_2 + d)}, \\
y' &= \frac{z' - \bar{z}'}{2i} = (ad - bc)\frac{y}{|cz + d|^2}.
\end{aligned}
$$

Therefore

$$
\begin{aligned}
\cosh(d(2',1')) &\equiv 1 + \frac{|z'_2 - z'_1|^2}{2y'_2 y'_1} = 1 + \frac{(ad - bc)^2 |z_2 - z_1|^2 |cz_1 + d|^2 |cz_2 + d|^2}{2|cz_1 + d|^2 |cz_2 + d|^2 y_1 y_2} \\
&= 1 + \frac{|z_2 - z_1|^2}{2y_2 y_1} = \cosh(d(2,1))
\end{aligned}
$$

and the distance remains invariant under the transformations (94). The existence of such a big group of transformations is a consequence of the constant curvature of our surface.

The transformation (94) is invariant under multiplication of the coefficients a, b, c, and d by the same number. Therefore, our transformations depend only on three parameters exactly as it was for the Euclidean plane. It is convenient to represent the transformation (94) as a matrix

$$M = \begin{pmatrix} a & b \\ c & d \end{pmatrix},\tag{95}$$

and to fix these parameters by fixing the determinant of the matrix. Two cases are possible. If $ad - bc > 0$ then by rescaling one can normalise the determinant to 1

$$\det M \equiv da - bc = 1.\tag{96}$$

In this case the transformation (94) transforms the upper-half plane into itself as it is clear from the relation

$$y' = (ad - bc)\frac{y}{|cz + d|^2}.\tag{97}$$

If $ad - bc < 0$ it is convenient to normalise the determinant to -1. The transformations (94) with $ad - bc = -1$ are orientation-reversing transformations. They naturally appear in billiard problems but we shall not discuss them here.

From now on we shall consider fractional transformations (94) normalised in such a way that $ad - bc = 1$ and to each of them we shall associate a matrix (96) with the determinant equal to 1. This connection between isometries of the upper-half plane and matrices is not just a formal one. Let us perform two transformations with two different sets of parameters. The first

$$z' = \frac{a_1 z + b_1}{c_1 z + d_1},$$

and the second one:

$$z'' = \frac{a_2 z' + b_2}{c_2 z' + d_2} = \frac{a_2(a_1 z + b_1) + b_2(c_1 + d_1)}{c_2(a_1 + b_1) + d_2(c_1 z + d_1)} = \frac{a_3 z + b_3}{c_3 z + d_3},$$

where

$$
\begin{aligned}
a_3 &= a_2 a_1 + b_2 c_1, & b_3 &= a_2 b_1 + b_2 d_1, \\
c_3 &= c_2 a_1 + d_2 c_1, & d_3 &= c_2 b_1 + d_2 d_1.
\end{aligned}
$$

It is easy to verify that

$$
\begin{pmatrix} a_3 & b_3 \\ c_3 & d_3 \end{pmatrix} = \begin{pmatrix} a_2 & b_2 \\ c_2 & d_2 \end{pmatrix} \begin{pmatrix} a_1 & b_1 \\ c_1 & d_1 \end{pmatrix},
$$

This means that the result of successive transformations can be obtained by the multiplication (from left to right) of the corresponding matrices. Mathematically, the group of isometries of the surface of constant negative curvature is the same as the projective special linear group of matrices

$$PSL(2, R) = SL(2, R)/\{\pm I\}.$$

$L(2, R)$ here denotes the (linear) group of 2×2 matrices with real coefficients, S, 'special', denotes that the determinant of these matrices equals 1 and P, 'projective', reminds that the changing the sign of all elements of a matrix leads to the same transformation. This isomorphism of isometries and groups of matrices is very useful in applications.

The Laplace-Beltrami operator

Having defined the isometries the next question is what is an invariant differential operator. This operator is the analogue of the usual Laplace operator and it is called the Laplace-Beltrami operator. A simple observation shows that it has the form

$$\tilde\Delta = y^2 \left(\frac{\partial^2}{\partial x^2} + \frac{\partial^2}{\partial y^2} \right). \tag{98}$$

As in the previous section, it is convenient to consider the Laplace-Beltrami operator in complex coordinates $z = x + iy$ in order to check its invariance under isometries (94). Then

$$\tilde\Delta = y^2 \frac{\partial^2}{4 \partial z \partial \bar z}.$$

However

$$\frac{\partial}{\partial z} f \left(\frac{az + b}{cz + d} \right) = \frac{1}{(cz + d)^2} \frac{\partial f(z')}{\partial z'}.$$

and

$$y' = \frac{y}{|cz + d|^2}.$$

Therefore

$$\tilde{\Delta} f(z', \bar{z}') = y^2 \frac{\partial^2}{4 \partial z \partial \bar{z}} f\left(\frac{az + b}{cz + d}, \frac{a\bar{z} + b}{c\bar{z} + d}\right)$$

$$= \frac{y^2}{(cz + d)^2 (c\bar{z} + d)^2} \frac{\partial^2}{\partial z' \partial \bar{z}'} f(z', \bar{z}') = y'^2 \frac{\partial^2}{4 \partial z' \partial \bar{z}'} f,$$

which is exactly the condition of invariance of $\tilde{\Delta}$ under isometries.

Now practically all constructions that we have used on the Euclidean plane can be translated to our constant negative curvature surface. As on the plane the motion on the whole upper half plane is not very interesting and one has to find an analogue of the torus. The torus was the result of the factorisation of a free motion by a discrete group of isometries. One can build a finite surface on the hyperbolic plane exactly in the same way, by factorising the upper half-plane by the action of discrete groups of (fractional) isometries. The first step in this direction is the construction of a discrete group. As was discussed above, the discreteness of a group means that there is a (finite) neighbourhood of every point of our space such that the results of the action of all group transformations on our point (except the identity) will be outside this neighbourhood. The images of a point cannot approach each other too closely. A crude analogy of this notion is the one-dimensional group of transformations of the unit circle into itself. Here the group of transformations consists of all transformations of the following form

$$z \to g(n)z, \quad \text{and} \quad g(n) = \exp(2\pi i \alpha n),$$

where α is a constant and $n = 0, \pm 1, \pm 2, \ldots$. If α is a rational number M/N, $g(n)$ can take only a finite number of values such that $(g(n))^N = 1$ and the corresponding group is obviously discrete. If instead α is an irrational number then the images of any point cover the whole circle uniformly and the group is not discrete. The important mathematical fact is that on the upper half plane there exists an infinite number of different discrete groups. This rich mathematical structure is one of the reasons of the strong mathematical interest to these problems. Instead of presenting the proof of this statement we consider an important example of discrete group.

Let us consider a group of 2×2 matrices with unit determinant but whose entries are integer. This group is called the modular group and it is one of the most investigated groups. Since all the elements are integers this group is evidently a discrete one. It is easy to show that this group is generated by two transformations: the translation $T : z \to z + 1$ and the inversion $S : z \to -1/z$. In the matrix representation T corresponds to

$$\begin{pmatrix} 1 & 1 \\ 0 & 1 \end{pmatrix}$$

and S to

$$\begin{pmatrix} 0 & 1 \\ -1 & 0 \end{pmatrix}.$$

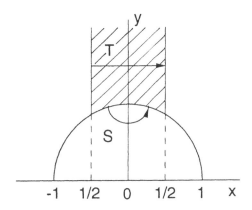

Figure 8. *Fundamental domain of the modular group.*

Following the construction of the fundamental domain for the torus model, one can perform analogous steps for the modular domain; it turns out that the fundamental domain for the modular group can be chosen as in Figure 8. The lines indicate the boundary of the fundamental domain identified by the action of the generators. Geometrically the fundamental domain of the modular group is the triangle with angles $(\pi/3, 0, \pi/3)$. Two sides of this triangle go to infinity performing a cusp. However, the area of this triangle is finite and equal to $\pi/3$ as it can be calculated from a general formula valid for any geodesics triangle with angles α, β γ

$$\text{Area} = \pi - \alpha - \beta - \gamma.$$

In general the fundamental domain of a discrete group has a shape of a polygon built by segments of geodesics. Group generators act as transformations which identify the corresponding sides of the polygon. Under certain consistency conditions this construction defines a discrete group of isometries. In particular, the area of a smooth compact surface generated by a discrete group should equal $4\pi(g - 1)$ where g is an integer called the genus of a surface. g is a topological characteristic of a surface and is equal to the number of handles. For the sphere $g = 0$, for the torus $g = 1$.

From now on we assume that we have a discrete group G with corresponding matrices $M \in SL(2, R)$

$$M = \begin{pmatrix} a & b \\ c & d \end{pmatrix},$$

a, b, c, and d are real and $ad - bc = 1$. Any such matrix can be characterised by point of the upper half-plane which does not move under the corresponding transformation. The equation of such fixed points read

$$z = \frac{az + b}{cz + d}$$

or

$$cz^2 + (d - a)z - b = 0. \tag{99}$$

Its solutions have the form

$$z = \frac{a - d \pm \sqrt{t^2 - 4}}{2c},$$

where $t \equiv \mathrm{Tr} M = a + d$ is the trace of the matrix M. Three different cases are possible:

- $|t| > 2$, these transformations and corresponding matrices are called hyperbolic.

- $|t| = 2$, these isometries are called parabolic.

- $|t| < 2$, the corresponding isometries are called elliptic.

If $|t| > 2$ Equation (99) has two real solutions which correspond to two fixed points on the real axis. There is one and only one geodesics connecting these two points. Its equation has the form

$$c(x^2 + y^2) + (d - a)x - b = 0. \tag{100}$$

It is possible to check that this geodesic is invariant under the transformation

$$z \to z' = \frac{az + b}{cz + d}$$

induced by matrix (95). This means that if one takes a point on this line then its image under this transformation will also belong to it.

If $|t| = 2$ there is only one real fixed point and in the case $|t| < 2$ one has two complex fixed points but only one of them belongs to the upper half-plane. The corresponding isometry is just a rotation around this point.

In the modular group all three types of matrices are present. Examples:

$$\text{hyperbolic:} \qquad M = \begin{pmatrix} 2 & 1 \\ 1 & 1 \end{pmatrix}$$

$$\text{parabolic:} \qquad M = \begin{pmatrix} 1 & 1 \\ 0 & 1 \end{pmatrix}$$

$$\text{elliptic:} \qquad M = \begin{pmatrix} 1 & -1 \\ 0 & 1 \end{pmatrix}$$

For a discrete group with a compact fundamental domain all group matrices have to be hyperbolic.

Each discrete group defines a surface of constant negative curvature by identifying points of the upper half-plane which are different under the action of group transformations. This means that all points of the form

$$z' \equiv g(z) = \frac{az + b}{cz + d} \tag{101}$$

for all $g \in G$ are considered as one point. The classical motion now is the motion with unit velocity on geodesics (*i.e.* semi-circles perpendicular to the real axis) inside the fundamental domain; when the trajectory hits a boundary of the fundamental domain it appears from the opposite side as prescribed by the boundary identification.

It is interesting to ask what are the classical periodic orbits of such motions. As for the torus model in the previous section, one can associate to each group matrix a periodic orbit as a geodesic which remains invariant under the corresponding transformation: the length of this periodic orbit should be computed as the hyperbolic distance between a point z whose coordinates obey Equation (100) and its image $z' = (az + b)/(cz + d)$. Using Equation (93) one concludes that

$$\cosh(l_p) = 1 + \frac{|z - z'|^2}{2yy'}.$$

But $y' = y/|cz + d|^2$ and

$$z - \frac{az + b}{cz + d} = \frac{c(x + iy)^2 - (d - a)(x + iy) - b}{(cz + d)^2} = y \frac{-2cy + i(d - a + 2cx)}{(cz + d)^2}.$$

Here we have used Equation (100). Therefore

$$
\begin{aligned}
\cosh(l_p) &= 1 + \frac{1}{2} |-2cy + i(d - a + 2cx)|^2 \\
&= 1 + \frac{1}{2} \left(4c^2 y^2 + (d - a)^2 + 4cx(d - a) + 4c^2 x^2 \right) \\
&= 1 + \frac{1}{2} \left(4c(c(x^2 + y^2) + (d - a)x) + (d - a)^2 \right) \\
&= 1 + \frac{1}{2} \left(4bc + (d - a)^2 \right) = \frac{(a + d)^2}{2} - 1.
\end{aligned}
$$

Note that the length of the periodic orbit depends only on the trace of the corresponding matrix. It is convenient to calculate not $\cosh(l_p)$ but $\cosh(l_p/2)$. The previous equation gives

$$2\cosh\left(\frac{l_p}{2}\right) = |\mathrm{Tr} M|. \tag{102}$$

To each group matrix one can associate a periodic orbit but each periodic orbit corresponds to infinite many matrices. The reason for this is the following. All points $g(z)$ where g is a transformation from our group should be considered as one point. Therefore if we have a periodic orbit which goes from point z to point Mz all trajectories connecting points $g(z)$ and $Mg(z)$ reduce to one orbit. This is equivalent to the condition

$$M' = S^{-1} M S, \tag{103}$$

where the matrix S belongs to our group and defines one and only one periodic orbit. These matrices form a the so-called class of conjugated matrices and periodic orbits of the classical motion are in one–to–one correspondence with classes of conjugated matrices. Of course, the length of periodic orbit does not change under the conjugation as it is evident from Equation (102) because $\mathrm{Tr}(AB) = \mathrm{Tr}(BA)$.

Quantum problems on a surface of negative curvature

Now we have all necessary ingredients to investigate quantum problems on constant negative curvature surfaces. As for the torus problem, the natural 'quantisation' consists of the problem of finding eigenvalues and eigenvectors of the invariant Laplace–Beltrami operator

$$-y^2 \left(\frac{\partial^2}{\partial x^2} + \frac{\partial^2}{\partial y^2} \right) \Psi_n(x,y) = E_n \Psi_n(x,y). \tag{104}$$

For the class of functions invariant under all transformations from a given discrete group G we have

$$\Psi_n(x',y') = \Psi_n(x,y), \tag{105}$$

where $z' = x' + iy'$ is connected to $z = x + iy$ by

$$z' = \frac{az+b}{cz+d}$$

for all transformations from the group G. In the mathematical literature such functions are called automorphic functions (automorphic with respect to a given group).

Later on we shall be interested in eigenfunctions of the Laplace-Beltrami operator (98) and its normalisation will be important. The correct orthogonal conditions for equations of the type (98) can be obtained by the following method. Consider two different eigenfunctions of Equation (98) with different eigenvalues

$$-\tilde{\Delta}\Psi_1 = E_1\Psi_1, \qquad -\tilde{\Delta}\Psi_2 = E_2\Psi_2, \tag{106}$$

multiply the first equation by $\bar{\Psi}_2$ and the complex conjugate of the second one by Ψ_1 and subtract them

$$\bar{\Psi}_2\tilde{\Delta}\Psi_1 - \tilde{\Delta}\bar{\Psi}_2\Psi_1 = (E_1 - E_2)\bar{\Psi}_2\Psi_1.$$

However,

$$\int_V (\bar{\Psi}_2\tilde{\Delta}\Psi_1 - \tilde{\Delta}\bar{\Psi}_2\Psi_1)\frac{dx\,dy}{y^2} = \int_V \nabla(\bar{\Psi}_2\nabla\Psi_1 - \nabla\bar{\Psi}_2\Psi_1)dx\,dy$$

$$= \int_B d\boldsymbol{\sigma} \cdot \mathbf{J},$$

where $\mathbf{J} = \bar{\Psi}_2\nabla\Psi_1 - \nabla\bar{\Psi}_2\Psi_1$ and the integration in the last integral is taken over the boundary of our region. We are interested on eigenvalues with periodic boundary conditions (105), *i.e.* we want the current $\mathbf{J}$ to take the same value on the opposite sides of the fundamental domain of our group identified by a group generator. Therefore

$$\int_{B_D} \mathbf{J} \cdot d\boldsymbol{\sigma} = 0$$

for any two eigenvalues of the 'quantum' problem (104), (105). From Equation (106) we obtain that the following orthogonality conditions can be imposed on the eigenfunctions

$$\int_D \bar{\Psi}_n(x)\Psi_m(x)d\mu = \delta_{nm}, \tag{107}$$

where the integration is taken over the fundamental domain of the discrete group and

$$d\mu = \frac{dx\,dy}{y^2}$$

is the invariant measure of the upper half–plane. The invariance of this measure means that it does not change under any fractional transformation. In order to derive the trace formula for these models we shall proceed exactly as in the case of the torus. The two main steps are the following. (1) We need to compute the exact Green function for the free motion on the whole upper half–plane. (2) We have to use the fact that the Green function automorphic with respect to a discrete group is a sum over all images of the source point under the group transformations.

The free Green function obeys, by definition, the equation

$$(\tilde{\Delta} + E)G_E^0(\mathbf{x}, \mathbf{x}') = \delta(\mathbf{x} - \mathbf{x}'). \tag{108}$$

The only difference with the plane case is that now $\tilde{\Delta}$ is not the usual Laplace operator but the Laplace-Beltrami operator

$$\tilde{\Delta} = y^2\left(\frac{\partial^2}{\partial x^2} + \frac{\partial^2}{\partial y^2}\right).$$

Similarly to the case on the plane, the free Green function $G_E^0(\mathbf{x}, \mathbf{x}')$ has to depend only on the (hyperbolic) distance between the points $\mathbf{x}$ and $\mathbf{x}'$. The latter is given by Equation (93). Therefore one can try to find the solution of Equation (108) in the form

$$G_E^0(\mathbf{x}, \mathbf{x}') = G(u)$$

where

$$u = \frac{(x - x')^2 + (y - y')^2}{yy'}.$$

One obtains

$$\frac{\partial G}{\partial x} = G'\frac{\partial u}{\partial x},$$

$$\frac{\partial^2 G}{\partial x^2} = G''\left(\frac{\partial u}{\partial x}\right)^2 + G'\frac{\partial^2 u}{\partial x^2},$$

and

$$\tilde{\Delta}G = y^2\left[\left(\left(\frac{\partial u}{\partial x}\right)^2 + \left(\frac{\partial u}{\partial y}\right)^2\right)G'' + \left(\frac{\partial^2 u}{\partial x^2} + \frac{\partial^2 u}{\partial y^2}\right)G'\right].$$

However,

$$\frac{\partial u}{\partial x} = 2\frac{x - x'}{yy'},$$

$$\frac{\partial^2 u}{\partial x^2} = \frac{2}{yy'},$$

$$\frac{\partial u}{\partial y} = 2\frac{y-y'}{yy'} - \frac{u}{y},$$

$$\frac{\partial^2 u}{\partial y^2} = \frac{2}{y^2} + \frac{2u}{y^2} - 2\frac{y-y'}{y^2 y'},$$

$$\left(\frac{\partial u}{\partial x}\right)^2 + \left(\frac{\partial u}{\partial x}\right)^2 = \frac{1}{y^2}(u^2 + 4u),$$

$$\frac{\partial^2 u}{\partial x^2} + \frac{\partial^2 u}{\partial y^2} = \frac{1}{y^2}(2u + 4).$$

Finally the equation for $G(u)$ takes the form:

$$(u^2 + 4u)G'' + (2u + 4)G' + EG = \delta(\mathbf{x} - \mathbf{x}').$$

If $\mathbf{x} \neq \mathbf{x}'$, $G(u)$ should obey the equation

$$(u^2 + 4u)G'' + (2u + 4)G' - l(l+1)G = 0, \tag{109}$$

where instead of E we have introduced the parameter l by using the relation

$$E = -l(l+1). \tag{110}$$

Standard notation (see below) is

$$E = \frac{1}{4} + k^2, \tag{111}$$

where k is called the momentum and

$$l = -\frac{1}{2} + ik.$$

It is convenient to change the variable u to $y = 1 + u/2$. Then $u = 2(y - 1)$ and the equation for $G(y)$ takes the form:

$$(1 - y^2)\frac{d^2 G}{dy^2} - 2y\frac{dG}{dy} + l(l+1)G = 0. \tag{112}$$

This is the equation of the Legendre function (see page 998 of [1]). As in the case of the torus the $i\epsilon$ prescription fixes the unique solution

$$G(y) = aQ_{-\frac{1}{2}+ik}(y), \tag{113}$$

where $Q_l(y)$ is the Legendre function of the second kind. When $l \to \infty$, Q_l has the following asymptotics (see page 1004 of [1])

$$Q_{-\frac{1}{2}+ik}(\cosh(d)) \to \sqrt{\frac{\pi}{2k\sinh d}} \exp\left(-i\left(kd + \frac{\pi}{4}\right)\right).$$

This function allows for the exact integral representation (see page 1002 of [1])

$$Q_{-\frac{1}{2}+ik}(\cosh(d)) = \frac{1}{\sqrt{2}}\int_d^\infty dr \frac{\exp(-ikr)}{\sqrt{\cosh(d) - \cosh(r)}}. \tag{114}$$

When $d \to 0$

$$Q_{-\frac{1}{2}+ik}(\cosh(d)) \to -\log d.$$

The asymptotics behaviour fixes the coefficient a in Equation (113) and finally one obtains that the free Green function of the Laplace-Beltrami operator has the following form

$$G_E^0(\mathbf{x}, \mathbf{x}') = -\frac{1}{2\pi} Q_{-\frac{1}{2}+ik}(\cosh(d(\mathbf{x}, \mathbf{x}'))), \tag{115}$$

where $d(\mathbf{x}, \mathbf{x}')$ is the hyperbolic distance between points $\mathbf{x}$ and $\mathbf{x}'$ given by

$$\cosh(d(\mathbf{x}, \mathbf{x}')) = 1 + \frac{|\mathbf{x} - \mathbf{x}'|^2}{2yy'}$$

and k is connected to the energy by the relation $E = 1/4 + k^2$.

As above, the exact Green function can be written as follows

$$G_E(\mathbf{x}, \mathbf{x}') = \sum_g G_E^0(\mathbf{x}, g(\mathbf{x}')), \tag{116}$$

where the summation is taken over all group transformations (or group matrices) and $g(z)$ is a fractional transformation

$$g(z) = \frac{az + b}{cz + d},$$

with the corresponding a, b, c, and d. The density of eigenvalues is related to this Green function by Equation (65)

$$d(E) = -\frac{1}{\pi} \int_D \text{Im}(G_E(\mathbf{x}, \mathbf{x})) d\mu. \tag{117}$$

Explicitly

$$d(E) = \frac{1}{2\pi^2} \sum_g \int_D Q_{-\frac{1}{2}+ik} \left(\cosh d(\mathbf{x}, g(\mathbf{x}))\right) \frac{dx\, dy}{y^2}. \tag{118}$$

Using the integral representation of the Legendre function (114) one obtains an explicit formula for $d(E)$

$$d(E) = \frac{1}{2\sqrt{2}\pi^2} \sum_g \int_D \left(\int_{d(z, g(z))}^{\infty} dr \frac{\sin kr}{\sqrt{\cosh d(z, g(z)) - \cosh r}} \right) \frac{dx\, dy}{y^2}. \tag{119}$$

Here the summation is performed over all group transformations or (which is the same) over all group matrices and D is the fundamental domain of a given discrete group.

Let us calculate first the contribution to the level density which we denoted by $\bar{d}(E)$ from the identity transformation. In this case $g(z) = z$ and $d(z, g(z)) = 0$. Then

$$\begin{aligned}
\bar{d}(E) &= \frac{1}{2\sqrt{2}\pi^2} \int_D \frac{dx\, dy}{y^2} \int_0^{\infty} dr \frac{\sin kr}{\sqrt{1 - \cosh(r)}} \\
&= \frac{\mu(D)}{(2\pi)^2} \int_0^{\infty} \frac{\sin kr}{\sinh(r/2)} dr,
\end{aligned}$$

where

$$\mu(D) = \int_D \frac{dx\,dy}{y^2}$$

is the (hyperbolic) area of the fundamental domain. This last integral can be easily computed (see page 344 in [1])

$$\int_0^\infty \frac{\sin kr}{\sinh(r/2)} dr = \pi \tanh(\pi k)$$

and the final expression for the smooth part of the level density takes the form

$$\bar{d}(E) = \frac{\mu(D)}{4\pi} \tanh(\pi k). \tag{120}$$

When $k \to \infty$, expression (120) tends to $\mu(D)/4\pi$ in exactly the same way as the corresponding term for the torus case.

The most difficult (and important) step is the calculation of the contribution from the non–trivial fractional transformations. Let us divide all group matrices into classes of conjugated elements. This means that we consider a matrix g and all matrices of the form

$$g' = SgS^{-1}, \tag{121}$$

where S is any matrix of the group, as belonging to one class. Note that two classes either have no common elements or they coincide. This follows from the fact that if

$$S_1 g_1 S_1^{-1} = S_2 g_2 S_2^{-1}$$

then

$$g_2 = S_3 g_1 S_3^{-1}$$

where $S_3 = S_1^{-1} S_2$. However, if S_1 and S_2 belong to G, S_3 also is matrix from G and g_2 belongs to the same class as g_1. Therefore all group matrices can be split into classes of mutually non–conjugated elements. One can replace the sum over all group matrices in Equation (116) by the sum over classes of conjugated matrices. Let g be a representative element of one class. The contribution of this class to the level density can be written as follows

$$\sum_S \int_D f(d(z, SgS^{-1}(z)) d\mu \tag{122}$$

where the integration is taken over the fundamental domain of our group and the summation is done over all group matrices S provided there is no double counting in the representation (121). This means that the set of matrices S should not contain matrices such that

$$S_1 g S_1^{-1} = S_2 g S_2^{-1}$$

From this relation, it follows that the matrix $S_3 = S^{-1} S_2$ commutes with g

$$S_3 g = g S_3.$$

Let us denote the set of matrices commuting with a given matrix g by S_g. This set, evidently, forms a sub–group of our initial group G because if s_1 and s_2 commute with g, the same happens to their product. In order to ensure the unique decomposition of group matrices into non–overlapping classes of conjugated elements (121) the summation in Equation (122) has to be performed over a group of matrices S where only one matrix can be represented as

$$S_2 = sS_1, \tag{123}$$

where s belongs to S_g. This is equivalent to the statement that one sums over all group matrices but matrices connected by Equation (123) are considered as one matrix only. It is similar to the factorisation of the plane by the action of a group and mathematically such set of matrices is denoted by G/S_g.

The distance between two points remains invariant under any fractional transformation of both arguments

$$d(z, z') = d(S(z), S(z')),$$

and

$$d(z, g(z)) = d(S(z), Sg(z)) = d(y, SgS^{-1}(y)),$$

where $y = S(z)$. These relations give

$$\int_D f(d(y, SgS^{-1}(y)))d\mu = \int_{S^{-1}(D)} f(d(z, g(z)))d\mu,$$

where the last integral is taken over the image of the fundamental domain under the transformation S^{-1}. Therefore

$$\sum_S \int_D f(d(z, SgS^{-1}(z)))d\mu = \sum_S \int_{S^{-1}(D)} f(d(z, g(z)))d\mu.$$

For different S, images $S^{-1}(D)$ are different and do not overlap. The integrand does not depend on S and

$$\sum_S \int_D f(d(z, SgS^{-1}(z)))d\mu = \int_{D_g} f(d(z, g(z)))d\mu,$$

where

$$D_g = \sum_S S^{-1}(D). \tag{124}$$

The sum of images $S^{-1}(D)$ over all S is, evidently, the whole upper half–plane. However, we do not need to sum over all S but only over S factorised by the action of the group of matrices commuting with a fixed matrix g. D_g will thus be a smaller region. To find it we have to investigate the structure of matrices commuting with g. Since any matrix g can be written as a positive power of a certain primitive element g_0 belonging to our group G

$$g = g_0^n$$

it is almost evident that matrices which commute with a non–degenerate matrix g are precisely the group of matrices generated by g_0. This is a cyclic abelian group consisting of all (positive, negative or zero) powers of g_0

$$S_g = g_0^m, \qquad m = 0, \pm 1, \pm 2, \ldots \qquad (125)$$

Like any discrete group, this cyclic abelian group has its fundamental domain FD_g.

The last step is to prove that D_g defined in Equation (124) is exactly the fundamental domain. This is doen as follows. Firstly, the sum over all $S^{-1}(D)$ without the factorisation over S_g equals the whole upper half-plane. Secondly, no two points in D_g are connected by a transformation from S_g because if $x = sys^{-1}$ for a certain s from S_g and x, y belong to D_g, then there are two transformations S_1, S_2 from G/S_g such that $x = S_1 x'$, $y = S_2 y'$ where points x', y' are from the fundamental domain of the group G. But then $x' = S_1^{-1} s S_2 y'$. The matrix $S_1^{-1} s S_2$ belongs to G and by definition of the fundamental domain D this equality is possible only if $S_1^{-1} s S_2 = 1$ and $S_1 = s S_2$ which contradicts the definition of G/S_g.

These considerations prove that the contribution of one class of conjugated matrices can be written in the form

$$\sum_{G/S_g} \int_D f(d(z, SzS^{-1}(z))) d\mu = \int_{FD_g} f(d(z, g(z))), \qquad (126)$$

where in the left hand side the integration is taken over the fundamental domain of the whole group G and the summation is done over matrices from G factorised by the subgroup of matrices S_g which commutes with a fixed matrix g. In the right hand side there is no summation but the integration is taken over the fundamental domain of the subgroup S_g. This formula allows for the explicit calculation of the contribution of one class of conjugated matrices in the trace formula (119). Using Equations (119) and (126)

$$d(E) = \bar{d}(E) + \sum_g d_g(E), \qquad (127)$$

where

$$d_g(E) = \frac{1}{2\sqrt{2}\pi^2} \int_{FD_g} d\mu \int_{d(z,g(z))}^{\infty} dr \frac{\sin kr}{\sqrt{\cosh(d(z, g(z))) - \cosh(r)}}, \qquad (128)$$

and the summation is performed over all classes of conjugated matrices. Here $d(z, g(z))$ is the distance between point z and $g(z)$ and g is any representative element of a conjugated class.

Let g be an hyperbolic matrix $g = g_0^n$ with g_0 belonging to G. One can transform g_0 to a diagonal form by a suitable matrix B

$$B g_0 B^{-1} = \begin{pmatrix} \lambda_0 & 0 \\ 0 & \lambda_0^{-1} \end{pmatrix}, \qquad (129)$$

where for hyperbolic matrices λ_0 is real and $\lambda_0 > 1$. By using the same transformation matrix, g can be transformed to

$$B g B^{-1} = \begin{pmatrix} \lambda & 0 \\ 0 & \lambda^{-1} \end{pmatrix}, \qquad (130)$$

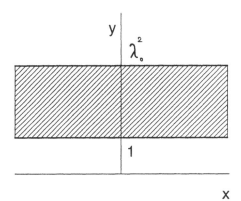

Figure 9. *Fundamental domain of the group of multiplication.*

where $\lambda = \lambda_0^n$. Of course, matrix B in general does not belong to our group G. However it is evident that Equation (128) remains valid for any form of the matrix g so that one can use g in the diagonal form (130). In this case the transformation

$$g(z) = \lambda^2 z \tag{131}$$

is just the multiplication and

$$\cosh(d(z, g(z))) = 1 + \frac{(\lambda^2 - 1)^2}{2\lambda^2} \left(\frac{x^2 + y^2}{y^2} \right).$$

The group of matrices which commute with g is the group generated by a multiplication

$$S_g : (\lambda_0^{2m}, \quad m = 0, \pm 1, \pm 2, \ldots).$$

It is quite evident that its fundamental domain can be chosen as displayed in Figure 9. Since λ_0 is real, the transformation $z' = \lambda_0^2 z$ gives $y' = \lambda_0^2 y$ and the interval $1 < y < \lambda_0^2$ is the fundamental domain.

Now Equation (128) can be rewritten as

$$d_g(E) = \int_{-\infty}^{+\infty} dx \int_1^{\lambda_0^2} dy\, F \left(\frac{(\lambda^2 - 1)^2 (x^2 + y^2)}{\lambda^2 y^2} \right).$$

Introducing the new variable $\zeta = xy$ one obtains

$$d_g(E) = \int_1^{\lambda_0^2} \frac{dy}{y} \int_{-\infty}^{+\infty} d\zeta\, F \left(\frac{(\lambda^2 - 1)^2 (1 + \zeta^2)}{\lambda^2} \right) = \log \lambda_0^2 \int_{-\infty}^{+\infty} d\zeta\, F \left(\frac{(\lambda^2 - 1)^2 (1 + \zeta^2)}{\lambda^2} \right).$$

Denoting

$$u = \frac{(\lambda^2 - 1)^2}{\lambda^2} (1 + \zeta^2)$$

one obtains

$$d_g(E) = \frac{\log \lambda_0^2}{\sqrt{u_0}} \int_{u_0}^{\infty} du \frac{F(u)}{\sqrt{u - u_0}},$$

where

$$u_0 = \frac{(\lambda^2 - 1)^2}{\lambda^2} = \lambda + \frac{1}{\lambda} - 2. \tag{132}$$

u is related to the distance d by

$$\cosh(d) = 1 + u/2$$

while from Equation (128)

$$F(d) = \frac{1}{2\sqrt{2}\pi^2} \int_d^{\infty} \frac{\sin kr}{\sqrt{\cosh(d) - \cosh(r)}} dr.$$

We introduce a new variable τ which is related to r as u is related with d

$$\cosh(r(\tau)) = 1 + \frac{\tau}{2}, \qquad \frac{dr}{d\tau} = \frac{1}{\sqrt{\tau^2 + 4\tau}}. \tag{133}$$

We obtain

$$F(u) = \frac{1}{2\pi^2} \int_u^{\infty} \frac{\sin kr(\tau)}{\sqrt{(\tau - u)(\tau^2 + 4\tau)}} d\tau,$$

and

$$d_g(E) = \frac{\log \lambda_0^2}{2\pi^2 \sqrt{u_0}} f(u_0),$$

where

$$f(w) = \int_w^{\infty} \frac{du}{\sqrt{u - w}} \int_u^{\infty} \frac{\sin kr(\tau)}{\sqrt{(\tau - u)(\tau^2 + 4\tau)}} d\tau.$$

Changing the order of integrations one obtains

$$f(w) = \int_w^{\infty} \frac{\sin kr(\tau) d\tau}{\sqrt{\tau^2 + 4\tau}} \int_w^{\tau} \frac{du}{\sqrt{(u - w)(\tau - u)}}.$$

Note that the last integral is half of the contribution from the pole at infinity

$$\int_w^{\tau} \frac{du}{\sqrt{(u - w)(\tau - u)}} = \pi,$$

and so one concludes that

$$f(w) = \pi \int_w^{\infty} \frac{\sin(kr(\tau)) d\tau}{\sqrt{\tau^2 + 4\tau}} = \pi \int_{l_p}^{\infty} \sin(kr) dr = \frac{\pi}{k} \cos(kl_p),$$

provided that one adds a small imaginary part to k. Here l_p is the minimal value of $r(\tau)$ (see (133)) corresponding to minimal value of $\tau = u_0$ where u_0 is defined in Equation (132)

$$\cosh(l_p) = 1 + \frac{1}{2}(\lambda + \frac{1}{\lambda} - 2) = \frac{1}{2}(\lambda + \frac{1}{\lambda}).$$

In other words

$$2\cosh(l_p) = \text{Tr}g.$$

But this is exactly the formula for the length of a periodic orbit associated with the class of conjugated matrices with a representative element g.

These calculations show that the contribution to the level density of one class of conjugated matrices is equal to

$$d_g(E) = \frac{\log \lambda_0^2}{2\pi k \sqrt{\lambda + \lambda^{-1} - 2}} \cos k l_p = \frac{l_{pp}}{4\pi k \sinh(l_p/2)} \cos(k l_p), \tag{134}$$

where l_p is the length of the periodic orbit and l_{pp} is the length of a primitive periodic orbit $l_p = n l_{pp}$ where n is integer.

The full trace formula is the sum over all conjugated classes or, equivalently, the sum over all periodic orbits, each contribution being given by Equation (134). Denoting the length of primitive periodic orbit by l_{pp} and explicitly summing over all occurrencies one obtains

$$d(E) = \frac{\mu(D)}{4\pi} \tanh(\pi k) + \sum_{pp} \frac{l_{pp}}{4\pi k} \sum_{n=1}^{\infty} \frac{\cos(k l_{pp} n)}{\sinh(l_{pp} n/2)}, \tag{135}$$

where the first sum is taken over all primitive periodic orbits ('pp') while the second one is a summation over all occurrencies of one primitive trajectory. k is the momentum connected with the energy by $E = 1/4 + k^2$. To avoid problems with convergence one should multiply both sides of this equation by a suitable even function $h(k)$ and integrate over E. (Note that the function $h(k)$ is any analytical function in a region $|\text{Im}(k)| < 1/2 + \delta$ such that $h(-k) = h(k)$ and $|h(k)| < A(1 + |k|)^{-2-\delta}$ for certain A and $\delta > 0$).

For the left hand side of (135) one obtains

$$\int d(E) h(k) dE = \int \sum_n \delta(E - E_n) h(k) dE = \sum_n \int \delta(k_n^2 - k^2) 2k dk = \sum_n h(k_n).$$

For the right hand side one instead obtains

$$\int h(k) \frac{\cos kl}{4\pi k} 2k dk = \frac{1}{2\pi} \int_{-\infty}^{+\infty} h(k) e^{-ikl} dk.$$

Finally if we define $g(l)$ to be the Fourier transform of $h(k)$

$$g(l) = \frac{1}{2} \int_{-\infty}^{+\infty} h(k) e^{-ilk} dk,$$

we obtain the famous Selberg trace formula:

$$\sum_n h(k_n) = \frac{\mu(D)}{2\pi} \int_{-\infty}^{+\infty} h(k)k\tanh(\pi k)dk + \sum_{\text{pp}} l_{\text{pp}} \sum_{n=1}^{\infty} \frac{g(l_{\text{pp}}n)}{2\sinh(l_p n/2)}. \tag{136}$$

This equation relates the eigenvalues of the Laplace–Beltrami operator for functions automorphic with respect to a discrete group (and with compact fundamental domain) to the classical periodic orbits. From a theoretical point of view this formula is very important. It is the only *exact* trace formula for strongly chaotic systems. More general Gutzwiller trace formulae [3] are only asymptotic formulae valid in the limit $E \to \infty$, only. The Selberg trace formula is based on a solid theoretical ground where different hypotheses can be checked mathematically. We often try to extract from (136) results which (hopefully) can be generalised to more generic systems.

One of the most important applications of this trace formula is the introduction of the Selberg Zeta function. Let us choose the function

$$h(k) = \frac{1}{k^2 + \alpha^2}$$

in Equation (136). Then

$$g(l) = \frac{1}{2\pi} \int_{-\infty}^{+\infty} \frac{\exp(-ikl)}{k^2 + \alpha^2} dk = \frac{1}{2\alpha} e^{-\alpha|l|}.$$

One cannot use this $h(k)$ in the trace formula because it does not decrease fast enough to ensure the convergence. To avoid this problem let

$$h(k) = \frac{1}{k^2 + \alpha^2} - \frac{1}{k^2 + \beta^2}.$$

Then

$$g(l) = \frac{1}{2\alpha} e^{-\alpha|l|} - \frac{1}{2\beta} e^{-\beta|l|},$$

and the Selberg trace formula gives

$$\sum_n \left(\frac{1}{k_n^2 + \alpha^2} - \frac{1}{k_n^2 + \beta^2} \right) = \frac{\mu(D)}{4\pi} \int_{-\infty}^{+\infty} k\tanh(\pi k)\left(\frac{1}{k^2 + \alpha^2} - \frac{1}{k^2 + \beta^2} \right) \tag{137}$$
$$+ \sum_{\text{pp}} \sum_{n=1}^{\infty} \frac{l_{\text{pp}}}{z} 2\sinh(l_{\text{pp}}n/2)\left(\frac{1}{2\alpha} e^{-\alpha l_{\text{pp}}} - \frac{1}{2\beta} e^{-\beta l_{\text{pp}}} \right).$$

Let us introduce formally the following function which is called the Selberg Zeta function

$$Z(s) = \prod_{\text{pp}} \prod_{m=0}^{\infty} (1 - e^{-l_{\text{pp}}(s+m)}), \tag{138}$$

One obtains

$$
\frac{1}{Z}\frac{dZ}{ds} = \sum_{pp}\sum_{m=0}^{\infty}\frac{l_{pp}e^{-l_{pp}(s+m)}}{1-e^{-l_{pp}(s+m)}}
$$

$$
= \sum_{pp}l_{pp}\sum_{n=1}^{\infty}\sum_{m=0}^{\infty}e^{-l_{pp}(s+m)n} = \sum_{pp}l_{pp}\sum_{n=1}^{\infty}\frac{e^{-l_{pp}sn}}{1-e^{-nl_{pp}}}
$$

$$
= \sum_{pp}l_{pp}\sum_{n=1}^{\infty}\frac{1}{2\sinh(l_{pp}n/2)}e^{-l_{pp}n(s-1/2)}.
$$

By substituting $\alpha = s - 1/2$ and $\beta = s' - 1/2$ in Equation (138) one obtains

$$
\sum_{n}\left(\frac{1}{k_n^2+(s-1/2)^2} - \frac{1}{k_n^2+(s'-1/2)^2}\right) =
$$

$$
\int_{-\infty}^{+\infty}k\tanh(\pi k)\left(\frac{1}{k^2+(s-1/2)^2} - \frac{1}{k^2+(s'-1/2)^2}\right)dk
$$

$$
+ \frac{1}{2s-1}\frac{Z'(s)}{Z(s)} - \frac{1}{2s'-1}\frac{Z'(s')}{Z(s')}. \tag{139}
$$

The integral in the right hand side can be computed by taking into account all pole terms in the upper half-plane of the complex variable k (the contribution from the pole at infinity cancels in the difference)

$$
\int_{-\infty}^{+\infty}k\tanh(\pi k)\left(\frac{1}{k^2+(s-1/2)^2} - \frac{1}{k^2+(s'-1/2)^2}\right)dk = f(s) - f(s'), \tag{140}
$$

where $f(s)$ is the sum of the residues from the pole at $k = i(s-1/2)$ coming from the pole term in Equation (140) and from poles at $k = i(n+1/2)(n = 0,1,2,\ldots)$, related to the poles of $\tanh(\pi k)$. It leads to

$$
f(s) = 2\pi i\left(\frac{1}{2}\tanh\left(i\pi\left(s-\frac{1}{2}\right)\right) + \frac{1}{\pi}\sum_{n=0}^{\infty}\frac{i(n+1/2)}{(s-1/2)^2-(n+1/2)^2}\right)
$$

$$
= \pi\cot(\pi s) - \sum_{n=1}^{\infty}\frac{1}{s-n} + \sum_{n=0}^{\infty}\frac{1}{s+n}.
$$

However,

$$
\cot(\pi s) = \frac{1}{\pi s} + \frac{2s}{\pi}\sum_{n=1}^{\infty}\frac{1}{s^2-n^2} = \frac{1}{\pi}\left(\sum_{n=1}^{\infty}\frac{1}{s-n} + \sum_{n=0}^{\infty}\frac{1}{s+n}\right),
$$

and finally

$$
f(s) = 2\sum_{n=0}^{\infty}\frac{1}{s+n}.
$$

Using these relations one obtains

$$
\frac{1}{2s-1}\frac{Z'(s)}{Z(s)} = \frac{1}{2s'-1}\frac{Z'(s')}{Z(s')} - \frac{\mu(D)}{2\pi}\left(\sum_{n=0}^{\infty}\frac{1}{s+n} - \sum_{n=0}^{\infty}\frac{1}{s'+n}\right)
$$

$$
+ \sum_{n}\left(\frac{1}{k_n^2+(s-1/2)^2} - \frac{1}{k_n^2+(s'-1/2)^2}\right). \tag{141}
$$

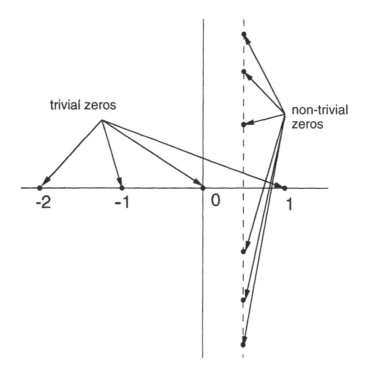

Figure 10. *Zeros of the Selberg Zeta function.*

The right hand side has poles at special values of $s = s_k$. Therefore the same poles should be present in the left hand side as well. This means that the Selberg Zeta function should have a zero (or a pole) in $s = s_k$. If

$$Z(s) \to (s - s_k)^{\nu_k} \quad \text{as } s \to s_k,$$

then

$$\frac{Z'(s)}{Z(s)} \to \frac{\nu_k}{s - s_k},$$

where ν_k is the multiplicity of the zero (or of the pole if $\nu_k < 0$). Combining all pole terms one concludes that the Selberg Zeta function has two different sets of zeros. The first set consists of zeros $s = 1/2 \pm i k_n$, called non–trivial zeros, which come from the eigenvalues of the Laplace–Beltrami operator. The second set comes from the $E = 0$ eigenvalue and from the contribution of the smooth term. These zeros are called trivial zeros and they are located at points $s = -m$ $(m = 1, 2, \ldots)$ with multiplicity $\nu_m = (2m + 1)\mu(D)/2\pi$, at point $s = 0$ with multiplicity $\nu_0 = \mu(D)/2\pi + 1$, and a simple zero at $s = 1$. These multiplicities are, in fact, integers if one remembers that the area of the compact fundamental domain is $\mu(D) = 4\pi(g - 1)$ where g is the genus of our surface.

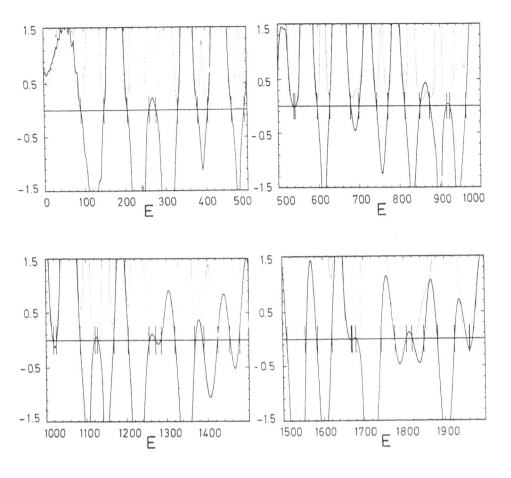

Figure 11. *Selberg Zeta function for the modular billiard.*

Schematically the structure of the zeros of the Selberg Zeta function is represented in Figure 10 where we have assumed that there are no small eigenvalues with $0 < E < 1/4$. The most important property of the Selberg Zeta function is that it has zeros at $s = 1/2 \pm ik$, *i.e.* exactly at the eigenvalues of our quantum problem. It is this property that makes Zeta functions the best means of computation of quantum properties from classical mechanics.

Let us set $s' = 1-s$ in Equation (139). In the left hand side the sum over eigenvalues cancels while $f(s) - f(1-s) = 2\pi \cot(\pi s)$. We then conclude that

$$\frac{1}{2s-1}\left(\frac{Z'(s)}{Z(s)} + \frac{Z'(1-s)}{Z(1-s)}\right) = -\frac{\mu(D)}{2}\cot(\pi s).$$

Setting

$$Z(s) = \phi(s)Z(1-s),\tag{142}$$

we obtain

$$\frac{\phi'(s)}{\phi(s)} = -\mu(D)\left(s - \frac{1}{2}\right)\cot(\pi s)$$

with $\phi(1/2) = 1$. This gives

$$\phi(s) = \exp\left(\mu(D)\int_0^{s-1/2} u\tan(\pi u)\,du\right).\tag{143}$$

This equation is called the functional equation for the Selberg Zeta function. It defines the analytical continuation of the Selberg Zeta function in the whole complex plane.

The Selberg trace formula and the Selberg Zeta function are the only known exact relations between quantum problems on constant negative curvature surfaces and quantities calculated from classical mechanics. They are used (explicitly or implicitly) in practically all theoretical works on quantum chaos problems. We shall not even try to discuss them but only present in Figure 11, as an illustration, a few pictures of the Selberg Zeta function taken from Reference [4] and computed by using a special method of regularisation based on all classical primitive periodic orbits up to the length 12 and their occurrencies. The evaluation of the Selberg Zeta function was not done for the kind of problems discussed here but for a billiard with Dirichlet boundary conditions on the triangle $(\pi/2, \pi/3, 0)$, *i.e.* half of the fundamental domain of the modular group. Ticks on these pictures show the position of the exact energy levels computed by a direct numerical method. One clearly sees how well the zeros of the Zeta function reproduce the energy levels.

Bibliography

These lectures are only an introduction to a very broad subject. We provide here a short list of references containing detailed material relevant to the topics discussed above.

- Reference [5] is an immediate continuation of these lectures. It is strongly recommended for readers interested in this subject.

- Reference [6] contains a good collection of papers on different aspects of quantum chaos problems.

- Reference [7] gives a simple discussion of different problems related to models on surfaces of constant negative curvature.

- For a mathematical introduction to the Selberg trace formula and the Selberg Zeta function see Reference [8]; detailed calculations of different types of Selberg trace formulae can be found in References [9, 10]

References

[1] I S Gradshtein and I M Pyzhik I M, *Table of Integrals, Series, and Products*, Fourth Edition (Academic Press, New York and London, 1965)

[2] M V Berry and M Tabor, *Proc R Soc Lond A*, **349**, 101, (1976)

[3] M C Gutzwiller, *Chaos in Classical Mechanics* (Springer–Verlag, New York, 1990)

[4] E B Bogomolny and C Schmit, *Nonlinearity*, **6**, 523 (1993)

[5] N L Balazs and A Voros, *Phys Rep* , **143**, 109 (1986)

[6] Les Houches Summer School, 1989, *Chaos and Quantum Physics*, edited by M J Giannoni, A Voros, and J Zinn-Justin (Elsevier, New York, 1991)

[7] A Terras, *Harmonic Analysis on Symmetric Spaces and Applications* (Springer, 1979)

[8] D Hejhal, *Duke Math J* , **43**, 441 (1976)

[9] D Hejhal, *The Selberg Trace Formula Part 1, Lecture Notes in Mathematics*, **548**, (Springer, Berlin, 1976)

[10] D Hejhal, *The Selberg Trace Formula Part 2, Lecture Notes in Mathematics*, **1001**, (Springer, Berlin, 1983)

Cavity Quantum Electrodynamics

M Brune and S Haroche

Laboratoire Kastler-Brossel
Paris

Cavity QED is the domain of quantum optics which studies the behaviour of single atoms and photons coupled together in an electromagnetic resonator. During the last ten years, various spectacular effects have been observed on such systems, either on the atom's or on the field's properties. The presence of a cavity around an atom can produce an alteration of the atomic radiative rates which has been observed in several experiments (spontaneous emission enhancement and inhibition [1]). Atomic energy level shifts due to the coupling with the cavity have also been studied [2]. Symmetrically, photons can be manipulated via their interaction with individual atoms crossing the cavity. Non-classical field states can be generated in this way in micromaser [3] or microlaser [4] systems. Experiments in which single atom index effects are observable can be performed. Due to recent progresses in the technology of high-Q cavities and in atomic beam manipulation, photons could be continuously observed in a cavity and counted non-destructively, in a way quite similar to the counting and manipulation of material particles [5, 6]. New 'Einstein-Podolsky-Rosen' (EPR) situations [7], involving atoms correlated at macroscopic distances via their interaction with a cavity field could also be studied [8]. Mesoscopic field coherences could be generated in high Q cavities [6, 9], which would display some of the properties discussed by Schrödinger in his famous cat paradox [10]. For example, non-local superpositions of fields occupying simultaneously two cavities could be prepared and detected [9] with possible applications to particle teleportation [11, 12], quantum cryptography or quantum computation [13, 14]. All these experiments would constitute stringent tests of quantum measurement theory.

The aim of these lectures is to give an overview of the physical principles involved in cavity QED as well as a description of the experiments presently carried out at ENS in this field. Since many review articles have already been devoted to this domain of quantum optics, we restrict the present analysis to a qualitative physical discussion and refer the reader for more details to other tutorial references [15, 16, 17, 18].

1 The basic cavity QED system: a two-level atom coupled to a single field mode

Ideally, cavity QED deals with two-level atoms coupled resonantly or nearly resonantly to a single cavity mode. This is of course a simplification of real situations, in which atoms have many levels and the cavity sustains more than one mode. This situation can however be nearly ideally realised by coupling very excited atoms to a high Q superconducting cavity with a mode matching or nearly matching a transition between two adjacent Rydberg levels $|e\rangle$ and $|g\rangle$. The strong resonant coupling of the atomic dipole to this cavity mode overcomes the coupling to all the other non resonant field modes or atomic transitions, which can be disregarded to a good approximation. This approximation becomes excellent if the atom is prepared in a special kind of Rydberg state, called 'circular' [19, 20, 21], in which the atom's angular momentum takes its maximum value. In a circular Rydberg state, spontaneous decay of the atom to more bound states is negligible and the two-level model becomes quite accurate. We restrict our discussion to Rydberg atom-superconducting cavity systems. Other cavity QED situations, involving optical cavities of various kinds and transitions between strongly bound electronic states [22, 23], are also being investigated in various laboratories.

The Hamiltonian $H = H_{at} + H_{cav} + V$ of the Rydberg atom-cavity system in its simplest version (Jaynes Cummings model [24]), can be expressed as the sum of three terms representing respectively the two-level atom Hamiltonian H_{at}, the free cavity mode Hamiltonian H_{cav} and the atom field coupling V, which can be written as:

$$H_{at} = \frac{\hbar\omega_0}{2}\left[b^\dagger b - bb^\dagger\right] \tag{1}$$

$$H_{cav} = \frac{\hbar\omega}{2}\left[a^\dagger a + aa^\dagger\right] \tag{2}$$

$$V = \hbar\Omega\left[ab^\dagger + a^\dagger b\right] \tag{3}$$

In these expressions, ω and ω_0 are the field and atomic transition angular frequencies, a and $a^\dagger$ the photon annihilation and creation operators in the field mode , $b = |g\rangle\langle e|$ and $b^\dagger = |e\rangle\langle g|$ are the atomic lowering and raising operators, and 2Ω is the rate at which the atom and the empty cavity exchange a single photon at exact resonance (vacuum Rabi frequency [15]).

The frequencies ω_0 and ω (typically in the ten to hundred GHz range) may differ by a small detuning δ (equal to zero at exact resonance, but in any case very small compared to ω; δ is typically in the kHz or MHz range). The Rabi frequency 2Ω, typically of the order of 10 to 100kHz, is very small compared to ω and ω_0. When it is also small compared to δ, the atom-cavity coupling is said to be non-resonant. Finally the atomic dipole and the cavity field relax with rates $\kappa_{at} = 1/t_{at}$ and $\kappa_{cav} = 1/t_{cav}$ where t_{at} and t_{cav} are the atom and field relaxation times respectively. Typically t_{at} falls in the 10^{-2} s range for circular Rydberg atoms with $n = 50$ [20]. The cavity mode damping rate κ_{cav} is equal to ω/Q where Q is the mode quality factor. Superconducting cavities at subKelvin temperatures may have Q factors in the 10^8 to 10^{10} range, corresponding to t_{cav} of the order of 10^{-1} to 10^{-3} s. The strong coupling regime in Cavity QED is defined by the conditions $\Omega \gg \kappa_{at}, \kappa_{cav}$, which are largely satisfied in circular Rydberg atom-superconducting cavity systems. These conditions mean that the atom-field coupling can give rise to appreciable effects within a time scale shorter than the atom and field relaxation times, an obvious requirement for the observation of atom-field coherent effects. We will assume in the following that relaxation rates can be in first approximation neglected for a qualitative description of the atom-field system evolution.

2   Resonant cavity QED effects

The system of equations (1–3) describes formally the coupling of a fermionic system (the two-level atom with b and $b^\dagger$ obeying anticommutation rules) and a bosonic system (the field whose a and $a^\dagger$ operators obey canonic harmonic oscillator commutation rules). For weak excitations, it is possible to neglect the difference between the boson and the fermion character, so that the atom and the field behave quite like two coupled ordinary oscillators, exchanging their excitation at resonance at rate 2Ω. For this reason, Cavity QED bears a strong analogy with the physics of coupled pendulums [15].

Many experiments can be envisaged with such a system. We will focus in this section on the resonant case and describe situations where the atom and the field can exchange energy. We will first consider experiments involving the coupling of a single atom to the cavity mode, then analyse situations in which a train of atoms interact successively with the cavity field. We will finally consider situations where several cavities are coupled to the same beam of atoms. The point of view adopted here is theoretical. We will describe ideal systems, in which relaxation is completely negligible and the atomic detection efficiency is perfect. Such idealisations correspond to 'gedanken experiments'.

2.1   A single atom in a cavity and the EPR paradox

Let us first consider the simple and ideal situation sketched in Figure 1. An atom in the upper level $|e\rangle$, prepared in 'circularisation box' CB, is introduced at time $t = 0$ in a resonant and initially empty cavity C. The system evolves as the linear superposition

$$|\Psi\rangle = \cos \Omega t |e, 0\rangle - i \sin \Omega t |g, 1\rangle \tag{4}$$

where, in each ket, the first symbol refers to the atom's state and the second to the photon number. The amplitudes associated to the upper and lower atomic states oscillate as a function of time, revealing the 'vacuum Rabi oscillation' effect. This oscillation is quite analogous to the reversible energy exchange between two identical coupled oscillators. The photon number (0 or 1) is strongly correlated to the atom's state (e or g). If the atom leaves the cavity at some time t, the system evolution is of course frozen at that time, and, provided that the atom and field relaxations can be neglected, the entanglement between field and matter can persist and lead to non-local correlations.

In this kind of Cavity QED experiment, all the information is provided by measuring the Rydberg atom's energy by selective field ionisation, after the atom has left the cavity [15]. The atom enters a detector D in which a variable electric field is applied. The thresholds for ionisation of levels e and g are reached at two different places inside

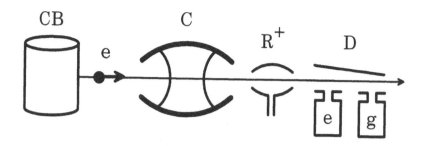

Figure 1. *Scheme of Rydberg atom resonant Cavity* QED *setup. The atom, prepared in level e in box CB crosses the cavity C before reaching the detector D. A microwave pulse applied between C and D in zone R_+ mixes levels e and g, making it possible to detect linear superpositions of these two states.*

the detector, leading to a spatially resolved ionisation process. From the position at which the resulting ion or electron is detected, one can infer the energy state (e or g) of the atom. When the atom's energy is measured, the field inside the cavity collapses into a Fock state (0 or 1 photon).

By a simple modification of the design, one should be able to detect a non diagonal atomic observable, with eigenstates of the form $c_e|e\rangle + c_g|g\rangle$. This can be achieved by mixing between the cavity and the detector the two levels e and g with the help of an auxiliary microwave pulse applied in zone R_+. The effect of this pulse amounts to performing a rotation on the pseudospin equivalent to the two-level atom system and the combination of R_+ and D is equivalent to a non diagonal observable detector. If the eigenstate $c_e|e\rangle + c_g|g\rangle$ is detected, the field in the cavity collapses into a linear superposition of Fock states, $a|0\rangle + b|1\rangle$, which can be easily deduced from equation 4.

We recognise here the ingredients of the Einstein-Podolsky-Rosen paradox [7]. The atom and the field in the cavity get entangled by their interaction. This entanglement survives the system separation. One subsystem (here the field) collapses into a state which depends upon the result of the measurement performed on the other part (here the atom), even if these two parts are far apart from each other when this measurement is performed. The state into which this collapse occurs depends upon the kind of measurement one decides to perform (here by adjusting the microwave parameters in R_+). This decision can even be made after the systems have ceased to interact (one can change these parameters while the atom is flying from C to R_+, thus realising a 'delayed choice' experiment). The analysis of such an experiment is incompatible with classical probability theories based on hidden variables and the actual realisation of such experiments would provide new tests of Bell's inequality violations.

2.2   A train of atoms to a resonant cavity: an ideal micromaser

A micromaser [3] is a device in which Rydberg atoms, initially prepared in level e, are successively sent across an initially empty cavity and detected one by one after they leave the cavity. In this device, the post selection zone R_+ is usually left inactive and one merely detects the atom's energy state (e or g) in D.

The atom considered in the previous subsection can be considered as the first atom crossing the maser cavity. If it is detected in level e, the cavity remains in its vacuum state after the interaction. It contains one photon if the atom is found in g. The initial state of the second atom plus cavity system is thus either $|e, 0\rangle$ or $|e, 1\rangle$, depending upon the outcome of the first atom's measurement. In the first case, the second atom merely reproduces the initial situation. At some point, after a finite number p of atoms has crossed the cavity, the photon number will eventually jump from zero to one and the $(p + 1)$th atom plus cavity system will turn out to be initially in the $|e, 1\rangle$ state. This state will evolve as a linear superposition of $|e, 1\rangle$ and $|g, 2\rangle$ and, depending upon the outcome of the measurement of that atom, the field will be left in state 1 or 2 and so on. Quite generally, if the atom number q is introduced in a cavity containing already N photons, the state of the atom plus field system at the time t when the atom leaves the cavity is described by:

$$|\Psi_q\rangle = \cos(\Omega\sqrt{N+1}\ t)|e, N\rangle - i\sin(\Omega\sqrt{N+1}\ t)|g, N+1\rangle. \tag{5}$$

Note that there is here again a Rabi oscillation between two quantum states $|e, N\rangle$ and $|g, N + 1\rangle$, but that the rate of oscillation is proportional to $\sqrt{N+1}$, i.e. to the amplitude of the field in the cavity. We see from this simple analysis that the probability for the photon number to increase from N to $N + 1$ depends upon N. The photon number in the cavity increases as a discrete random variable, and the whole process can be easily simulated by a Monte Carlo calculation [25].

If one starts the process again, a new random walk of the photon number will be obtained. The state of the field reached after a given number of atoms will be different from one realisation to the next and an ensemble of realisations will exhibit in general photon number fluctuations at a given time. A statistical ensemble of a large number of field realisations will be described by a density matrix whose diagonal elements in the Fock state representation correspond to the probabilities of finding a given photon number at a given time. Repeating the Monte Carlo calculation of the field evolution many times provides a very illustrative and simple way of constructing this density matrix.

In this ideal maser model, the number of photons could in principle increase without limit, since field relaxation is completely neglected. A photon number upperbound is however eventually reached if the condition $\Omega\sqrt{N_0 + 1}\ t = k\pi$ is fulfilled for two integer values N_0 and k. Such a condition can be achieved by adjusting the atom's velocity across the cavity and hence the atom cavity interaction time t. In this case, the photon number cannot exceed the value N_0. When this value is reached, the next atom has indeed a zero probability of ending up in level g. In fact, the atom undergoes an integer number of Rabi oscillations and exits the cavity in the excited state e, without imparting any energy to the field. The field remains in a 'trapping state', a Fock state with N_0 photons [26, 27]. Note that this limit is certainly reached for all field realisations, so that the equilz,ibrium state attained by the field is in this case the pure state $|N_0\rangle$ and not a statistical mixture.

In real micromasers, field relaxation, blackbody radiation effects and imperfect atomic detection complicate the above analysis. A complete treatment of the micromaser can be found in several references [28, 29]. The model we have developed here provides a simple example of random walk calculation in a quantum optics situation. Experiments realising closely such an ideal maser situation should be possible in the near future with circular Rydberg atoms crossing high Q cavities in a time t very short compared to t_{cav} and t_{at}.

Note finally that micromasers operating on two-photon transitions by emitting pairs of microwave quanta inside a very high Q cavity have also been studied theoretically [30] and realised experimentally [31]. We will not consider them here.

2.3   A single atom and two cavities: non-local photon states

An interesting situation could arise when a resonant atom, initially excited in level e, is sent across two (or more) cavities and its energy is detected downstream, after it has emerged from the last resonator. If the atom is detected in level g, there is no way to know in which cavity the atom has been de-excited, and as a result, the field must end up in a linear superposition of states corresponding respectively to a single photon stored in each of the cavities [32]. Such delocalised photon states, which have not yet been experimentally observed, would display intriguing features of quantum theory and would be useful to implement, for example, teleportation of particles between the cavities [12].

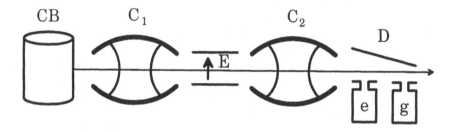

Figure 2. *Set up for the study of photon states delocalised between two cavities C_1 and C_2. The electric field E between C_1 and C_2 makes it possible to tune the phase of the field linear superposition.*

Many different situations can be envisaged, in which one could vary the strength of the atom field coupling independently in the different cavities and obtain field superpositions with adjustable amplitudes. In order to be specific, let us consider the situation depicted in Figure 2. A single atom, prepared in level e, is sent across two initially empty cavities C_1 and C_2, exactly resonant with the $e{\rightarrow}g$ transition. The Rabi couplings with the cavities are respectively Ω_1 and Ω_2 and the times T_1 and T_2 the atom takes to cross each cavity are adjusted so that $\Omega_1 T_1 = \pi/4$ and $\Omega_2 T_2 = \pi/2$. These conditions can be met by adjusting the atom's velocity and having the atom cross the two cavities along trajectories where the integrated product of the field amplitudes by the atomic pathlength across the cavity differ by a factor two. When the atom is between C_1 and C_2, the atom plus field system is in the state:

$$|\Psi_1\rangle = (1/\sqrt{2})(|e;0,0\rangle - i|g;1,0\rangle) \tag{6}$$

the three symbols in each ket referring to the state of the atom and of the field in C_1 and C_2 respectively. If the atom enters C_2 in state e, it must emit a photon in the second cavity, while it leaves of course the second cavity empty if it enters it in state g. As a result, the final atom-field state is:

$$|\Psi_2\rangle = (1/\sqrt{2})(|g;0,1\rangle + |g;1,0\rangle) \tag{7}$$

The atom then emerges from C_2 in level g and the field is left in a linear superposition, with equal probabilities, of finding the photon in either cavity. The phase between the two probability amplitudes could also be adjusted, by performing a small variant on this Gedanken experiment. Assume that between the two cavities the atom is subjected to a small electric field E which Stark-shifts by different amounts the levels e and g (see fig 2). As a result, the two components of the atom+ field state $|\Psi_1\rangle$ are dephased by an angle ϕ before the atom enters C_2 and the final field state becomes:

$$|\Psi_{C_1 C_2}\rangle = (1/\sqrt{2})(e^{i\phi}|1,0\rangle + |0,1\rangle), \tag{8}$$

where ϕ can be adjusted by varying the electric field between the two cavities.

Such field superpositions are different from the mixture described by the density operator:

$$\rho = (1/2)(|1,0\rangle\langle 1,0| + |0,1\rangle\langle 0,1|) \tag{9}$$

which corresponds to a classical situation where there is a mere statistical uncertainty about the location of the photon. The field described by equation (8) displays a coherence which could lead to interesting interference effects. Suppose that a second 'probe' atom, prepared in level g, is sent across a cavity system in which a field has been prepared in the state $|\Psi_{C_1 C_2}\rangle$. Let us call $a_1 e^{i\phi_1}$ the probability amplitude that the atom would be excited by a photon in state $|1,0\rangle$ and $a_2 e^{i\phi_2}$ the probability amplitude that the atom would be excited by the photon in state $|0,1\rangle$ (these amplitudes can be computed easily from the evolution equations of the system). The probability $p_{2:e}$ that the 'probe' atom absorbs the delocalised photon and ends up in level e appears as the square of the sum of these two amplitudes, the first one being affected by the phase factor $e^{i\phi}$:

$$p_{2:e} = |a_1 e^{i(\phi_1+\phi)} + a_2 e^{i\phi_2}|^2 = a_1^2 + a_2^2 + 2a_1 a_2 \cos(\phi + \phi_1 - \phi_2). \tag{10}$$

This probability thus presents an interference term, with an oscillatory behaviour when the phase ϕ is swept by changing the electric field amplitude between C_1 and C_2. If the field were in the classical mixture described by equation (9) instead, the probability would be $a_1^2 + a_2^2$, appearing as the mere sum of the probabilities that the photon has been absorbed from the two cavities, without interference. The presence of the interference term in equation (10) is the signature of a nonclassical situation, which could be demonstrated by performing a statistical experiment. One should send repeatedly pairs of atoms across the cavities, the first one preparing the field coherence and the second one, acting as a probe, being detected downstream. The cavities should be allowed to relax down to the vacuum state between the launch of two successive atom pairs, but no relaxation should be allowed to occur between the two atoms of a given pair. The statistics from a large set of data obtained for various values of the dephasing ϕ should reconstruct the result of equation (10).

The above discussion shows that various 'quantum games' can be played with an atom-cavity system at exact resonance. One could witness the collapse of the field wave function when the atom is detected, perform nonlocal correlation experiments which would be as many tests of Bell's inequalities and more generally manipulate non classical fields in one or several cavities correlated to each other by a beam of atoms. The operation of micromasers has already shown evidence of some of these effects, but the experiments performed so far with ordinary (non circular) Rydberg atoms are more complicated to interpret than the simple 'gedanken' situations envisioned above (relaxation effects are usually non negligible in these experiments). Simpler and more dramatic illustrations of these ideas would require the use of circular Rydberg atoms (experiments are in progress). At resonance, the atoms play a double role in these Cavity QED experiments. First, they act as source and absorber for the field. Second, they are the probes of the field, since the system evolution is witnessed by atomic detection. This double role, typical of micromaser physics, is a source of complexity in the system analysis. By making the atom-cavity interaction non-resonant, it is possible to separate these two roles and to turn the atom into a 'non-perturbing' measuring probe for the cavity photons (quantum non-demolition experiments). The photons must then be introduced in the cavity by some 'external' means (coupling of the cavity to a classical microwave source for example). We discuss in the next section the very interesting experiments which could be performed on the atom-cavity system in the non-resonant case.

3   The dispersive regime in cavity QED

If the cavity and the atomic transition are slightly mistuned, with a frequency difference $\delta = \omega - \omega_0$, any exchange of energy between atom and field is made impossible. The atom-field coupling becomes purely dispersive. The interaction produces a mere dephasing of the field (index effect of the atom crossing the cavity) and also dephases the atom's wave function by an angle depending upon the number of photons in the cavity and of the quantum state of the atom. More precisely, the non-resonant atom field interaction can be described by the effective Hamiltonian:

$$H_{\text{eff}} = (\hbar\omega_0/2)[b^\dagger b - bb^\dagger] + (\hbar\omega/2)[a^\dagger a + aa^\dagger] - (\hbar\Omega^2/\delta)[aa^\dagger b^\dagger b - bb^\dagger a^\dagger a] \qquad (11)$$

whose eigenstates are of the form $|e, N\rangle$ and $|g, N\rangle$ with the energies [15, 16]:

$$
\begin{aligned}
E_{e,N} &= (\hbar/2)[\omega_0 + (2N + 1)\omega] - (\hbar\Omega^2/\delta)(N + 1) \\
E_{g,N} &= (\hbar/2)[-\omega_0 + (2N + 1)\omega] + (\hbar\Omega^2/\delta)N \ .
\end{aligned}
\qquad (12)
$$

The last terms in the right hand sides of equation (12) describe the light shifts induced by the non-resonant field containing N photons on the atom in level e or g. These shifts vary linearly with the field intensity (N) and are proportional to the square of the atom field coupling. They are inversely proportional to the atom field detuning (dispersive effect). They have opposite signs in levels e and g. Note finally that, for $N = 0$, there is a residual shift effect of the $|e, 0\rangle$ state (equivalent to a cavity induced Lamb shift).

3.1 Inverse Stern-Gerlach effect.

The energy shift per photon $\hbar\Omega^2/\delta$ varies spatially, reflecting the variations of the field mode amplitude in the cavity. It thus appears as a 'potential energy per photon' for the atom moving across the cavity. Although small, this potential energy might be not negligible. It corresponds typically to shifts of a few kHz per photon, of the order of the kinetic energy associated to temperatures of about a microKelvin. The forces deriving from this potential might thus delay slow atoms crossing the cavity by an amount depending upon N [33]. It could also trap very slow atoms inside the cavity [34] or reflect these atoms at cavity entrance [35]. The photon dependent delay would lead to a potentially observable 'inverse Stern Gerlach effect' [16].

Assume that an atom in level e or g crosses a cavity containing a field in a superposition of Fock states of the form $|\Psi_{\text{field}}\rangle = \sum_N c_N |N\rangle$. The state of the total atom-field system can then be expressed as:

$$|\Psi\rangle = \sum_N c_N \int dx\, \phi_{i,N}(x)|x,i;N\rangle, \tag{13}$$

where i stands for e or g and $\phi_{i,N}(x)$ is the atomic wave function of an atom in level i crossing the cavity when it contains N photons. This equation describes an entanglement between the atomic position and the state of the field in the cavity. If the wavelets corresponding to different N values are spatially resolved after the atom has crossed C, the atomic detection should result in a collapse of the field into a Fock state. Note the analogy with the usual Stern-Gerlach effect in which particles with spin are sent across a classical magnet deflecting by different angles the various spin states. Here, the cavity plays the role of a 'quantised magnet' and the atomic trajectories are non-locally correlated to the state of this 'magnet'. Such an inverse Stern-Gerlach experiment is yet of the 'Gedanken' type. The difficulty in realising it is to avoid field relaxation during the relatively long time required for the slow atoms to cross the cavity.

3.2 Atomic phase shift induced by a cavity field and photon number QND measurement

Let us assume now that the field is in a energy eigenstate $|N\rangle$ and that the atom is injected in C in a linear superposition $c_e|e\rangle + c_g|g\rangle$. This can be realised by submitting the atom, initially in level e, to a microwave pulse coherently admixing levels e and g before it enters in C (this pulse is produced in zone R_- before the cavity (see Figure 3). After the atom has crossed C, the state of the atom field system is:

$$|\Psi\rangle = c_e \exp(i\varepsilon(N+1)T)|e,N\rangle + c_g \exp(-i\varepsilon NT)|g,N\rangle, \tag{14}$$

where T is the atom-cavity crossing time and $\varepsilon = \overline{\Omega}^2/\delta$ is the frequency shift per photon averaged over the atom's trajectory across C. We have adopted here the interaction representation which neglects the phase evolution produced by the 'unperturbed' Hamiltonian $H_{\text{at}} + H_{\text{cav}}$ and we consider only the phase shifts produced by the atom-cavity interaction. The phase distorsion between the two parts of the system state can be measured by Ramsey interferometry. After the atom has left the cavity, it could be subjected to a second microwave pulse in zone R_+ which mixes again e and g (see Figure 3). The probability that the atom has finally been transferred from e to g by the combined action of R_+ and R_- is then determined by detecting the atom's energy downstream. This probability oscillates as a function of the microwave frequency applied in R_+ and R_- (Ramsey fringes [36]). The position of the fringes depends upon the additional dephasing produced by the field in C, the fringe phase shift being equal to $\varepsilon(2N+1)$.

The Ramsey fringe detection corresponds to an atomic interference effect. The atom can follow two distinct 'paths' across the cavity, one in which it is in state e, the other in state g. The fringes result from the interference of the amplitudes associated to these two paths. If the cavity contains N photons, the two path amplitudes undergo different phase shifts and the fringe pattern is accordingly shifted, in a way quite analogous to the shift of the fringe pattern in a Young double slit experiment when a dispersive plate is inserted behind one of the two slits [37].

Note that the measurement of the Ramsey fringe shift amounts to a determination of the photon number in the cavity. Such a measurement does not change the field's energy, since the atom field interaction in C is purely dispersive. This is thus an inherently 'quantum non-demolition' method (QND)[5, 6, 38].

We have analysed elsewhere in details how the Ramsey method could be adapted to measure the field photon number when the initial field state is in a superposition of Fock state [5, 6]. We have shown how the detection of a train of atoms crossing the cavity results in the final collapse of the field into a Fock state, in what could be the prototype of an ideal quantum measurement. The experiment performing this ideal QND measurement, which is presently in progress at ENS, is described in the next section.

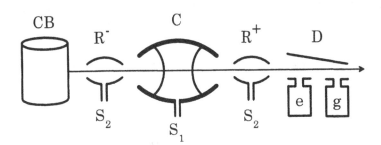

Figure 3. *Set up for the detection of cavity induced atomic phase shifts by Ramsey interferometry. The atomic transition off-resonant with the cavity mode in C (fed by the microwave source S_1) and resonant with the field applied in the Ramsey zones R_- and R_+ (source S_2).*

3.3   Dispersive generation of field mesoscopic coherences

The set up depicted in Figure 3, or a simple variant of this set up, could also be used to generate and study non-classical fields which are in quantum superpositions of states differing by some macroscopic parameter (phase, amplitude or location in space). Let us consider now explicitly the situation where both the initial field and the initial atom states are coherent superpositions. More explicitly we will assume that the field is in a Glauber coherent state [39] with complex amplitude α, the field amplitudes being $c_N = \exp(-|\alpha|^2/2)\alpha^N/\sqrt{N!}$. Such a field can be produced by coupling C to a classical source of current which is switched off immediately before the atom is sent across the apparatus. The state of the 'atom plus field' system after the atom leaves the cavity is obtained by mere superposition of terms analogous to those in equation (14):

$$
\begin{aligned}
|\Psi\rangle &= c_e e^{i\varepsilon t}|e\rangle \left\{ \sum_N c_N \exp[i\varepsilon NT]|N\rangle \right\} + c_g|g\rangle \left\{ \sum_N c_N \exp[-i\varepsilon NT]|N\rangle \right\} \\
&= c_e e^{i\varepsilon t}|e, \alpha e^{i\varepsilon T}\rangle + c_g|g, \alpha e^{-i\varepsilon T}\rangle.
\end{aligned}
\tag{15}
$$

This expression shows that the atom in $|e\rangle$ (resp in $|g\rangle$) dephases the field in the cavity by the angle $+\varepsilon T$ (resp $-\varepsilon T$). This dephasing corresponds to an atomic index effect. If the atom is in a linear superposition of the two levels, this phase shift results in an entanglement of the system after the interaction, the internal state of the atom being correlated to the phase of the field in the cavity. Note that this entanglement is of a quite different nature than the one discussed in the resonant case (Section 2 above). Here the number of photons in C cannot be changed by the atom-field interaction and the entanglement results from a purely dispersive phase shift distorsion of the wave function, different for each photon number and atomic state.

If the microwave zone R_+ is left inactive, measuring the atom's state in the detector D 'collapses' the phase of the field to a single value, leaving the field in either the state $|\alpha \exp(i\varepsilon T)\rangle$ or the state $|\alpha \exp(-i\varepsilon T)\rangle$, a rather trivial result. A very interesting situation arises however if a $\pi/2$ microwave pulse mixing levels $|e\rangle$ and $|g\rangle$ is applied on the atoms in R_+. Then, the atom+field state immediately after the atom leaves R_+ becomes:

$$
|\Psi_3\rangle = c_e(e^{i\varepsilon T}/\sqrt{2}) \left([|e, \alpha e^{i\varepsilon T}\rangle + |g, \alpha e^{i\varepsilon T}\rangle \right) + (c_g/\sqrt{2}) \left(|g, \alpha e^{-i\varepsilon T}\rangle - |e, \alpha e^{-i\varepsilon T}\rangle \right)
\tag{16}
$$

and the subsequent detection of the atom in level $|g\rangle$ or $|e\rangle$ results in the collapse of the field into one of the two states:

$$
|\Phi^{\pm}\rangle = c_e e^{i\varepsilon T}|\alpha e^{i\varepsilon T}\rangle \pm c_g|\alpha e^{-i\varepsilon T}\rangle.
\tag{17}
$$

These are linear superposition of field states with different classical phases which have been dubbed 'Schrödinger cat states' of the field [6, 40].

We recognise here again, now in a dispersive context, the ingredients of the Einstein-Podolsky-Rosen paradox. One subsystem (here the field) collapses into a state which depends upon the result of the measurement performed on the other part (here the atom), even if these two parts are far apart from each other when this measurement is performed.

For sake of definiteness, let us take the atom's parameters to be $c_e = -i/\sqrt{2}$, $c_g = 1/\sqrt{2}$, $\varepsilon T = \pi/2$. Equation (17) then becomes:

$$
|\Phi^{\pm}_{\text{phase}}\rangle = (1/\sqrt{2})(|\beta\rangle \pm |-\beta\rangle),
\tag{18}
$$

with $\beta = i\alpha$. The field in the cavity is then prepared in a linear superposition of two coherent field states with opposite phases. These particular superpositions are even (respectively odd) photon number states when the sign in Equation 18 is $+$ (respectively $-$). Shrödinger cat states have been studied extensively in theoretical papers [6, 41]. Cavity QED provides for the first time a practical mean to generate and detect them.

We have described elsewhere a possible experiment with these states [42]. It is of course essential to perform a measurement probing directly the coherent character of the superposition and enabling us to distinguish between a field described by equation (18) and a mere statistical mixture of the $|\beta\rangle$ and $|-\beta\rangle$ field states. As for the non-local photon state case discussed in subsection 2.3, an elegant way to achieve this is to send a second atom after the first one has been detected and to measure the probability of detecting this second atom in $|e\rangle$ or $|g\rangle$. This probability presents an interference term between two probability amplitudes, one associated to each of the $|\beta\rangle$ and $|-\beta\rangle$ states. This interference is constructive for the probability of detecting the second atom in the same state as the first one, making this conditional probability equal to 1 and destructive for the probability of the second atom to be detected in a state different from the first one, making this conditional probability equal to zero. If the field in C is instead in a statistical mixture, the interference vanishes and both conditional probabilities level off to $1/2$.

We have neglected in the analysis so far the relaxation of the field in the cavity. Dissipative processes have a strong effect on these quantum superpositions. In a time of the order of $t_{\text{cav}}/\overline{N}$, where $\overline{N} = |\alpha|^2$ is the average number of photons in the coherent field, they evolve into a classical statistical mixture [6]. We thus expect the conditional probability of detecting the first and the second atom in the same quantum state to be a function of the delay T_{12} between the two atomic detections. For short delays (*i.e.* $T_{12} \ll t_{\text{cav}}/\overline{N}$), this probability should be close to 1. on the other hand for large delays ($t_{\text{cav}} \gg T_{12} \gg t_{\text{cav}}/\overline{N}$), it should take the value $1/2$.

The continuous change of this conditional probability from 1 to $1/2$ as T_{12} is increased should be a direct evidence of the 'Schrödinger cat's' decoherence, a physical process which is at the heart of the quantum measurement process [43]. It is instructive to see the phase of the field in C as a kind of 'needle' pointing in two possible directions, each direction being correlated to one of the two Rydberg states $|e\rangle$ or $|g\rangle$ of the first atom crossing C. This 'needle' remains for some time in a quantum superposition of its two possible classical positions, but in the end it chooses one or the other (when the quantum superposition has evolved, due to the field dissipative coupling to its environment, into a statistical mixture). For 'small needles' ($\overline{N} \simeq 1$), this decoherence occurs in the relatively long time t_{cav}. For 'long needles' ($\overline{N} = 10$–100), the decoherence becomes much faster. Since we can adjust the intensity of the field initially present in the cavity from small to large values of N, such an experiment would enable us to explore the fuzzy boundary between the quantum world (where 'small needles' are, at least for some time, quantum objects existing in several possible states susceptible to create interference effects), and the classical world (where 'large needles' decohere into mutually exclusive states much faster than they can be observed).

Other kinds of cats can be generated with simple variants of this cavity QED set up. Instead of preparing a coherent field inside the cavity prior to the first atom injection, it is possible to employ the atom itself as a kind of 'quantum switch' governing the flow of the field inside the cavity [9]. The cavity must then be connected to a classical source slightly mistuned, so that, in the absence of an atom, it cannot feed any field inside C. The atomic parameters are then adjusted so that an atom crossing C in level $|e\rangle$ provides exactly the mode frequency shift required to tune it into resonance with the source. On the other hand, the atom in level $|g\rangle$ leaves the cavity and the source mistuned. We take again here advantage of the single atom index effect, the atom behaving as a kind of dispersive 'plunger' tuning C in and out of resonance with the source. Such a device allows us to prepare 'amplitude cat states' of the form $|\Phi^{\pm}_{\text{amplitude}}\rangle = (1/\sqrt{2})(|\alpha\rangle \pm |0\rangle)$. whose coherence can also be tested by sending a second atom across the system and measuring the conditional probability that both atoms end up being detected in the same state.

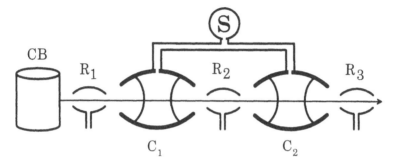

Figure 4. *Set up for the preparation of a delocalised coherent state between two cavities.*

The quantum switch can also be used to generate non local 'Schrödinger cat' states in two identical cavities C_1 and C_2 [9] (see Figure 4). The two cavities are now coupled symmetrically to a slightly mistuned source and a single atom is sent across both cavities. A microwave zone R_1 in front of C_1 prepares again the atom in a linear symmetrical superposition of $|e\rangle$ and $|g\rangle$ states, realising the quantum switch device. A π microwave pulse applied in zone R_2 turns $|e\rangle$ into $|g\rangle$ between C_1 and C_2 and exchanges the open and closed states of the switch. The two levels $|e\rangle$ and $|g\rangle$ are finally mixed again in the downstream zone R_3 before the atom is detected. In this way, one can generate the superposition state

$$|\Phi^{\pm}_{\text{non local}}\rangle = (1/\sqrt{2})(|\alpha; 0\rangle \pm |0; \alpha\rangle), \tag{19}$$

which represents a non local field, with equal (or opposite) probability amplitudes for the coherent field to be in the first or in the second cavity (the first and the second symbol in each ket refer to the field in C_1 and C_2 respectively). Note the analogy between Equations 19 and 8. The non-locality involved by Equation (19) is in a sense 'stranger' than the one described by Equation (8). Quantum mechanics tells us that a single particle (here a photon) can be delocalised between two (or several) positions, which makes Equation (8) somewhat 'familiar'. The non-locality it describes is the same as the one produced by a beamsplitter in optics, which can send a single photon in two different directions with equal probabilities. The non-locality described by Equation (19) involves on the other hand a quasi-classical object, a field which can contain a relatively large number of quanta. Such a situation, which cannot be produced by a beamsplitter-like device, is much more 'exotic' than the one described by Equation (8).

4 Ramsey interferometry and trapped photons

The cavity QED effects presented in the previous sections were described in the context of the simple Jaynes-Cummings model of a single two level atom interacting with a single cavity mode. In real experiments, we have to deal with more complex situations and many experimental tricks have to be developed in order to 'force' real atoms and fields to behave according to the Jaynes Cummings picture. In this section, we describe the dispersive cavity QED experiments presently carried out at Ecole Normale Supérieure [44] which are a first step towards the actual realisation of several of the gedanken experiments described above.

All these experiments rely on the extensive use of non-local atom-field correlations appearing during the interaction of the atom and field systems. These correlations are essential to get informations about the field by performing various atomic measurements after the atom has left the cavity. Practically, this means that one has to give a particular care to minimise atomic and field damping. Atomic relaxation has to be negligible during the atomic flight time across the apparatus. Since the characterisation of non-classical properties of the field rely on the detection of correlations between atoms interacting successively with the field, cavity damping must be negligible over even longer periods of times. The preparation of strong atom field correlations during the finite atom-cavity interaction time t_{int} also requires a large enough vacuum Rabi frequency Ω so that $\Omega t_{int} \simeq \pi$.

As mentioned above, circular Rydberg atoms coupled to a single resonant mode of a superconducting microwave cavity is a system meeting all these criteria. Other systems using optical cavities or trapped ions are also interesting but their presentation is out of the scope of these notes. The first experiment we have performed which combines circular atoms and a superconducting cavity has been the detection by Ramsey interferometry of very small fields, down to the vacuum in a single cavity mode [44].

4.1 Properties and preparation of circular Rydberg atoms

Circular Rydberg atoms [19, 20, 21] are atoms in which one electron is placed on a very excited energy level with a large value of the principal quantum number n (typically $n \simeq 50$ in our case) and where the angular momentum takes its maximum value: $\ell = m = n - 1$. The wavefunction of such an electron is a torus located around a circle centred on the atom's core. The radius of the 'orbit' is of the order of $a_0 n^2$. This gives also the typical size of the electric dipole of these atoms. With $n = 50$, it is about 2500 times larger than the dipole of weakly excited levels. Such large dipoles provide a very strong coupling with the cavity field, even if it contains only one photon.

The main advantage of circular Rydberg atoms over 'usual' Rydberg atoms (low value of ℓ and m) is a much longer lifetime. It is related to the smaller acceleration of the electron on a circular orbit as compared to the very elliptic orbits of non-circular states. The radiated field being proportional to this acceleration, it is much weaker with circular Rydberg states. As a result, the lifetime of circular states scales as n^5, and is of the order of 30ms for $n = 50$ (it is only 80μs for the 50f level). For a thermal beam at 300m/s, the flight time before decay of circular states is of the order of 10m. It ensures a completely negligible damping at the scale of the experimental setup (dimension $\simeq$ 25cm).

By applying a static electric field to circular atoms, it is also possible to make them a nearly perfect approximation of a two level system. Figure 5 shows the Stark diagram of the high angular momentum states of the $n = 50$ and 51 manifolds. The circular states correspond to the isolated levels at the right of the triangularly shaped level structures, the other states representing elliptical states. Most of the degeneracy with elliptic states is removed by a linear Stark effect of 100 MHz/(V/cm). Suppose now that the cavity is tuned on the 50c ('c' meaning circular) to 51c transition at $\omega_0/2\pi = 51.099$GHz. Applying the usual selection rules for electric dipole transitions and neglecting non-resonant processes, the only couplings to be considered are the two transitions $\Delta|m| = 1$ starting from the 50c level (shown by arrows). The ratio of the coupling strengths of these two transitions being 1/75, one can neglect the $\Delta m = -1$ coupling and approximate the circular atom undergoing transitions between the 50 and 51 manifold as a closed two-level system. The matrix element of a linear projection of the electric dipole between n and $n+1$ circular levels is $d_n = n^2/2$ (a.u). This two level approximation holds as long as the atom cavity detuning δ and the Rabi frequency Ω are much less than the linear Stark shifts. This condition is fulfilled in our experiments.

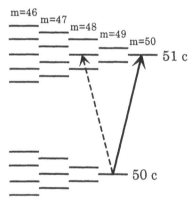

Figure 5. *Stark diagram of Rydberg states in the $n = 51$ and $n = 50$ manifolds. Each column corresponds to a given value of m. Only the last values $m \geq 47$ are displayed. The transitions quasi resonant with the cavity mode from the circular $n = 50$ state are indicated by arrows. The solid line arrow corresponds to a matrix element 75 time larger than the dashed one.*

A DC field E_0 is applied on the atoms along their trajectory in the setup. This field plays an essential role. By defining a quantisation axis, it keeps the circular orbit horizontal and prevents it from wobbling and being deformed by the coupling to elliptic states, produced by unavoidable stray electric or magnetic fields. The presence of this directing field is especially important inside the cavity in order to keep a maximal coupling with the linear polarization of the microwave cavity field. The requirement of such a directing field inside C excludes the use of 'traditional' micromaser cavities. In these closed cylindrical superconducting boxes no directing field can be applied and the atoms, submitted to stray fields, are no longer 'circular' at cavity exit. We thus have developedȷ new 'open' microwave cavities which are presented in the next section.

The preparation of the $n = 51$ circular Rydberg atoms consists in two main stages: laser excitation of the $n = 51f$ Rydberg state, and transfer of angular momentum by adiabatic absorption of low energy rf photons. The method is adapted from the original method of Hulet and Kleppner [19]. It has been described in detail elsewhere [20] and we will only recall its principle here.

A thermal beam of Rubidium is excited by three properly polarized diode lasers in the 51f, $m = 2$ Rydberg state via the $5p_{3/2}$ and $5d_{5/2}$ levels. A static electric field E, parallel to the quantisation axis, is then switched on adiabatically, transferring the atomic population in one of the lowest energy levels of the $m = 2$ manifold. This level is connected to the circular state by 48 degenerate $\Delta m = +1$ transitions. These transitions are induced by a resonant radiofrequency electric field polarized perpendicular to E. A small magnetic field B, parallel to E, removes the degeneracy between 'good' $\Delta m = +1$ and 'bad' $\Delta m = -1$ transitions which would populate unwanted 'elliptical' states. The transfer efficiency to the circular state is optimised with the help of an adiabatic transfer method generalising the well known adiabatic inversion of a spin $1/2$.

The method is time resolved. It produces up to 300 circular atoms per pulse in the 51 state at a maximal rate of 15000 pulses per second. The circularisation region is enclosed in a superconducting box so that the magnetic field B does not leak in the other parts of the setup.

4.2   The microwave cavity

The superconducting cavities developed for this experiment consist of two spherical mirrors in front of each other like in an usual optical Fabry-Perot resonator. The mirror curvature radius is 40mm, the distance L between mirrors is 27.54mm and the mirror diameter is 50 mm. The one centimetre spacing between the two mirrors ensures that, even at cavity entrance, the atoms are far enough from metallic surfaces so that stray fields are negligible and the atomic coherences are preserved during cavity crossing. The open structure of this cavity makes it less immune than closed cavities to diffraction losses produced by the residual roughness of the mirror surfaces. The quality factor of the first pair of mirrors we used was limited to 8.0×10^5. With better mirrors, we are now able to reach Q values of 2.0×10^8 and further improvements are in view. A very small hole is drilled at the centre of each mirror. It allows us to monitor the cavity transmission with a vector network analyser (AB millimetre, Paris) and to determine the resonance frequencies and quality factors. One hole can also be connected to a microwave source via calibrated attenuators in order to inject a field with a controlled amplitude inside the cavity.

The cavity sustains two TE_{900} standing wave gaussian modes presenting 9 antinodes between the mirrors and a gaussian transverse profile along the atomic trajectory, which is perpendicular to the cavity axis (mode waist $w = 5.96$mm). The two modes, with linear polarizations orthogonal to the cavity axis, have slightly different frequencies due to a small mirror ellipticity. Their angular frequencies (ω_{c1} and ω_{c2}) are tuned very close to the 50c $\rightarrow$ 51c transition by using a micrometer screw and a piezo stack. A DC field is applied between the mirrors so that the circular atomic orbit remains perpendicular to the cavity axis Oz.

The coupling of an atom to a linearly polarized cavity mode is: $\Omega = d_n\, E_{\text{vac}}(z,r)/\hbar$ with

$$E_{\text{vac}}(z,r) = E_{\text{vac}}(0,0)\exp(-r^2/w^2)\cos(i\pi\omega_0 z/c).$$

The vacuum field at cavity centre is $E_{\text{vac}}(0,0) = (\hbar\omega_0/2\epsilon_0 V)^{1/2}$ where ϵ_0 is the vacuum permitivity and $V = \pi w^2 L/4$ the effective cavity volume (0.7 cm^3). The maximum coupling at cavity centre is $\Omega(0,0) = 25$kHz.

4.3   The experimental setup

The setup is sketched on Figure 3. A thermal beam of Rubidium atoms is excited in the $n = 51$ circular state in the 'circularisation box' CB. The atomic orbit is in the horizontal plane. The atoms then cross three microwave cavities (R_-, C and R_+) before being detected in D on the right side. While crossing the superconducting cavity C, the atoms experience dispersive phase shifts induced by non-resonant coupling. The microwave source S_1 is used to inject calibrated fields in C. The shifts are probed by the Ramsey interferometer made of two low Q cavities R_- and R_+. These cavities are fed by the same microwave source S_2 with frequency ω_r. They are tuned on resonance on the 51c $\rightarrow$ 50c transition and the S_2 field amplitude is adjusted so that it induces $\pi/2$ pulses. The whole setup, except the oven, is enclosed in a shield cooled at 1.45K by a ^{4}He cryostat.

Atomic populations in states 50c and 51c are measured successively in the detector D by switching the ionisation field E_i between the ionisation thresholds of these two levels (136 and 148V/cm respectively). The resulting electrons are accelerated and counted by an electron multiplier. Ramsey fringes are recorded by monitoring the 51c to 50c transfer rate as a function of ω_r.

The fringe pattern shown on Figure 6 has been monitored with monokinetic atoms ($v = 295$m/s). The circular state excitation being pulsed, the atomic velocity was selected by gating the detection. The selectivity is $\delta v/v = 1.5\%$. The microwave fields in R_- and R_+ were also gated in 10μs pulses. The recording corresponds to the detection of $3.0{\times}10^5$ atoms in 20 min. It shows a 3.2kHz fringe spacing equal to the inverse of the atomic transit time between R_- and R_+.

The observation of these fringes requires a good control of the static electric and magnetic fields. The electric field E_0 preserving circular atoms in the Ramsey setup region was 0.3V/cm. It was produced by a set of gold coated electrodes. A particular care was taken to compensate the contact potential between these electrodes and the niobium mirrors. The laboratory magnetic field was cancelled by external coils. Its fluctuations were reduced with a mu-metal cylinder around the cryostat and a super-conducting indium coating of the 1.45K thermal shield. An upper bound of transverse inhomogeneities of the electric and magnetic fields can be deduced from the value of the fringe contrast. These fluctuations are below 3mV/cm and 400μG. The observed fringe pattern drift is less than 25Hz/day, which sets the upper limit of residual magnetic field fluctuations to 20μG/day.

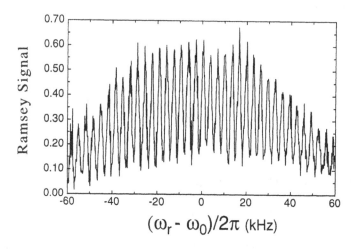

$$(\omega_r - \omega_0)/2\pi \ (\text{kHz})$$

Figure 6. *Population transfer signal between circular Rydberg levels $n = 51$ and $n = 50$ as a function of the frequency $\omega_r/2\pi$ applied in the microwave zones R_- and R_+. The cavity is detuned by $\delta/2\pi = 150$ kHz from the atomic transition. Ramsey fringes are clearly exhibited (from Reference [44]).*

4.4   Measurement of cavity field induced phase shifts

The cavity induced phase shifts are measured by monitoring the fringe shift while a small coherent field is injected at frequency ω_{c1}. About three fringes are recorded at the centre of the fringe pattern and we get the fringe phase by fitting a sine function to the experimental points (figure 7a). The signal to noise limited accuracy of the fit is 25Hz (0.05 radian phase uncertainty). The two signals presented on Figure 7a correspond to injected fields with zero and one photon on average. The atom cavity detuning has the value $\delta_1/2\pi = (\omega_{c1} - \omega_0)/2\pi = 150$kHz. Figure 7b shows the linear dependence of the observed shifts as a function of the field intensity. For a given field intensity, the fringe translation is found to be inversely proportional to δ_1, demonstrating the dispersive character of this effect.

As shown in Section 3.4, the single mode cavity induced fringe frequency shift is the product of the spatial average of the atomic energy shifts along the atomic trajectory times the effective cavity crossing time:

$$\Delta(N) = (N + 1/2)(\overline{\Omega}^2/\delta_1)(2\pi w/c), \tag{20}$$

where N is the cavity photon number. In this preliminary experiment, the cavity damping time (2μs, corresponding to a Q value of 8.0×10^5) was about ten times shorter than the atom cavity interaction time t_{int} (20μs), which entails that N fluctuated during t_{int}. The observed phase shifts was thus obtained by replacing N by its average $\overline{N}$ in equation (20). This shift is defined without adjustable parameter and can be used to calibrate the average intensity of the measured fields of Figure 7. The mean $\overline{\Omega}^2$ is obtained by averaging $\Omega^2(r, z)$ over the 1mm beam diameter. The measured 0.7mm vertical shift of the atomic beam with respect to the cavity centre results in a reduced

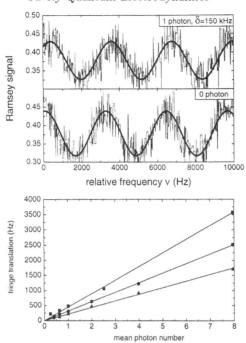

Figure 7. *(a) Ramsey fringe signal (central part of the spectrum in Figure 6 when the cavity contains zero photon (lower trace) or one photon on average (upper trace). The translation of the fringe pattern can reveal the dispersive light shifts produced by subphoton fields. (b) Fringe shift as a function of the average photon number in the cavity for three values of the atom-cavity detuning (δ=110, 150 and 220kHz —from Reference [44]).*

rms atom-field coupling $\overline{\Omega}/2\pi = 17(3)$kHz. This corresponds to an observed fringe shift of 315Hz/photon.

4.5 Observation of 'vacuum' shifts

A non vanishing residual shift $\Delta(0)$ is predicted in Equation (20) when no field is injected in the cavity mode $(\overline{N} = 0)$. It originates from the cavity induced 'Lamb shift' of the upper state. Its magnitude corresponds to the usual 'half-photon' energy of the vacuum field fluctuations. We have measured this shift by monitoring the fringe translation as a function of atom-cavity detuning δ_1 without injecting any field in the cavity (see Figure 8). The measured data (open circles) are in a very good agreement with the theoretical prediction (dotted line) taking into account the contributions of the two cavity modes and the residual thermal field in the cavity. Resonant atoms which thermalise with the cavity field during their crossing time were used as field thermometer. A field temperature of 1.73K, a little bit higher than the 1.45K setup temperature was deduced from the measured resonant atomic transfer rates, corresponding to an average number of thermal photons $N_{th} = 0.32(0.02)$ photons. The full circle points on Figure 8 are obtained by subtracting the effect of the thermal field. They clearly demonstrate the sensitivity of this experiment to vacuum fluctuations.

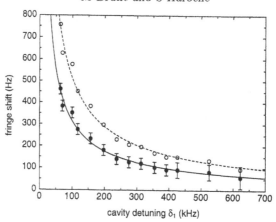

Figure 8. *Fringe frequency shifts as a function of atom-cavity detuning with no injected field in C (experiment: open circles; theory: dashed line). The cavity vacuum shift, obtained by subtracting the thermal radiation induced light shift, is shown by full circles (experiment) and solid line (theory) (from Ref.[44]).*

Agreement between theory and measurement is obtained by fitting the coupling $\overline{\Omega}/2\pi$ at 16.5(0.5)kHz. This value is in good agreement with, and more precise than the direct geometric determination (17 (3)kHz). It calibrates the observed one photon phase shift of Figure 7 to be 0.62(0.04) radian. The 25 Hz fringe position resolution results in a sensitivity to $\overline{N}$ better than 0.1 photon.

The sensitivity of Ramsey interferometry with circular Rydberg atoms has thus been demonstrated down to the vacuum field. These experiments were performed using a relatively 'low Q' cavity , whose damping time was much shorter than the atom-cavity interaction time. With new and better polished mirrors, we have now achieved a photon lifetime of $650\mu s$. This is long enough to ensure that non-local atom-field quantum correlations will survive until atomic detection, making it possible to implement the QND measurement method presented previously. The next step will consist in the preparation and characterisation of mesoscopic coherences of non-classical fields such as the quantum superposition of two quasi-classical fields with opposite phases. This will require another factor ten increase of the cavity damping time as well as a better control of the atomic beam parameters. Using a reduction of the beam divergence by transverse cooling, we are preparing a new beam setup with a very high atomic flux at 100m/s. A Doppler velocity selective excitation will also allow us to use monokinetic atoms in order to control actively the atom-cavity interaction time.

The marriage of Ramsey interferometry with dispersive cavity QED effects thus promises to open new fascinating perspectives for the manipulation of quantum fields in a cavity. At a fundamental level, as discussed above, these experiments will provide new tests of quantum mechanics and quantum measurement theory. Whether or not applications will follow is an open question. Recently, a lot of interest has been raised by the possibility of using an experimental scheme very close to the one described above to implement a logic gate for quantum computing [13, 14]. Realising such gates and trying to combine them to perform actual computing based on the manipulation of quantum mechanical amplitudes is a formidable challenge for experimentalists. It requires indeed to be able to keep quantum amplitudes alive on time scales much longer than the ones necessary to perform the experiments discussed above. Even if this remains very far from present possibilities, Cavity QED has had the merit of stimulating a flurry of activity among computer scientists and theoretical quantum opticians, which might lead to a better understanding of what quantum computers could do. The other obvious field where cavity QED could be applied is in the domain of microscopic integrated optics. Cavity QED effects have been demonstrated with very small structures of various kinds (microscopic 'thresholdless' lasers made of quantum wells [22], microspherical cavities [45, 46]). In this domain, the atomic and field relaxation times are much shorter than in the Rydberg atom-microwave cavity systems. Use of these devices for quantum computation in the optical domain seems thus even less likely, but other applications such as the generation or control of non-classical optical fields by microlasers are possible.

References

[1] P Goy, J M Raimond, M Gross and S Haroche *Phys Rev Lett* , **50**, 1903 (1983); G Gabrielse and H Dehmelt *Phys Rev Lett* , **55**, 67 (1985); R G Hulet, E S Hilfer and D Kleppner *Phys Rev Lett* , **55**, 2137 (1985); W Jhe, A Anderson, E A Hinds, D Meschede, L Moi and S Haroche *Phys Rev Lett* , **58**, 666 (1987); F De Martini, G Innocenti, G R Jacobovitz and P Mataloni *Phys Rev Lett* , **59**, 2955 (1987)

[2] D J Heinzen and M S Feld *Phys Rev Lett* , **59**, 2623 (1987)

[3] D Meschede, H Walther and G Muller *Phys Rev Lett*, **54**, 551 (1985); G Rempe, H Walther and N Klein *Phys Rev Lett* , **58**, 353 (1987)

[4] K An, J J Childs, R R Dasari, M S Feld, *Phys Rev Lett* , **73**, 3375 (1994)

[5] M Brune, S Haroche, V Lefhvre, J Raimond, N Zagury, *Phys Rev Lett* , **65**, 976 (1990)

[6] M Brune, S Haroche, J Raimond, L Davidovich, N Zagury, *Phys Rev* , **A 45** (1992) p 5193

[7] A Einstein, B Podolsky and N Rosen, *Phys Rev* , **47**, 777 (1935)

[8] S Haroche in Proceedings of 'Fundamental Problems in Quantum Theory', D Greenberger editor, to be published by the New York Academy of Sciences (1995)

[9] L Davidovich, A Maali, M Brune, J Raimond, S Haroche *Phys Rev Lett* , **71**, 2360 (1993)

[10] E Schrödinger, *Naturwissenschaften*, **23**, 807 (1935), **23**, 823 (1935); **23**, 844 (1935) (English translation by J D Trimmer, Proc Am Phys Soc **124**, 3235 (1980))

[11] C H Bennet, G Brassard, C Cripeau, R Josza, A Peres and W Wootters, *Phys Rev lett* , **70**, 1895 (1993)

[12] L Davidovich, A Maali, M Brune, J Raimond, S Haroche, *Phys Rev* , **A50**, R895 (1994)

[13] A K Ekert, Proceedings of the 14th International Atomic Physics Conference, C Wieman and D Wineland editors (1994);

[14] T Sleator, H Weinfurter, Proceedings of 'Fundamental Problems in Quantum Theory', D Greenberger (ed), to be published by the New York Academy of Sciences (1995)

[15] S Haroche 'Cavity Quantum Electrodynamics' in 'Systhmes fondamentaux en Optique Quantique', les Houches session LIII J Dalibard, J M Raimond et J Zinn Justin editeurs, Elsevier Science Publishers (1992)

[16] S Haroche and J M Raimond 'Manipulation of non classical field states by atom interferometry' in 'Cavity Quantum Electrodynamics', supplement of Advances in Atom and Molec Physics, Berman editor, Academic Press (1994)

[17] S Haroche and D Kleppner *Physics Today*, **42**, 24 (1989)

[18] S Haroche and J M Raimond, *Scientific American*, April 1993

[19] R G Hulet and D Kleppner, *Phys Rev Lett* , **51**, 1430 (1983)

[20] P Nussenzveig, F Bernardot, M Brune, J Hare, J M Raimond, S Haroche and W Gawlik *Phys Rev* **A48**, 3991 (1993)

[21] R J Brecha, G Raithel, C Wagner and H Walther, *Opt Comm* **102**, 257 (1993)

[22] Y Yamamoto, S Machida, K Igeta and Y Horikoshi in Coherence and Quantum Optics VI, L Mandel, E Wolf and J H Eberly editors (Plenum, New York, 1990)

[23] J Kimble in in 'Systhmes fondamentaux en Optique Quantique', les Houches LIII J Dalibard, J M Raimond et J Zinn Justin editeurs, Elsevier Science Publishers (1992)

[24] E T Jaynes and F W Cummings, Proc IEEE **51**, 89 (1963)

[25] P Meystre and E M Wright, *Phys Rev* **A37**, 2524 (1988)

[26] P Filipowicz, J Javanainen and P Meystre, *J Opt Soc Am* **B3**, 906 (1986)

[27] P Meystre, G Rempe and H Walther, *Opt Lett* **13**, 1078 (1988)

[28] P Filipowicz, J Javanainen and P Meystre, *Phys Rev* **A34**, 3077

[29] L A Lugiato, M O Scully and H Walther, *Phys Rev* **A36** ,740 (1987)

[30] L Davidovich, J M Raimond, M Brune and S Haroche, *Phys Rev* **A36**, 3771 (1987)

[31] M Brune, J Raimond, P Goy, L Davidovich, S Haroche, *Phys Rev Lett* , **59**, 1899 (1987)

[32] P Meystre in Progress in Optics XXX,, edited by E Wolf (Elsevier Science 1992)

[33] D Ivanov and T A B Kennedy, *Phys Rev* **A47**, 566 (1993)

[34] S Haroche, M Brune and J M Raimond,*Euro Phys Lett* , **14**, 19 (1991)

[35] B G Englert, J Schwinger, A O Barut and M O Scully, *Euro Phys Lett* , **14**, 20 (1991)

[36] N Ramsey, Molecular Beams, Oxford University Press, New York (1985)

[37] S Haroche, M Brune and J M Raimond, *Appl Phys B* , **54** (1992) 355

[38] V B Braginsky, Y I Vorontsov and F Y Khalili, Zh Eksp Teor Fiz **73**, 1340 (1977) (*Sov Phys JETP*, **46**, 705 (1977)); W G Unruh, *Phys Rev* D **18**, 1764 (1978); V B Braginsky and F Y Khalili, Zh Eksp Teor Fiz **78**, 1712 (1980) (*Sov Phys JETP*, **51**, 859 (1980))

[39] R J Glauber, *Phys Rev* , **130**, 2529 (1963) and **131**, 2766 (1963)

[40] B Yurke and D Stoler, *Phys Rev Lett* , **57**, 13 (1986)

[41] I A Malkin and V I Man'ko, Dynamical symmetries and coherent states of quantum systems, Nauka, Moscow (1979); J Jansky and A V Vinogradov, *Phys Rev Lett* , **64**, 2771 (1990); W Schleich, M Pernigo and F Lekien, *Phys Rev* , **A44**, 2172 (1991); V Buzek, H Moya-Cessa, P L Knight and S D L Phoenix, *Phys Rev* , **A45**, 8190 (1992)

[42] S Haroche, M Brune, J Raimond, L Davidovich 'Mesoscopic Quantum Coherences in Cavity QED' in 'Fundamentals of Quantum Optics III', Ehlotzky (ed), Springer Verlag 1993

[43] J Von Neumann, Die Mathematische Grundlagen der Quantenmechanik (Springer Verlag, Berlin 1932); K Hepp, *Helv Phys Acta*, **45**, 237 (1972); J S Bell, Helv Phys Acta **48**, 93 (1975); Quantum Theory and Measurement edited by J A Wheeler and W Zurek (Princeton Univ Press, Princeton, 1983); W Zurek, *Phys Rev* **D24**,1516 (1986); F Haake and D Waals, *Phys Rev* , **A36**, 730 (1986); W Zurek, *Phys Today*, **44** No 10, 36 (1991)

[44] M Brune, P Nussenzveig, F Schmidt-Kaler, F Bernardot, A Maali, J M Raimond and S Haroche, *Phys Rev Lett* , **72**, 3339 (1994)

[45] A J Campillo, J D Eversole and H B Lin, *Phys Rev Lett* , **67**, 437 (1991)

[46] F Treussart, J Hare, L Collot, V Lefevre, D Weiss, V Sandoghdar, J M Raimond and S Haroche, *Opt Lett* , **19**, 1651 (1994)

Coherent interactions of two trapped ions

R G DeVoe

IBM Almaden Research Center, San Jose

1 Introduction

Recent advances in laser cooling and trapping have made possible a new kind of microscopic atomic physics experiment in which each atom is detected individually and in which the interatomic distances are known and controlled on a sub-micron scale. One example of this is the observation [1–8] of 'ion crystals' in radio frequency quadrupole traps (Paul traps), in which clouds of trapped ions have been shown to condense into essentially stationary ordered arrays when laser cooled to a few milli-kelvin degrees. With proper design, the ions can be localised to less than $\lambda/2\pi$, where λ is the wavelength of the cooling laser light. One can then perform a microscopically resolved experiment by studying the coherent interactions of the ions as the ion-ion distance is varied by changing the trap parameters. This can be contrasted to a *macroscopic* experiment using an atomic beam or a gas cell in which the coherent signals from a large number of atoms are averaged over a statistical distribution of initial positions, velocities, and states of excitation. The requirement that the signals survive this average restricts the kinds of interactions that can be studied and also leads to phenomena such as inhomogeneous broadening, hole-burning, and coherent directional effects which can obscure the fundamental interactions. By studying only two trapped atoms, in which all the initial conditions are known and under the control of the experimenter, new aspects of atomic interactions may be revealed. For example, the superradiance signal studied below changes sign as the atomic separation varies by $\lambda/2$ and therefore averages to zero in conventional macroscopic experiment.

Here we study the spontaneous emission of two trapped atoms separated by a distance $R \approx \lambda$. At these distances there is a substantial probability that a photon emitted by one atom will be absorbed by the other. Dicke first showed [9, 10] in 1954 that the correct quantum mechanical approach to this problem is to treat the system of two coupled two-level atoms as a single 4-level system in which the exchange of energy between

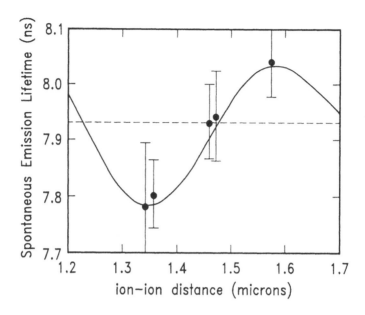

Figure 1. *Experimental result showing the* $\sin(kR)/kR$ *variation of the spontaneous emission lifetime of a two-ion crystal of* $^{138}Ba^+$ *versus theory.*

the atoms leads to enhanced and inhibited spontaneous decay rates which he called superradiant and subradiant respectively. Dicke's theory shows that a pair of two-level systems coupled to a common radiation field interact in a way which is analogous to how two spin 1/2 angular momenta couple to form a spin 1 system. This is because the atomic raising and lowering operators responsible for spontaneous emission obey commutation relations which are identical (up to a numerical factor) to those of an angular momentum. In his theory superradiance and subradiance are associated with 'triplet' $|1,0\rangle$ and 'singlet' $|0,0\rangle$ states of the two-atom system, where these terms are used in an abstract sense and do not refer to physical angular momenta.

The purpose of this paper is to describe the theory of superradiant and subradiant spontaneous emission in sufficient detail to understand current experiments. A recent experimental result [11] is presented in Figure 1 which shows oscillations in the spontaneous emission rate of a two-ion crystal of $^{138}Ba^+$ where the decay rate varies by $\pm 1\%$ as the ion-ion distance is changed by $\pm \lambda/4$ around a nominal value of 1470nm. Particular attention is given to forming a simple physical picture of these oscillations and of the singlet and triplet states. We begin with a review of the spontaneous emission of one atom (Section 2), then apply Dicke's theory to two trapped atoms (Section 3), compare to the results of a master equation (Section 4), discuss a simple physical picture of singlet and triplet states in both wave function and density matrix approaches (Section 5), describe a classical model of the decay rate oscillations (Section 6), and finally show how the experimental signal depends on the excitation conditions (Section 7). The apparatus necessary to observe such a signal is described elsewhere [12–15].

2 Spontaneous emission of a single atom.

We first review the spontaneous emission of one atom to establish the method and notation. The Hamiltonian of a two-level atom coupled to a quantised electromagnetic field can be written in the electric dipole approximation as

$$H = \hbar\omega_0 S_z - \mathbf{d}\cdot\mathbf{E}(\mathbf{r}) + \sum_i \hbar\omega_i \left[a_i^\dagger a_i + 1/2\right] \tag{1}$$

where the first term represents the atomic Hamiltonian H_A, the second term is the atom-field coupling H_{AF}, and the third is the field Hamiltonian H_F. The atom is located at $\mathbf{r}$ and has ground state a and excited state b with $E_b - E_a = \hbar\omega_0$. The electric dipole term contains the atomic dipole moment operator

$$\mathbf{d} = \mathbf{d}_{ab} \begin{pmatrix} 0 & 1 \\ 1 & 0 \end{pmatrix} = \mathbf{d}_{ab}\left(S_+ + S_-\right) \tag{2}$$

(where $S_\pm$ are defined below) and the electric field operator

$$\mathbf{E}(\mathbf{r}) = i\sum_i \sqrt{\frac{2\pi\hbar\omega_i}{V}}\boldsymbol{\epsilon}_i \left[a_i \exp(i\mathbf{k}_i\cdot\mathbf{r}) - a_i^\dagger \exp(-i\mathbf{k}_i\cdot\mathbf{r})\right] \tag{3}$$

where the i-th mode of the field has wave-vector $\mathbf{k}_i$, polarisation $\boldsymbol{\epsilon}_i$, creation and destruction operators $a_i^\dagger$ and a_i, and quantisation volume V. The interaction term can be written as

$$\mathbf{d}\cdot\mathbf{E} = (S_+ + S_-)\sum_i \left(g_i(\mathbf{r})a_i + g_i^*(\mathbf{r})a_i^\dagger\right) \tag{4}$$

where the coupling constants g_i are

$$g_i(\mathbf{r}) = i(\mathbf{d}_{ab}\cdot\boldsymbol{\epsilon}_i)\sqrt{\frac{2\pi\hbar\omega_i}{V}} \exp(i\mathbf{k}_i\cdot\mathbf{r}). \tag{5}$$

The atomic population difference operator S_z and raising and lowering operators $S_\pm$ are defined by

$$S_z = \frac{1}{2}\left(|b\rangle\langle b| - |a\rangle\langle a|\right) \tag{6}$$

$$S_+ = |b\rangle\langle a| \qquad S_- = |a\rangle\langle b|. \tag{7}$$

It is easy to verify directly that these operators obey commutation relations identical (except for a factor of $\hbar$) to those of an angular momentum

$$[S_x, S_y] = iS_z \tag{8}$$

and cyclic permutations, where $S_\pm = S_x \pm iS_y$. By dropping the energy nonconserving terms $S_+a^\dagger$ and S_-a the one-atom Hamiltonian becomes

$$H = \hbar\omega_0 S_z + \sum_i \left[g_i(\mathbf{r})S_+a_i + g_i^*(\mathbf{r})S_-a_i^\dagger\right] + \sum_i \hbar\omega_i \left[a_i^\dagger a_i + 1/2\right]. \tag{9}$$

This form is useful for later comparison to the two-atom case.

The decay rate from a given initial state i can be calculated from Fermi's Golden Rule

$$\Gamma_i = \frac{2\pi}{\hbar} \sum_f |\langle f| - \mathbf{d} \cdot \mathbf{E}(\mathbf{r})|i\rangle|^2 \, \delta(\hbar\omega_f - \hbar\omega_0) \,, \tag{10}$$

where the initial state $|i\rangle = |b; 0\rangle$ contains the atom in the excited state with no photons in the field and the final state $|f\rangle = |a; \mathbf{k}_j \epsilon_j\rangle$ contains the atom in the ground state having emitted a photon of wave-vector $\mathbf{k}_j$ and polarisation ϵ_j. Evaluating the matrix element gives

$$\Gamma_i = \frac{2\pi}{\hbar} \sum_f |g_i^*|^2 \, \delta(\hbar\omega_f - \hbar\omega_0). \tag{11}$$

The sum over final states reduces to an integral over solid angle using the density of states

$$\frac{d\rho}{d\Omega} = \frac{V\omega^2}{(2\pi c)^3 \hbar} \tag{12}$$

so that Equation (11) becomes

$$\Gamma_i = \frac{k^3}{2\pi\hbar} \int \left[|\mathbf{d}_{ab} \cdot \epsilon_1|^2 + |\mathbf{d}_{ab} \cdot \epsilon_2|^2 \right] d\Omega. \tag{13}$$

Here ϵ_1 and ϵ_2 are perpendicular to $\mathbf{k}$ and to each other and can be chosen, for example, to be $\mathbf{e}_\theta$ and $\mathbf{e}_\phi$ in spherical polar coordinates. For a $\Delta m = 0$ transition the polarisation factor in Equation (13) is $d^2 \sin^2 \theta$ and for a $\Delta m = \pm 1$ transition it is $d^2(1 + \cos^2 \theta)$. Evaluating these integrals gives decay rates of $4d^2k^3/3\hbar$ and $8d^2k^3/3\hbar$ for $\Delta m = 0$ and $\Delta m = 1$ transitions respectively, showing the 1:2 intensity ratio. We now show how these decay rates are affected by the presence of a second atom approximately λ away.

3   Dicke's superradiance theory for two atoms.

The Hamiltonian for two identical atoms interacting with a common electromagnetic field may be written as

$$H = \hbar\omega_0 S_{1,z} + \hbar\omega_0 S_{2,z} - \mathbf{d}_1 \cdot \mathbf{E}(\mathbf{r}_1) - \mathbf{d}_2 \cdot \mathbf{E}(\mathbf{r}_2) + \sum_i \hbar\omega_i \left[a_i^\dagger a_i + 1/2 \right] \tag{14}$$

where atoms 1 and 2 are coupled to the electromagnetic field at $\mathbf{r}_1$ and $\mathbf{r}_2$ respectively. Using Equations (6–8) this may be simplified for arbitrary $\mathbf{r}_1$ and $\mathbf{r}_2$ to

$$\begin{aligned}
H = \quad & \hbar\omega_0(S_{1,z} + S_{2,z}) \\
& + \sum_i \left[(g_i(\mathbf{r}_1)S_{1,+} + g_i(\mathbf{r}_2)S_{2,+})a_i + (g_i^*(\mathbf{r}_1)S_{1,-} + g_i^*(\mathbf{r}_2)S_{2,-})a_i^\dagger \right] \\
& + \sum_i \hbar\omega_i a_i^\dagger a_i + 1/2 \,.
\end{aligned} \tag{15}$$

Note that the coefficients $g_i(\mathbf{r}_1)$ and $g_i(\mathbf{r}_2)$ defined in Equation (5) differ only by a phase factor $\exp(i\mathbf{k}_i \cdot \mathbf{R})$. Consider now the case where the atoms are less than one wavelength apart so that $g_i(\mathbf{r}_1) \approx g_i(\mathbf{r}_2)$, but not so close that the energy levels are perturbed. Then Equation (16) simplifies to

$$H = \hbar\omega_0 R_z + \sum_i \left[g_i R_+ a_i + g_i^* R_- a_i^\dagger \right] + \sum_i \hbar\omega_i \left[a_i^\dagger a_i + 1/2 \right] \tag{16}$$

where the R operators are defined via $R_z = S_{1,z} + S_{2,z}$ and $R_\pm = S_{1,\pm} + S_{2,\pm}$. As first noted by Dicke [9], Equation (16) for two atoms is identical in form to the one-atom Hamiltonian Equation (9), so that by determining the matrix elements and eigenfunctions of R_z and $R_\pm$ we will be able to directly calculate the two-atom spontaneous emission rates using Fermi's Golden Rule.

In Dicke's theory the basis vectors of the two-atom system are chosen to be eigenfunctions of R_z and of $\mathbf{R}^2$. Since $\mathbf{R}, \mathbf{S}_1$ and $\mathbf{S}_2$ all obey commutation relations of the form of Equation (6), the basis vectors may be found in the same way as for the addition of angular momentum. Recall that in the addition of two angular momenta $\mathbf{J}_1$ and $\mathbf{J}_2$ to form a sum $\mathbf{J}$ there are two different representations possible, the $|j_1 m_1 j_2 m_2\rangle$ basis and the $|j^2, j_1^2, j_2^2, j_z\rangle$ basis. The first set can be called a direct product or independent atom representation since the two-atom wave-functions are direct (or tensor) products of the one atom functions, here denoted by primes $1' - 4'$

$$\begin{aligned} |1'\rangle &= |b_1 b_2\rangle & |2'\rangle &= |a_1 b_2\rangle \\ |3'\rangle &= |b_1 a_2\rangle & |4'\rangle &= |a_1 a_2\rangle \end{aligned} \tag{17}$$

which results from the direct product of column vectors

$$\begin{pmatrix} b_1 \\ a_1 \end{pmatrix} \otimes \begin{pmatrix} b_2 \\ a_2 \end{pmatrix}. \tag{18}$$

These wave-functions are simultaneous eigenfunctions of the one atom operators S_1^2, $S_{1,z}$, S_2^2 and $S_{2,z}$. It is useful in such calculations to express the operators in terms of Pauli matrices, where for example $S_{1,z} = \sigma_z \otimes I$ where I is a unit matrix in the space of the second atom. The second representation has basis vectors which are simultaneous eigenfunctions of R^2 and R_z. This may be called the collective or Dicke representation and is defined (unprimed) as

$$\begin{aligned} |1\rangle &= |1,1\rangle = |b_1 b_2\rangle & |2\rangle &= |1,0\rangle = \frac{1}{\sqrt{2}} \left[|a_1 b_2\rangle + |b_1 a_2\rangle \right] \\ |3\rangle &= |0,0\rangle = \frac{1}{\sqrt{2}} \left[|a_1 b_2\rangle - |b_1 a_2\rangle \right] & |4\rangle &= |1,-1\rangle = |a_1 a_2\rangle . \end{aligned} \tag{19}$$

See Figure 2. For convenience we also use the notation $|+\rangle \equiv |1,0\rangle$ for the symmetric triplet state and $|-\rangle \equiv |0,0\rangle$ for the antisymmetric singlet state. The two representations are connected by a unitary transformation U, whose elements are the Clebsch-Gordon coefficients. U is found by inspection to be

$$U = \begin{pmatrix} 1 & 0 & 0 & 0 \\ 0 & 1/\sqrt{(2)} & 1/\sqrt{2} & 0 \\ 0 & 1/\sqrt{(2)} & -1/\sqrt{2} & 0 \\ 0 & 0 & 0 & 1 \end{pmatrix} \tag{20}$$

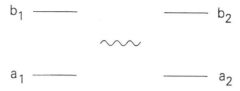

$$|\pm\rangle = \frac{1}{\sqrt{2}}\left(|a_1 b_2\rangle \pm |b_1 a_2\rangle\right)$$

Figure 2. *(a) Two two-level atoms coupled to a common radiation field (b) form a single four-level system with triplet and singlet states* $|\pm\rangle$.

which we will use later in the density matrix theory.

Following Dicke we compute the matrix elements for the spontaneous emission from the triplet and singlet states $|\pm\rangle$ to the ground state $|a_1 a_2\rangle$ which take the form of

$$g_i^* \langle a_1 a_2 | R_- | \pm \rangle \langle k_j \epsilon_j | a_i^\dagger | 0 \rangle . \tag{21}$$

Recall from the theory of angular momentum that the ladder operators $J_\pm = J_x \pm i J_y$ transform a state $|j, m\rangle$ into $|j, m \pm 1\rangle$ via

$$J_\pm |j, m\rangle = \hbar\sqrt{(j \mp m)(j \pm m + 1)}|j, m \pm 1\rangle \tag{22}$$

so that

$$J_- |1, 1\rangle = \sqrt{2}\hbar |1, 0\rangle \qquad J_- |1, 0\rangle = \sqrt{2}\hbar |1, -1\rangle . \tag{23}$$

The same relations apply to the atomic operators $R_\pm$ without $\hbar$. Note also that $R_\pm |0, 0\rangle = 0$ and that in general ladder operators do not connect states of different j. Hence, the singlet state $|-\rangle$ does not couple to the other three states and

$$\langle 1, -1 | R_- | + \rangle = \langle + | R_- | 1, 1 \rangle = \sqrt{2}. \tag{24}$$

Because of the similarity of the two-atom Hamiltonian Equation (16) to the one-atom result of Equation (9), the calculation of the Fermi's Golden Rule decay rates proceeds just as for one atom. Since the decay rates are proportional to the square of the matrix elements

$$\Gamma_+ = 2\Gamma \qquad \Gamma_- = 0 \tag{25}$$

where $\Gamma_\pm$ are rates for decays which have the states $|\pm\rangle$ as initial or final states. When the two-atom system is prepared in the triplet state, it decays at twice the rate of a single atom, or has half the lifetime, and is therefore called superradiant. The singlet state is metastable in this approximation and is called subradiant.

When the two atoms are separated by an arbitrary distance R the coupling constants $g_i(\mathbf{r}_1)$ and $g_i(\mathbf{r}_2)$ differ by a phase factor $\exp(i\mathbf{k}\cdot\mathbf{R})$ so that the atomic operators $S_{1,\pm}$ and $S_{2,\pm}$ can no longer be combined into a collective operator $R_\pm$. Then the nonzero matrix elements for the decay of the $|\pm\rangle$ states in Fermi's Golden Rule are then [16]

$$g_i^*(\mathbf{r}_1)\left[\langle a_1 a_2|S_{1,-}|\pm\rangle + \langle a_1 a_2|S_{2,-}|\pm\rangle \exp(i\mathbf{k}_i\cdot\mathbf{R})\right]\langle \mathbf{k}_j\epsilon_j|a_i^\dagger|0\rangle \tag{26}$$

and the decay rate becomes

$$\Gamma_\pm(\mathbf{R}) = \frac{2\pi}{\hbar}\sum_i |g_i|^2 \left[1 \pm \cos(\mathbf{k}_i\cdot\mathbf{R})\right]\delta(\hbar\omega_i - \hbar\omega_0). \tag{27}$$

Note that this is identical to the single atom result of Equation (11) except for the interference factor $1 \pm \cos(\mathbf{k}_i\cdot\mathbf{R})$. As in the single atom case, the sum over final states reduces to an integral over solid angle in which the interference term is multiplied by polarisation factors $\sin^2\theta$ or $(1 + \cos^2\theta)$ for $\Delta m = 0$ and $\Delta m = \pm 1$ cases respectively. These integrals are evaluated in the Appendix; for the case in which $\mathbf{R}$ is parallel to $\mathbf{z}$, the $\Delta m = \pm 1$ result is

$$\Gamma_\pm(\mathbf{R}) = \Gamma_0\left[1 \pm \frac{3\sin(kR)}{2kR} \pm \frac{3}{2}\left[\frac{\cos(kR)}{(kR)^2} - \frac{\sin(kR)}{(kR)^3}\right]\right] \tag{28}$$

while the $\Delta m = 0$ case gives

$$\Gamma_\pm(\mathbf{R}) = \Gamma_0\left[1 \pm 3\left[\frac{\cos(kR)}{(kR)^2} - \frac{\sin(kR)}{(kR)^3}\right]\right] \tag{29}$$

where Γ_0 is the single atom decay rate. In the region $R \geq \lambda$ the $\sin(kR)/kR$ term in the $\Delta m = \pm 1$ decay dominates since $kR \geq 10$. All other terms are of order $1/(kR)^2$. Equation (28) is plotted in Figure 3 for the special case of the $6P_{1/2}$ to $6S_{1/2}$ transition of the $^{138}Ba^+$ ion at 493nm as used in a recent experiment. Decay rate variations of as much as 10% are predicted for a two level system in the region near 1400nm. The $\Delta m = 0$ transition produces no visible signal on this scale and the disappearance of the signal on change of polarisation is an important signature for superradiance.

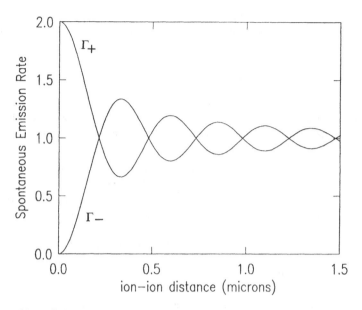

Figure 3. *Plot of the superradiant and subradiant decay rates $\Gamma_\pm$ for a $\Delta m = \pm 1$ transition from Equation (28).*

4   Master equation.

In the previous section we have assumed that the system was initially prepared in a pure singlet or triplet state. However, an experiment will in general prepare a partially coherent state described by a density matrix. A master equation is needed to describe spontaneous emission in the general case and to verify that the $|\pm\rangle$ states are diagonal during the decay. The master equation [17] for two stationary atoms separated by a distance R is given by

$$
\begin{aligned}
\frac{d\rho}{dt} = {} & -\frac{\Gamma}{2}\left[(S_{1,+}S_{1,-}+S_{2,+}S_{2,-})\rho+\rho(S_{1,+}S_{1,-}+S_{2,+}S_{2,-})-2S_{1,-}\rho S_{1,+}-2S_{2,-}\rho S_{2,+}\right] \\
& -\frac{\Gamma'}{2}\left[(S_{1,+}S_{2,-}+S_{2,+}S_{1,-})\rho+\rho(S_{1,+}S_{2,-}+S_{2,+}S_{1,-})-2S_{1,-}\rho S_{2,+}-2S_{2,-}\rho S_{1,+}\right]
\end{aligned}
$$
(30)

where the interaction Hamiltonian of Equation (16) has been treated as the perturbation in a derivation following Reference [18]. Here Γ is the single atom spontaneous emission rate and Γ' is defined as a Fermi's Golden Rule sum over final states

$$
\Gamma'(\mathbf{R}) = \frac{2\pi}{\hbar}\sum_i |g_i|^2 \cos(\mathbf{k}\cdot\mathbf{R})\delta(\hbar\omega_i - \hbar\omega_0)
$$
(31)

so that the superradiant and subradiant rates $\Gamma_\pm$ are given by $\Gamma_\pm = \Gamma\pm\Gamma'$, in agreement with Equation (27). Evaluation of Equation (30) yields an expression for the time rate of change of the density matrix

$$\frac{d\rho}{dt} = \begin{bmatrix} \Gamma_+\rho_{22} + \Gamma_-\rho_{33} & -\Gamma_+(\rho_{12} - 2\rho_{24})/2 & -\Gamma_-(\rho_{13} - 2\rho_{34})/2 & -\Gamma\rho_{14} \\ -\Gamma_+(\rho_{21} - 2\rho_{42})/2 & -\Gamma_+(\rho_{22} - \rho_{44}) & -\Gamma\rho_{23} & -(2\Gamma + \Gamma_+)\rho_{24}/2 \\ -\Gamma_-(\rho_{31} + 2\rho_{43})/2 & -\Gamma\rho_{32} & -\Gamma_-(\rho_{33} - \rho_{44}) & -(2\Gamma + \Gamma_-)\rho_{34}/2 \\ -\Gamma\rho_{41} & -(2\Gamma + \Gamma_+)\rho_{42}/2 & -(2\Gamma + \Gamma_-)\rho_{43}/2 & -2\Gamma\rho_{44} \end{bmatrix}.$$

Note that the populations and coherences are uncoupled from each other. The former may therefore be written in this representation as rate equations

$$\begin{aligned} \frac{d\rho_{11}}{dt} &= \Gamma_+\rho_+ + \Gamma_-\rho_- \\ \frac{d\rho_+}{dt} &= -\Gamma_+\rho_+ + \Gamma_+\rho_{44} \\ \frac{d\rho_-}{dt} &= -\Gamma_-\rho_- + \Gamma_-\rho_{44} \\ \frac{d\rho_{44}}{dt} &= -(\Gamma_+ + \Gamma_-)\rho_{44} \end{aligned} \tag{32}$$

which have the solution

$$\begin{aligned} \rho_{44}(t) &= \rho_{44}(0)\,\exp(-2\Gamma t) \\ \rho_+(t) &= \rho_+(0)\,\exp(-\Gamma_+ t) + \rho_{44}(0)\frac{\Gamma_+}{\Gamma_-}\left[\exp(-\Gamma_+ t) - \exp(-2\Gamma t)\right] \\ \rho_-(t) &= \rho_-(0)\,\exp(-\Gamma_- t) + \rho_{44}(0)\frac{\Gamma_-}{\Gamma_+}\left[\exp(-\Gamma_- t) - \exp(-2\Gamma t)\right] \end{aligned} \tag{33}$$

where we have abbreviated $\rho_{22} = \rho_+$ and $\rho_{33} = \rho_-$. These equations show that decays to or from the $|\pm\rangle$ states occur at rates $\Gamma_\pm$ as shown in Figure 2b. The total photon emission rate from all 4 transitions is

$$R = \rho_{44}\Gamma_+ + \rho_{44}\Gamma_- + \rho_+\Gamma_+ + \rho_-\Gamma_-. \tag{34}$$

5   Physical picture of singlet and triplet states.

Here we show that the triplet and singlet states represent states in which the dipole moments of the two atoms are either correlated or anticorrelated. This, together with the classical model in Section 6, gives a simple physical picture of why spontaneous emission is enhanced in the former and inhibited in the latter. Consider first a system in which both atoms 1 and 2 are in pure states with equal wave-functions

$$|\Psi_1\rangle = |\Psi_2\rangle = \begin{pmatrix} b \\ a \end{pmatrix}. \tag{35}$$

Since the two atoms' wave-functions are identical, so are all their physical observables, including the dipole moments. The probability amplitude of the $|\pm\rangle$ states, called $c_\pm$, can be calculated from Equations (19)

$$c_\pm = \frac{1}{\sqrt{2}}(ab \pm ab) \tag{36}$$

so that the populations of the singlet and triplet states are

$$|c_+|^2 = 2|ab|^2 \qquad |c_-|^2 = 0. \tag{37}$$

Hence identically prepared atoms populate the triplet state, and decay at a rate given by $\Gamma_+ = 2\Gamma$, while the singlet state is empty. On the other hand, when the atoms wave-functions are

$$|\Psi_1\rangle = \begin{pmatrix} b \\ a \end{pmatrix} \qquad |\Psi_2\rangle = \begin{pmatrix} -b \\ a \end{pmatrix} \tag{38}$$

so that the dipole moment of atom 2 is opposite that of atom 1, the triplet state is empty and the metastable singlet state is occupied with $|c_-|^2 = 2|ab|^2$.

The same point can be made more generally for mixed states by using the density matrix formalism. To do this we evaluate the 4×4 density matrix for the two-atom system in two different representations, the direct product (primed) representation of Equations (17) and the collective (unprimed) representation of Equations (19). They are related by the unitary transformation U of Equation (20), so that $\rho = U^{-1}\rho'U$ where $\rho' = \rho^1 \otimes \rho^2$ is the direct product of the two single atom, 2×2 density matrices ρ^1 and ρ^2. The populations of the collective states are then given by

$$\begin{aligned}
\rho_{11} &= \rho_{aa}^1 \rho_{aa}^2 \\
\rho_\pm &= \frac{1}{2}\left[\rho_{aa}^1\rho_{bb}^2 + \rho_{bb}^1\rho_{aa}^2 \pm (\rho_{ab}^1\rho_{ba}^2 + \rho_{ba}^1\rho_{ab}^2)\right] \\
\rho_{44} &= \rho_{bb}^1 \rho_{bb}^2 .
\end{aligned} \tag{39}$$

For pure states this reproduces the results of the wave-function calculation above.

Consider the case in which the two atoms have identical density matrices except for the phases of the off-diagonal elements $\rho_{ab}^1 = \rho_{ab}^2 \exp(i\phi)$. Then

$$\rho_\pm = \rho_{aa}\rho_{bb} \pm |\rho_{ab}|^2 \cos(\phi) \tag{40}$$

which shows again that the populations of the singlet and triplet states depend upon the relative phases of the dipole moments of the two atoms. When the atoms are in phase, the triplet state is preferentially populated and the decay is superradiant, while subradiant decay occurs when the atoms are out of phase.

6   Interference of two classical dipoles.

Some insight into the interference integral of Equation (27) and of the singlet and triplet states can be gained by considering the classical radiation of electromagnetic power by charged particles undergoing simple harmonic motion. From classical electromagnetic theory one may derive a radiative decay constant

$$\Gamma_{cl} = \frac{2e^2\omega_0^2}{3mc^3} \tag{41}$$

which represents the rate at which a simple harmonic oscillator with a mechanical energy of oscillation W radiates electromagnetic energy P, where $P = \Gamma_{cl}W$. Here the

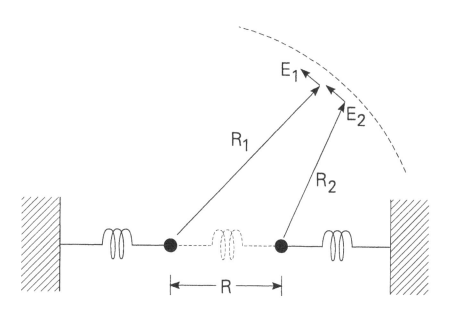

Figure 4. *Interference of the classical electromagnetic radiation of two particles of charge e and mass m undergoing coupled simple harmonic motion.*

particle has mass m and charge e and is suspended by a spring of force constant k so that the oscillation frequency $\omega_0 = \sqrt{k/m}$. This results from dividing the time-average of Larmor's formula for the power radiated by an accelerated charge

$$P = \frac{2e^2}{3c^3}|\dot{\mathbf{v}}|^2 \tag{42}$$

by the mechanical energy $W = m\omega_0^2 a^2/2$ where simple harmonic motion $z = a \cos \omega_0 t$ about an equilibrium point is assumed.

Now consider the power radiated by two identical charged particles suspended by springs as shown in Figure 4. The particles are assumed to be coupled by a very weak spring k' where $k' \ll k$. We will not attempt to calculate k' and its value is not important for what follows. It is well known that two such coupled particles oscillate in normal modes, one symmetric in which their displacement from equilibrium positions are equal, that is $\mathbf{z}_1 = \mathbf{z}_2 = \hat{z} a \cos(\omega_0 t)$ and one antisymmetric in which $\mathbf{z}_1 = -\mathbf{z}_2$. Because the particles motions are coherent, their electric fields will interfere and two modes will radiate different amounts of total power. The electric field [19] at position $\mathbf{R}$ radiated by an accelerated charge at the origin is given by

$$\mathbf{E}(\mathbf{R}) = \frac{e}{c^2 R} [\hat{n} \times [\hat{n} \times \dot{\mathbf{v}}]]_{\text{ret}} \tag{43}$$

where $\hat{n} = \mathbf{R}/R$ and $R = |\mathbf{R}|$. For the given simple harmonic motion along the z-axis this yields

$$\mathbf{E}(\mathbf{R}) = -\frac{ea\omega_0^2}{c^2 R} [\hat{n} \times (\hat{n} \times \hat{z}) \cos[\omega_0(t - R/c)]] . \tag{44}$$

The total radiated power is given by integrating the Poynting vector over a sphere at infinity

$$P = \frac{c}{4\pi} \int |\mathbf{E}_1 + \mathbf{E}|^2 d\Omega.$$ (45)

Consider first the case in which the particles are much closer together than a wavelength of light. Then for the symmetric mode $\mathbf{E}_1 = \mathbf{E}_2$ for all $d\Omega$ so that the fields are in-phase everywhere and the two-particle system radiates 4 times the power of a single particle. Hence the decay constant for the symmetric mode $\Gamma_+ = 2\Gamma_{cl}$. For the antisymmetric mode, $\mathbf{E}_1 = -\mathbf{E}_2$ and the fields cancel so that $\Gamma_- = 0$.

When the particles are at arbitrary separations $\mathbf{R}$ the path lengths R_1 and R_2 vary over the surface at infinity so that $\mathbf{E}_1$ and $\mathbf{E}_2$ are alternately in and out of phase. An angular integral over $d\Omega$ yields the radiated power for the symmetric and antisymmetric modes

$$P_\pm = \frac{e^2 a^2 \omega_0^4}{4\pi c^3} \int \sin^2\theta \left[\cos(\omega_0 t - \mathbf{k}\cdot\mathbf{R}_1) \pm \cos(\omega_0 t - \mathbf{k}\cdot\mathbf{R}_2)\right]^2 d\Omega$$ (46)

where $\mathbf{k} = \hat{n}\omega_0/c$ and the $\sin^2\theta$ factor is due to the polarisation, here assumed to be along the z axis. The time average of the quantity in brackets is equal to $2(1\pm\cos(\mathbf{k}\cdot\mathbf{R}))$ where $\mathbf{R} = \mathbf{R}_1 - \mathbf{R}_2$ so that the the time average $\langle P_\pm \rangle$ is

$$P_\pm = \frac{e^2 a^2 \omega_0^4}{c^3 4\pi} \int [1 \pm \cos(\mathbf{k}\cdot\mathbf{R})]\sin^2\theta d\Omega.$$ (47)

This equation contains the same angular integral as in the quantum mechanical case for a $\Delta m = 0$ transition, yielding a classical damping time for the symmetric and antisymmetric modes $\Gamma_\pm$

$$\Gamma_\pm = \Gamma_{cl}\left[1 \pm 3\left[\frac{\cos(kR)}{(kR)^2} - \frac{\sin(kR)}{(kR)^3}\right]\right]$$ (48)

by dividing the radiated power by the total mechanical energy $W = ma^2\omega_0^2$. This shows that the decay rate oscillations are due to phase or retardation factors that are common to both classical and quantum mechanical theory.

7   Experiments on superradiance and subradiance.

The simplest method in principle of observing superradiant and subradiant decay is to excite the two-atom system with a short intense pulse of resonant laser light at $t = 0$ so that the system may be described by a density matrix $\rho(0)$. The light is then turned off and $\rho(t)$ evolves according to Equation (33). A photomultiplier records the total photon emission rate, whose decay can be fitted to yield the rates $\Gamma_\pm(R)$ as the ion-ion distance is varied. A typical experiment is shown in Figure 5 and experimental details are described elsewhere [11–13]. The exciting laser pulses are assumed to be short and intense so the initial states of each atom are completely described by a Bloch angle Θ and the phases of the dipole moments $\phi_{1,2}$. In the geometry of Figure 5, the ions are crystallised in the radial plane of the trap, perpendicular to the z axis, so that

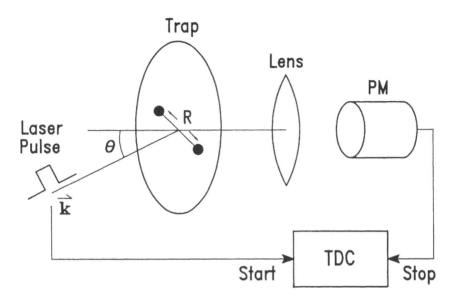

Figure 5. *Experimental arrangement. Two ions are trapped in in the radial plane of a microscopic ion trap, here shown schematically as a ring. The decay of the spontaneous emission is detected by a photomultiplier and recorded on a time-to-digital converter. The lifetime is extracted by fitting to an exponential.*

$\Delta\phi \equiv \phi_2 - \phi_1 = kR\sin\theta$. Coupling between the atoms can be ignored during the excitation process because it occurs in a time short compared to $1/\Gamma$. The density matrices at $t = 0$ are then given by the optical Bloch equations

$$\rho_{aa}^{1,2} = \cos^2(\Theta/2) \qquad \rho_{bb}^{1,2} = \sin^2(\Theta/2)$$
$$\rho_{ab}^{1,2} = \frac{\sin(\Theta)}{2}\exp(i\phi_{1,2}). \tag{49}$$

These form initial conditions for the collective system and $\rho(0)$ may be computed in the Dicke representation by using Equation (40), which yields

$$\rho_{11} = \cos^4(\Theta/2) \qquad \rho_{44} = \sin^4(\Theta/2)$$
$$\rho_{\pm} = \frac{1}{4}\sin^2(\Theta)\left[1 \pm \cos(\Delta\phi)\right]. \tag{50}$$

Consider first the case of weak coherent excitation in which $\Theta \ll 1$. In this limit the excited state population $\rho_{44} \to \Theta^4/16$ and may be neglected. One may then put all the excitation into either the triplet state $|+\rangle$ or the singlet state $|-\rangle$ by choosing $\Delta\phi = 0$ or π respectively. This can be accomplished experimentally either by changing the incidence angle θ or by changing R. More generally both $|\pm\rangle$ states will be excited and the total decay rate from Equation (34) is

$$W = \Gamma_+\rho_+ + \Gamma_-\rho_- \tag{51}$$

or

$$W = W_0 \sin^2(\Theta) \left[1 + \frac{3}{2} \frac{\sin(kR)}{kr} \cos(kR \sin \theta) + ... \right] \tag{52}$$

where W_0 is an experimental constant including the detection efficiency and repetition rate and higher order terms in $1/kR$ have been neglected. In a practical experiment level degeneracy and micromotion Doppler shifts will reduce the $3/2$ factor above by reducing the coherences ρ_{ab}. For comparison to the experiment in Figure 1 the $3/2$ factor has been replaced by 0.33, as described elsewhere.

Appendix.

Here we evaluate the angular integrals for $\Delta m = 0$ (Equation 28)

$$\frac{1}{4\pi} \int d\Omega \left[1 \pm \cos(\mathbf{k}\cdot\mathbf{R}) \right] \sin^2 \theta =$$
$$\frac{2}{3} \left[1 \pm \frac{3}{2} \left[(\hat{R}_x^2 + \hat{R}_y^2) \frac{\sin(kR)}{kR} - (2 - 3\hat{R}_x^2 - 3\hat{R}_y^2) \left[\frac{\cos(kR)}{(kR)^2} - \frac{\sin(kR)}{(kR)^3} \right] \right] \right] \tag{53}$$

and for $\Delta m = \pm 1$ (Equation 29)

$$\frac{1}{4\pi} \int d\Omega \left[1 \pm \cos(\mathbf{k}\cdot\mathbf{R}) \right] (1 + \cos^2 \theta) =$$
$$\frac{4}{3} \left[1 \pm \frac{3}{4} \left[(1 + \hat{R}_z^2) \frac{\sin(kR)}{kR} - (1 - 3\hat{R}_z^2) \left[\frac{\cos(kR)}{(kR)^2} - \frac{\sin(kR)}{(kR)^3} \right] \right] \right] . \tag{54}$$

The terms containing $\cos(\mathbf{k}\cdot\mathbf{R})$ may be evaluated by differentiating under the integral sign. In the first case, write

$$\sin^2 \theta = \frac{k_x^2 + k_y^2}{k^2} \tag{55}$$

so that the $\cos(\mathbf{k}\cdot\mathbf{R})$ term reduces to the real part of

$$-\frac{1}{4\pi k^2} \left(\frac{\partial^2}{\partial R_x^2} + \frac{\partial^2}{\partial R_y^2} \right) \int d\Omega \exp(i\mathbf{k}\cdot\mathbf{R}) . \tag{56}$$

Equation (54) involves a similar differentiations with respect to R_z.

References

[1] F Diedrich, E Peik, J M Chen, W Quint, H Walther, *Phys Rev Lett* , **59**, 2931 (1987)
[2] D J Wineland, J C Bergquist, W M Itano, J J Bollinger, and C H Manning, *Phys Rev Lett* , **59**, 2935 (1987)
[3] J Hoffnagle, R G DeVoe, L Reyna, R G Brewer, *Phys Rev Lett* , **61**, 255 (1988)
[4] R G Brewer, J Hoffnagle, and R G DeVoe, *Phys Rev Lett* , **65**, 2619 (1990)
[5] R G Brewer, J Hoffnagle, R G DeVoe, L Reyna, W Henshaw, *Nature*, **344**, 306 (1990)
[6] U Eichmann, J C Bergquist, J J Bollinger, W M Itano, and D J Wineland, *Phys Rev Lett* , **70**, 2359 (1993)

[7] R Bluemel, E Peik, W Quint, and H Walther, Springer Proceedings in Physics, Vol **41** (Springer Verlag, Berlin, 1989)

[8] M G Raizen, J M Gilligan, J C Bergquist, W M Itano, and D J Wineland, *Phys Rev A*, **70**, 6493 (1992)

[9] R H Dicke, *Phys Rev* , **93**, 99 (1954)

[10] M Gross and S Haroche, *Phys Rep* , **93**, 301 (1982)

[11] R G DeVoe and R G Brewer, Proceedings of the Twelfth International Conference on Laser Spectroscopy (World Scientific, Honk Kong, 1995)

[12] R G Brewer, R G DeVoe, and R Kallenbach, *Phys Rev A*, **46**, 4781 (1992)

[13] R G DeVoe and R G Brewer, *Opt Lett* , **19**, 1891 (1994)

[14] N Yu, W Nagourney, and H Dehmelt, *J Appl Phys* , **69**, 3779 (1991); N Yu, H Dehmelt, and W Nagourney, *Proc Natl Acad Sci* , **86**, 5672 (1989)

[15] C A Schrama, E Peik, W W Smith, and H Walther, *Opt Comm* , **101**, 32 (1993)

[16] E A Power, *J Chem Phys* , **46**, 4297 (1967)

[17] G S Agarwal, Sringer Tracts in Modern Physics, Vol **70** (Springer Verlag, Berlin, 1974)

[18] C Cohen-Tannoudji, J Dupont-Roc, and G Grynberg, *Atom-Photon Interactions* (Wiley, New York, 1993)

[19] J D Jackson, *Classical Electrodynamics* (Wiley, New York, 1975)

The two-photon decay of metastable atomic hydrogen and tests of Bell's inequality

A J Duncan

Atomic Physics Laboratory
University of Stirling, Scotland

1 Introduction

Since the early days of this century, the theoretical and experimental study of atomic hydrogen has served to elucidate our understanding and extend our knowledge of the fundamental properties and behaviour of atoms, and has provided an important testing ground for quantum theory. The states with principal quantum number $n = 2$ are and have been of special interest and importance, in particular with regard to the determination of the fine structure constant and measurement of the Lamb shift. It was, of course, the observations of the Lamb shift in 1947 and 1950 by Lamb and Retherford [1] which, by demonstrating the nondegeneracy of the $2^2S_{1/2}$ and $2^2P_{1/2}$ states, confirmed that the $2^2S_{1/2}$ state would be metastable in experimentally realisable situations, and showed that it should be possible to observe the two-photon emission which is the main mode of decay of this state. However, Göppert-Mayer [2] in 1931, in a paper which pioneered the field of multiphoton transitions, was the first to predict the possibility of the spontaneous two-photon decay process and, in 1940, Breit and Teller [3] applied this theory to the $2^2S_{1/2} - 1^2S_{1/2}$ transition in atomic hydrogen. Improved calculations of the characteristics of the two-photon decay process were carried out by Spitzer and Greenstein [4], Shapiro and Breit [5], Zon and Rapaport [6], Klarsfield [7] and Johnson [8]. Further refinements to the theory continue to be made for example by Goldman and Drake [9], Parpia and Johnson [10], Tung et al. [11], Florescu [12], Costescu [13] and Drake [14].

Although the existence of the Lamb shift indicated that it should be possible to observe the two-photon decay process, the process is second-order with a comparatively low transition probability and it was not until 1965 that Lipeles, Novick and

Tolk [15] made a successful measurement in singly ionised helium, further results being obtained by Artura, Tolk and Novick in 1969 [16]. Subsequently, the two-photon decay of hydrogen-like argon and sulphur ions was observed by Schmeider and Marrus [17] and Marrus and Schmeider [18] in a beam-foil-type experiment. In a similar experiment Cocke et al. [19] measured the lifetime of the $2^2 S_{1/2}$ state of hydrogen-like fluorine and oxygen and, about the same time Prior [20] measured the lifetime of the $2^2 S_{1/2}$ state of singly ionised helium using an ion-trapping technique, while Kocher et al. [21] and later Hinds et al. [22] did the same using a decay-in-flight method. In 1983, Gould and Marrus [23] carried out a further investigation of the two-photon decay process in hydrogen-like argon and derived a value for the Lamb shift by observing the quenching of the $2^2 S_{1/2}$ state in an electric field. Other closely related experiments of interest have been performed, for example that of Bannett and Freund [24] in which K-shell hole states in molybdenum were filled by $2S - 1S$ transitions with the accompanying emission of two X-ray photons, that of Braunlich et al. [25] in which singly stimulated emission from the $2^2 S_{1/2}$ state of atomic hydrogen was observed and that of Hippler [26] involving the observation of two-photon bremsstrahlung. More recently there has been considerable interest shown in the properties of the two photons produced in the process of parametric down conversion in which an incident (pump) photon can be considered to be split into two lower frequency (signal and idler) photons to form a highly correlated photon pair (Burnham and Weinberg [27]; Hong et al. [28]; Steinberg et al. [29]; Ou et al. [30]; Brendel et al. [31]).

Because of the long lifetime of the $2^2 S_{1/2}$ state, and the problems involved in producing a source of sufficient intensity, the experimental observation of the two-photon decay process in atomic hydrogen itself is very difficult. However, in 1975 the observation was carried out successfully at Stirling by O'Connell et al. [32] while almost simultaneously Kruger and Oed [33] at Tubingen published their results. These first experiments on atomic hydrogen concentrated on measurement of the lifetime of the metastable state as well as the angular correlation and spectral distribution of the two photons emitted in the decay process. However, considerable interest attaches to the measurement of the polarisation correlation of the two photons both as a sensitive test of the theory of the two-photon decay process itself and as a test of Bell's inequality [34] which allows a quantitative distinction to be made between, on the one hand, the predictions of quantum mechanics and, on the other, the predictions of local realistic or hidden variables theories. At Stirling, measurements of the linear and circular polarisation have been carried out and Bell's inequality put to the test [35, 36, 37, 38].

It is important to realise that experimental arrangements in which tests of Bell's inequality are possible are rare and represent the only known situations where, making an apparently reasonable additional assumption, one is compelled to choose between the validity of quantum mechanics and local realism. Nearly all previous tests have been made by observing the polarisation correlation either of the two rays that are emitted as a result of electron-positron annihilation or of the two photons emitted in succession in an atomic cascade when an excited atom decays to a state of lower energy via an intermediate state with a finite lifetime. Work in the field up to 1978 is reviewed in the articles by Pipkin [39] and Clauser and Shimony [40] and up to 1988 in the articles by Duncan [41] and Duncan and Kleinpoppen [42] which include reference to the important work of Aspect and his collaborators [43, 44, 45]). Unfortunately, because of the limited efficiency of detectors, all optical experiments to date to test Bell's inequality are subject to the criticism that the analysing polarisers used may themselves introduce a bias or enhancement effect into the probability of detection of a photon. To investigate such a possibility a series of experiments was carried out at Stirling in which a third element, for example a half-wave plate [36] or a a linear polariser [37] was inserted into the optical system.

The present article presents an account of work carried out on metastable atomic hydrogen to measure the polarisation correlation of the two photons produced in the decay process and to test Bell's inequality and ideas of enhancement. Before describing these experiments in detail, however, the theory of the two-photon decay will be discussed and an appropriate form of Bell's inequality will be derived.

2   Theory of the two-photon decay

For metastable atomic hydrogen [H($2S$)] electric dipole and electric quadrupole transitions from the $2^2S_{1/2}$ to $1^2S_{1/2}$ state are forbidden and, as a result, the H($2S$) state has a long lifetime of about $1/8$ second and decays primarily by the simultaneous emission of two photons as indicated in Figure 1.

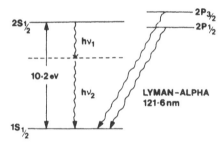

Figure 1. *Energy level diagram, neglecting hyperfine structure, for atomic hydrogen.*

Contributions to the decay of the $2^2S_{1/2}$ state are also possible by:

1. Single photon magnetic dipole transitions which only become significant, however, when relativistic effects are important, for example in high-Z hydrogenic ions. For hydrogen itself the lifetime of this type of transition is about 4×10^5 seconds [10] and can be neglected for most purposes.

2. A cascade involving the sequential emission of two photons through the $2^2P_{1/2}$ state which, because of the Lamb shift, lies slightly below the $2^2S_{1/2}$ state in energy. For hydrogen the associated lifetime is about 5×10^9 seconds [5] and hence this process can also be effectively neglected.

3. other types of two-photon transition. For example, the $2^2S_{1/2}$ state could decay by emitting two quadrupole photons [46] but the effect of such processes is negligible for the present purposes.

The theory of the two-photon decay, first put forward by Göppert-Mayer using second-order perturbation theory, can be developed by expanding the scattering operator S to second order in terms of the interaction Hamiltonian H_i between the atomic electron and the radiation field, the relevant part of the interaction Hamiltonian being given by $H_i = -(e/m)\mathbf{p}\cdot\mathbf{A}$ where e, m, and $\mathbf{p}$ are, respectively, the electron charge, mass, and canonical momentum, and $\mathbf{A}$ is the vector potential of the radiation field. This form for H_i gives rise to a matrix element describing the two-photon decay in the form

$$\frac{e^2}{m^2} \sum_j \frac{\langle\phi_f|\mathbf{p}\cdot\mathbf{A}|\phi_j\rangle\langle\phi_j|\mathbf{p}\cdot\mathbf{A}|\phi_i\rangle}{E_i - E_j}, \tag{1}$$

where $|\phi_i\rangle$, $|\phi_f\rangle$ represent, respectively, the initial and final states of the atom and radiation field together, and E_i is the energy of the initial state. The summation is to be taken over all allowed virtual intermediate states $|\phi_j\rangle$ of energy E_j which, in the present case, contain one more photon than the initial state $|\phi_i\rangle$. In terms of the electronic states $|\psi_i\rangle$, $|\psi_f\rangle$, and $|\psi_j\rangle$ of energies W_i, W_f, and W_j, respectively, when photons of energies $h\nu_1$ and $h\nu_2$ are emitted, it is possible to write

$$\begin{aligned}
|\phi_i\rangle &= |\psi_i\rangle|0\rangle \\
|\phi_j\rangle &= |\psi_j\rangle|\nu_1\rangle \text{ or } |\psi_f\rangle|\nu_2\rangle \\
|\phi_f\rangle &= \psi_f\rangle|\nu_1\rangle|\nu_2\rangle
\end{aligned} \tag{2}$$

where $|0\rangle$ represents the initial state of the radiation field, $|\nu_1\rangle$ ($|\nu_2\rangle$) its state when a photon of energy $h\nu_1$ ($h\nu_2$) has been added. The intermediate states $|\psi_j\rangle$ are, of course, the P states of atomic hydrogen. In terms of Feynman diagrams the process of two-photon emission can be represented as shown in Figure 2, which demonstrates the two possible ways the photons of energies $h\nu_1$ and $h\nu_2$ can be emitted.

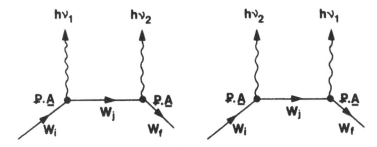

Figure 2. *Feynman diagrams for two-photon emission.*

If $\overline{\overline{A}}(\nu_1)d\nu_1$ denotes the probability per second for the spontaneous, simultaneous emission of two photons with one photon in the frequency range between ν_1 and $\nu_1 + d\nu_1$ irrespective of the directions of emissions or polarisation of the photons, then making the dipole approximation and changing from the dipole velocity to the dipole length form of the matrix element gives the result

$$\overline{\overline{A}}(\nu_1)d\nu_1 = \frac{1024\,\pi^6 e^4 \nu_1^3 \nu_2^3}{c^6} \left| \sum_j \left\{ \frac{\langle\psi_f|\mathbf{r}\cdot\hat{\mathbf{e}}_1|\psi_j\rangle\langle\psi_j|\mathbf{r}\cdot\hat{\mathbf{e}}_2|\psi_i\rangle}{W_i - W_j - h\nu_2} + \frac{\langle\psi_f|\mathbf{r}\cdot\hat{\mathbf{e}}_2|\psi_j\rangle\langle\psi_j|\mathbf{r}\cdot\hat{\mathbf{e}}_1|\psi_i\rangle}{W_i - W_j - h\nu_1} \right\} \right|^2_{\text{av}} d\nu_1 \tag{3}$$

where $\hat{e}_1$ and $\hat{e}_2$ represent unit vectors in the directions of polarisation of the two photons and the average is taken over the directions of emission and polarisation as indicated by the double bar over $A(\nu_1)$. The presence of the two terms in Equation (3) is the result of the fact that the emission of a complementary pair of photons of energies $h\nu_1$ and $h\nu_2$ can occur in either of the two ways shown in the Feynman diagrams in Figure 2, but it should be borne in mind that the two photons are in effect emitted simultaneously.

Finally, before leaving this discussion of the formal theory it should be noted that in the description of the two-photon decay it is customary practice to talk in terms of the directions of emission, the directions of polarisation, and the frequencies or energies of the photons. As the experiments to be described here forcefully demonstrate, the attribution of objective properties to photons in this way prior to an act of measurement, although semantically convenient, runs contrary to the fundamental concepts of quantum mechanics as expressed in the Copenhagen interpretation and, in some cases, can result in serious error.

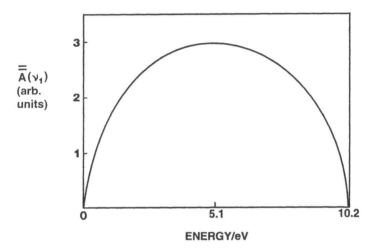

Figure 3. *Spectral distribution of the two photons emitted in the decay of H(2S).*

As a consequence of the above theoretical considerations the two-photon decay process is characterised by the following properties:

1. The energies $h\nu_1$, and $h\nu_2$ of the two complementary photons satisfy

$$h\nu_1 + h\nu_2 = 10.2\text{eV}. \qquad (4)$$

2. Each photon can have any energy from zero up to 10.2 eV, corresponding to a maximum frequency $\nu_0 = 2.47 \times 10^{15}$ Hz, with the spectral distribution shown in Figure 3. The corresponding allowed wavelengths for each photon range from infinity to 121.6nm with the maximum of the spectral distribution occurring at a wavelength of 243nm. Experimentally the fact that the spectral distribution has its maximum at a wavelength of 243nm is important, since it allows measurements

of photon polarisation to be made in air using simple lenses and pile-of-plates polarisers. The hydrogen decay is unique in this respect, all other hydrogenic two-photon decay processes and positronium annihilation producing photons of much shorter wavelength in regions of the spectrum where it is difficult to measure polarisation directly.

It is also interesting to note at this point that, for photon pairs for which the recoil momentum of the atom is less than the uncertainty in momentum resulting from localisation, the frequency component of the state vector representing the two photon radiation can be written in the form [47]

$$|\psi\rangle = \int a(\nu)|\nu\rangle_1 |\nu_0 - \nu\rangle_2 d\nu \tag{5}$$

where $\overline{\overline{A}}(\nu) = |a(\nu)|^2$.

3. Since the electric dipole operator er mediating the two-photon decay process is diagonal in the electronic and nuclear spin, fine and hyperfine structure plays no part in determining the decay process [3] or the properties, particularly the polarisation correlation properties, of the two photons emitted.

4. If $\hat{e}_1$ and $\hat{e}_2$ are unit linear polarisation vectors for photons of energies $h\nu_1$ and $h\nu_2$ detected at angle α relative to each other as shown in Figure 4, then it follows that the transition probability has a dependence on the polarisations of the photons given by

$$A(\nu_1) \propto |\hat{e}_1 \cdot \hat{e}_2|^2 = \cos^2\theta \tag{6}$$

where θ is the angle between $\hat{e}_1$ and $\hat{e}_2$, while averaged over polarisation

$$\overline{A}(\nu_1) \propto \langle|(|\hat{e}_1 \cdot \hat{e}_2|)^2|\rangle_{av} \propto (1 + \cos^2\alpha). \tag{7}$$

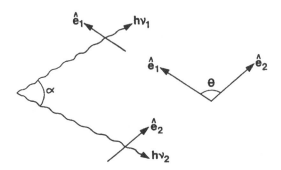

Figure 4. *Diagram to illustrate the angles involved in two-photon emission.*

When the two photons are detected in diametrically opposite directions, *i.e.* when $\alpha = \pi$, the polarisation part of the two-photon state vector takes a particularly interesting form, which can easily be derived from the requirements of conservation of angular momentum and parity. For example, in order to conserve angular momentum when the two photons are detected in opposite directions from a source whose angular momentum remains zero, it is necessary for the photons to have equal helicity. In terms of right-handed ($|R\rangle$) and left-handed ($|L\rangle$) helicity states it is thus expected that photon pairs will be found in states represented by $|R\rangle_1|R\rangle_2$, $|L\rangle_1|L\rangle_2$ or a superposition of these, where $|R\rangle_1$ denotes a photon of right-handed helicity propagating to the right, $|R\rangle_2$ a photon of right-handed helicity propagating to the left, and so on. In addition, since the initial and final states of the hydrogen atom both have even parity, the two-photon state vector must also possess even parity. Now, if P is the parity operator, $P|R\rangle_1 = |L\rangle_2$, $P|L\rangle_1 = |R\rangle_2$, etc., so that, to ensure even parity, the two-photon state vector must take the form

$$|\psi\rangle = 2^{-1/2}\left(|R\rangle_1|R\rangle_2 + |L\rangle_1|L\rangle_2\right). \tag{8}$$

Alternatively, assuming the photons are detected in the $+z$ and $-z$ directions, it is possible to describe the two-photon state vector in terms of the linear polarisation basis states $|x\rangle$ and $|y\rangle$ using the relations

$$\begin{aligned} |R\rangle_1 &= 2^{-1/2}\left(|x\rangle_1 + i|y\rangle_1\right), & |R\rangle_2 &= 2^{-1/2}\left(|x\rangle_2 - i|y\rangle_2\right) \\ |L\rangle_1 &= 2^{-1/2}\left(|x\rangle_1 - i|y\rangle_1\right), & |L\rangle_2 &= 2^{-1/2}\left(|x\rangle_2 + i|y\rangle_2\right). \end{aligned} \tag{9}$$

Substituting these relations into Equation (8) then gives

$$|\psi\rangle = 2^{-1/2}\left(|x\rangle_1|x\rangle_2 + |y\rangle_1|y\rangle_2\right) \tag{10}$$

as the appropriate state vector describing the polarisation properties of the two photons. The fact that this state vector is invariant in form with respect to rotation about the z axis plays an important role in the formulation of the Einstein-Podolsky-Rosen (EPR) argument (Einstein *et al.* [48]), as applied to photons, concerning the incompleteness of quantum mechanics and in the discussion of the nonlocality and lack of realism inherent in the quantum formalism.

It is interesting to note that both the frequency (Equation 5) and polarisation (Equation 10) components of the state vector are in what is now commonly referred to as an "entangled" form. Before any detection event takes place neither the polarisation nor frequency or energy of a single photon is defined or, indeed can be assigned any meaning. The entangled state vector represents the properties of the photon pair not single photons.

3   Bell's inequality

Experiments to test Bell's inequality usually involve placing a linear polariser and detector diametrically on each side of a source as shown in Figure 5. The form of the state vector given in Equations (5) and (10) then implies a strong correlation between the energies and polarisations of the two photons detected on either side of the source. For example, detection of a photon polarised, say, in the x direction on the left results in a collapse of the state vector to the form $|x\rangle_1|x\rangle_2$, and this implies that a photon polarised in the x direction will also be detected on the right. However, as has already been pointed out, since the form of the state vector in Equation (10) is invariant in form with respect to rotation about the z axis, the choice of x and y directions is quite

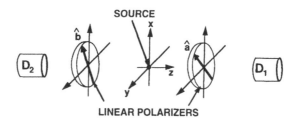

Figure 5. *Schematic diagram of the typical ideal experimental arrangement using two linear polarisers and two detectors D_1 and D_2. The linear polarisers are set with their transmission axes in the directions a and b.*

arbitrary. Thus, the polarisation of the photon, say on the right, can be considered to have been determined by the polarisation measurement chosen to be made on the other member of the pair on the left, despite the fact that the detection events are spatially separated in the relativistic sense. A similar type of argument can be applied in relation to the energy or frequency of the photons with interesting consequences [29].

The above discussion regarding polarisation, in fact, forms the essence of the Bohm [49, 50] and Bohm-Aharonov [51] versions of the Einstein-Podolsky-Rosen [48] argument against the completeness of quantum mechanics, for it implies that for every setting of the polariser on one side of the source, there exists an "element of reality" corresponding to the polarisation of the photon on the other side. That a photon can be linearly polarised in more than one direction at once is, of course, more than is allowed by quantum mechanics, and Einstein, Podolsky, and Rosen hoped that the situation might be resolved by the introduction of hidden variables, which would, at one and the same time, "complete" quantum mechanics and restore determinism to physics. However, the phenomenon is now seen more as a compelling example of the nonlocality inherent in nature as described by quantum mechanics, which, although it seems strange, is quite consistent with the Copenhagen interpretation in which the polarisation correlation must be considered a property of the photon pair and measuring apparatus taken as an indivisible whole. It is, on this view, not possible to regard individual photons as particles with objectively real properties whose values are independent of any measurements that may be made on them. It appears, therefore, that quantum mechanics presents a picture of the world that is neither local nor realistic in nature. Many people have found such a picture unsatisfactory, and attempts have been made to find local realistic theories (local hidden variables theories) to replace or supplement quantum mechanics. However, in 1964 Bell showed that, in a certain type of ideal experiment, local realistic theories were incapable of predicting the strong correlation given by quantum mechanics, and he set an upper bound on the strength of the correlation allowed by those theories, Subsequently, Clauser *et al.* [52] and, later, Clauser and Horne [53] extended the work of Bell so that it could be applied in actual experimental situations, where only a very small fraction of emitted photon pairs is detected, by introducing an additional assumption in either of the following two forms:

1. If a pair of photons emerges from two polarisers, the probability of their joint detection is independent of the orientation of the polarisers' axes.

2. For every atomic emission, the probability of a count with a polariser in place is not larger than with the polariser removed. The first form, due to Clauser *et al.* [52] can, of course be criticised on the grounds that it depends on the concept of emergence of individual photons from polarisers. The second form, due to Clauser and Horne [53] and not subject to this objection, was referred to by them as the no-enhancement assumption. Both forms are simply statements that the presence of the polarisers does not introduce any unsuspected bias or enhancement into the detection process for photon pairs.

Bell's inequality in the version suitable for use with the present series of experiments can be derived most easily following the method of Clauser and Horne [53]. Referring to Figure 5, consider a source emitting an ensemble of photon pairs, the state of each pair being described by a parameter λ (a hidden variable) of arbitrary complexity with a normalised probability density $\rho(\lambda)$. With reference to a pair described by λ, let the probability of detection on the right, when the corresponding polariser is orientated with its transmission axis in the direction $\hat{\mathbf{a}}$, be denoted by $p_1(\lambda, \hat{\mathbf{a}})$ with a similar definition for $p_2(\lambda, \hat{\mathbf{b}})$. The probability of joint detection of the two photons of a pair can be written $p_{12}(\lambda, \hat{\mathbf{a}}, \hat{\mathbf{b}})$. Then the ensemble probabilities $p_1(\hat{\mathbf{a}})$, $p_2(\hat{\mathbf{b}})$, and $p_{12}(\hat{\mathbf{a}}, \hat{\mathbf{b}})$ to detect a photon or a photon pair are given by

$$
\begin{aligned}
p_1(\hat{\mathbf{a}}) &= \int_\Gamma p_1(\lambda, \hat{\mathbf{a}})\rho(\lambda)d\lambda \\
p_2(\hat{\mathbf{b}}) &= \int_\Gamma p_2(\lambda, \hat{\mathbf{b}})\rho(\lambda)d\lambda \\
p_{12}(\hat{\mathbf{a}}, \hat{\mathbf{b}}) &= \int_\Gamma p_{12}(\lambda, \hat{\mathbf{a}}, \hat{\mathbf{b}})\rho(\lambda)d\lambda
\end{aligned}
\tag{11}
$$

where Γ is the space of the states λ.

It should be noted that, since λ only determines the probability of detection, the inequality to be derived here applies to inherently stochastic local realistic theories. It is the question of locality, not strict determinism, that is of crucial importance to the argument, and the assumption of locality is made, in this case, by demanding that

$$
p_{12}(\lambda, \hat{\mathbf{a}}, \hat{\mathbf{b}}) = p_1(\lambda, \hat{\mathbf{a}})p_2(\lambda, \hat{\mathbf{b}}).
\tag{12}
$$

The no-enhancement assumption is introduced, for every λ and for all orientations $\hat{\mathbf{a}}$ and $\hat{\mathbf{b}}$, in the form

$$
\begin{aligned}
0 &\leq p_1(\lambda, \hat{\mathbf{a}}) \leq p_1(\lambda, \infty) \leq 1 \\
0 &\leq p_2(\lambda, \hat{\mathbf{b}}) \leq p_2(\lambda, \infty) \leq 1
\end{aligned}
\tag{13}
$$

where the symbol ∞ is used to indicate the absence of a polariser. Now, as shown by Clauser and Horne [53], if x, x', y, y', X, and Y are real numbers such that $0 < x, x' < X$ and $0 < y, y' < Y$ then

$$
-XY \leq xy - xy' + x'y + x'y' - Yx' - Xy \leq 0.
\tag{14}
$$

Thus, if $\hat{\mathbf{a}}'$ and $\hat{\mathbf{b}}'$ are alternative settings of the polarisers, the above result can be used to write the inequality

$$
\begin{aligned}
-p_1(\lambda,\infty)p_2(\lambda,\infty) \;\leq\;\; & p_1(\lambda,\hat{\mathbf{a}})p_2(\lambda,\hat{\mathbf{b}}) - p_1(\lambda,\hat{\mathbf{a}}))p_2(\lambda,\hat{\mathbf{b}}') \\
& + p_1(\lambda,\hat{\mathbf{a}}')p_2(\lambda,\hat{\mathbf{b}}) + p_1(\lambda,\hat{\mathbf{a}}')p_2(\lambda,\hat{\mathbf{b}}') \\
& - p_2(\lambda,\infty)p_1(\lambda,\hat{\mathbf{a}}') - p_1(\lambda,\infty)p_2(\lambda,\hat{\mathbf{b}}) \\
\leq\;\; & 0
\end{aligned}
\tag{15}
$$

which, after multiplying by $\rho(\lambda)$ and integrating over λ, becomes

$$
\begin{aligned}
-p_{12}(\infty,\infty) \;\leq\;\; & p_{12}(\hat{\mathbf{a}},\hat{\mathbf{b}}) - p_{12}(\hat{\mathbf{a}},\hat{\mathbf{b}}') + p(\hat{\mathbf{a}}',\hat{\mathbf{b}}) + p_{12}(\hat{\mathbf{a}}',\hat{\mathbf{b}}') \\
& - p_{12}(\hat{\mathbf{a}}',\infty) - p_{12}(\infty,\hat{\mathbf{b}}) \\
\leq\;\; & 0.
\end{aligned}
\tag{16}
$$

This inequality, which, conveniently, only contains probabilities p_{12} for joint detection, can be simplified if the source possesses rotational symmetry about the propagation axis since, then, $p_{12}(\hat{\mathbf{a}}',\infty)$ and $p_{12}(\infty,\hat{\mathbf{b}})$ are independent of $\hat{\mathbf{a}}'$ and $\hat{\mathbf{b}}$, respectively, and $p_{12}(\hat{\mathbf{a}},\hat{\mathbf{b}}), p_{12}(\hat{\mathbf{a}},\hat{\mathbf{b}}')$, etc. depend only on the relative angle $(\hat{\mathbf{a}},\hat{\mathbf{b}}), (\hat{\mathbf{a}},\hat{\mathbf{b}}')$ etc. between the orientations of the transmission axes of the polarisers. By choosing the particular relative orientations $(\hat{\mathbf{a}},\hat{\mathbf{b}}) = (\hat{\mathbf{a}}',\hat{\mathbf{b}}) = (\hat{\mathbf{a}}',\hat{\mathbf{b}}') = \theta$, $(\hat{\mathbf{a}},\hat{\mathbf{b}}') = 3\theta$ for $\hat{\mathbf{a}}, \hat{\mathbf{a}}', \hat{\mathbf{b}}$, and $\hat{\mathbf{b}}'$, Inequality (16) can be expressed as

$$
-p_{12}(\infty,\infty) \leq 3p_{12}(\theta) - p_{12}(3\theta) - p_{12}(\hat{\mathbf{a}}',\infty) - p_{12}(\infty,\hat{\mathbf{b}}) \leq 0.
\tag{17}
$$

Experimentally, of course, probabilities are deduced from observations carried out over an ensemble of emissions. If $R(\hat{\mathbf{a}},\hat{\mathbf{b}})$ denotes the number or rate of joint detections (coincidence counts) with the transmission axes of the polarisers set in the directions $\hat{\mathbf{a}}$ and $\hat{\mathbf{b}}$, and R_0 the number or rate with both polarisers removed, then $p_{12}(\hat{\mathbf{a}},\hat{\mathbf{b}})/p_{12}(\infty,\infty) = R(\hat{\mathbf{a}},\hat{\mathbf{b}})/R_0$ with similar relations for the other probabilities. Inequality (17) then takes the form

$$
-1 \leq \left[3R(\theta) - R(3\theta) - R(\hat{\mathbf{a}}',\infty) - R(\infty,\hat{\mathbf{b}}) \right]/R_0 \leq 0.
\tag{18}
$$

Writing Inequality (18) for $\theta = 22.5°$ and $\theta = 67.5°$ and subtracting gives Bell's inequality in the form

$$
\eta = \left| \frac{R(22.5°) - R(67.5°)}{R_0} \right| \leq 0.25
\tag{19}
$$

originally derived by Freedman [54]. The Freedman form of the inequality has proved particularly useful from the experimental point of view since it only necessitates making three measurements.

In contrast, it follows directly from consideration of the state vector given in Equation (10), which predicts a $(1 + \cos 2\theta)/4$ variation in the ratio $R(\theta)/R_0$, that, in the ideal case, according to quantum mechanics, $\eta_{\mathrm{QM}} = 0.354$, in clear violation of the Freedman form of Bell's inequality. However, various factors act in practice to reduce η_{QM} and care must be taken experimentally to ensure that η_{QM} does not fall below 0.25. The main factors acting to reduce η_{QM} in this way are the transmission efficiencies of the polarisers and the finite solid collection angle subtended at the source by the lenses normally placed on either side of the source in front of the polarisers. If the

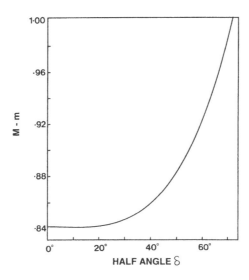

Figure 6. *Lower limit on $(M - m)$ as a function of the half-angle δ subtended by the collection optics at the source. In order for a meaningful test of Bell's inequality to be made, the experiment must be carried out in the region above the curve.*

transmission efficiencies of the polarisers (assumed identical) for light polarised parallel to and perpendicular to their transmission axes are M and m, respectively, and if the half-angle subtended at the source by each lens is δ, then it can be shown that according to quantum mechanics (Clauser *et al.* [52])

$$\frac{R(\theta)}{R_0} = \frac{1}{4}(M + m)^2 + \frac{1}{4}(M - m)^2 F(\delta) \cos 2\theta \qquad (20)$$

where θ is the relative angle between the transmission axes and $F(\delta)$ is a geometrical factor, depending on δ, which takes into account the effect of the non-colinear emission of photon pairs. Thus, on the basis of quantum mechanics, in a real situation , it follows that

$$\eta_{QM} = \frac{F(\delta)}{2\sqrt{2}}(M - m)^2 \qquad (21)$$

and, clearly, this result places constraints on the values of M, m and δ if a meaningful test of Bell's inequality is to be made. The requirement that $\eta_{QM} > 0.25$, in fact, sets a lower limit on the allowed value for $(M - m)$ which depends on δ in the way shown in Figure 6.

4   The two-photon apparatus

In the apparatus, shown schematically in Figure 7, a 1keV beam of metastable atomic deuterium [D($2S$)], of density about 10^4cm^{-3}, was produced by a charge exchange process in which deuterons extracted from a radio-frequency ion source passed through cesium vapour in a cell constructed according to a design by Bacal *et al.* [55, 56].

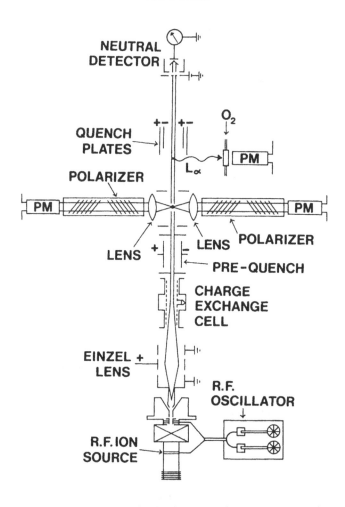

Figure 7. *Schematic diagram of the Stirling two-photon apparatus (not to scale).*

Deuterium was used rather than hydrogen since, for a given metastable density and hence two-photon signal, the noise generated by the interaction of the beam with the background gas was less in the former case. Also, although it was possible to produce higher beam densities at higher energies, it was found, in practice, that the best statistical accuracy in a given time could be achieved at an energy of 1keV.

After emerging from the charge exchange cell and being collimated down to about 10mm in diameter, the D(2S) beam passed through the electric field prequench plates which effectively allowed the metastable component of the beam to be switched on and off by making use of the effect of Stark mixing of the $2^2S_{1/2}$ and $2^2P_{1/2}$ states. In the same way, at the end of the apparatus the beam was totally quenched and the resulting Lyman-α signal used to normalise the two-photon coincidence signal. The Lyman-α signal itself was detected by means of a solar blind UV photomultiplier used together with an oxygen filter cell with lithium fluoride windows through which dry oxygen flowed at a constant rate. A series of experiments was carried out to check that the two-photon coincidence signal was indeed proportional to the Lyman-α signal over the range of conditions encountered in practice.

At right angles to the metastable beam the two-photon radiation was collected and collimated by two 50mm diameter lenses, each with a focal length of 43mm at a wavelength of 243nm. Each lens was placed close to 50mm from the beam, and subtended a half angle $\delta \simeq 23°$ at the beam. The position of the lenses was chosen so that, for a wavelength of 243nm, the centre of the beam formed an image on the 46-mm-diam cathodes of the photomultipliers, 53cm from the centre of the beam. By considering the design of the optical system it was estimated that only a 4mm diameter section of the beam acted as a source for two-photon radiation. For the linear polarisation correlation measurements, two high-transmission ultraviolet polarisers, each made from 12 plates polished flat to 2λ at 243nm and set nearly at Brewster's angle, were placed on either side of the source as shown in Figure 7. For other polarisation correlation measurements additional optical elements such as quarter-wave plates, half-wave plates, and linear polarisers could be inserted in the optical system as required. The lenses and plates were made from high-quality fused silica with a short wavelength cut off at 160nm. However, because of absorption in oxygen, the short wavelength cutoff occurred in practice at 185nm, which in turn implied a long wavelength cutoff at the complementary wavelength of 355nm. Hence, given that the quantum efficiency of the photomultipliers was fairly constant at about 20% over this range, all photons in the wavelength range 185–355nm were capable of contributing to the two-photon coincidence signal. Also, in order to compare the experimental results with theory it was necessary to know the values of the transmission efficiencies M and m for light polarised, respectively, parallel to and perpendicular to the transmission axes of the polarisers, and these quantities were measured carefully in a subsidiary experiment, making use of the 254nm line from a mercury lamp. In this experiment, using the apparatus as shown in Figure 7, the photomultiplier was removed from one arm and an almost parallel beam of polarised light was passed through the pile-of-plates polariser (aligned with its transmission axis parallel to the plane of polarisation of the light from the mercury source) to come to a focus at the position normally occupied by the centre of the atomic source. The light emerging from this focus was then analysed by the pile-of-plates polariser and photomultiplier in the other detection arm. After application of a small correction to take into account the slight wavelength dependence of the absorption in the fused silica between 185 and 200nm, it was found that for two of the polarisers $M = 0.908 \pm 0.013$ and $m = 0.0299 \pm 0.0020$, while, for a third polariser, constructed from plates provided by a different manufacturer, the values $M = 0.938 \pm 0.010$ and $m = 0.040 \pm 0.002$ were obtained.

The signal was detected on either side of the source by fast-rise-time photomultipliers, with bialkali cathodes and fused silica windows, which were specially selected for high gain, good single-photoelectron resolution, and good timing characteristics, and had a low dark count rate of about $10^2 s^{-1}$. The photomultiplier pulses were fed to the coincidence circuit previously described by O'Connell *et al.* [32] which consisted of the usual combination of constant fraction discriminators, a time-to-amplitude converter, and a multichannel analyser operated in the pulse-height analysis mode as shown in Figure 8.

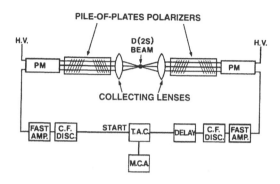

Figure 8. *Detection optics and electronics (not to scale).*

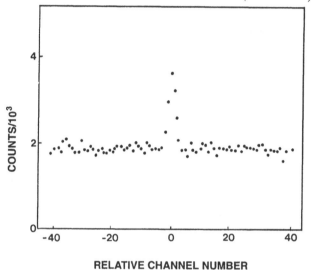

RELATIVE CHANNEL NUMBER

Figure 9. *A typical time-correlation spectrum (Perrie et al. [35]) after subtraction of the spectrum obtained with the metastable component of the beam quenched. Polariser plates are removed. Time delay per channel is 0.8 ns, total collection time 21.5 h. Singles rate with metastables present (quenched) is about $1.15 \times 10^4 s^{-1}$ $(0.85 \times 10^4 s^{-1})$. True two-photon coincidence rate is $490\ h^{-1}$.*

The time correlation spectra obtained in this way with the metastable atoms present and quenched were stored in separate segments of the multichannel analyser memory and subtracted at the end of each run, which normally lasted for about 22 hours. This procedure was adopted to ensure the elimination of any spurious true coincidence events due, for example, to cosmic rays and residual radioactivity in the apparatus. With the photomultiplier cathodes 1.06 m apart these events occurred at a rate of about $10^{-2}\ s^{-1}$, which decreased as the separation of the photomultipliers increased. A typical spectrum obtained in this way is shown in Figure 9, and, as expected for a simultaneous-emission process, the coincidence peak can be seen to be symmetrical. It should also be noted that the background signal resulting from the typical singles count rate of order

10^4 s^{-1} was due mainly to radiation produced by interaction of the atomic beam with background gas in the vacuum system at a pressure of about 2×10^{-7} Torr, and only very little (about 0.01%) to uncorrelated photons from the two-photon decay process itself.

5   Quantum mechanical predictions

In terms of spinors the state vector given by Equation (10) can be written as

$$|\psi\rangle = \frac{1}{\sqrt{2}} \left[\begin{pmatrix} 1 \\ 0 \end{pmatrix} \otimes \begin{pmatrix} 1 \\ 0 \end{pmatrix} + \begin{pmatrix} 0 \\ 1 \end{pmatrix} \otimes \begin{pmatrix} 0 \\ 1 \end{pmatrix} \right] \tag{22}$$

and the polarisation state can conveniently be represented in terms of the density matrix $\rho = |\psi\rangle\langle\psi|$ where here

$$\rho = \frac{1}{2} \begin{pmatrix} 1 & 0 & 0 & 1 \\ 0 & 0 & 0 & 0 \\ 0 & 0 & 0 & 0 \\ 1 & 0 & 0 & 1 \end{pmatrix}. \tag{23}$$

Optical elements whose action is represented by the matrix operators A and B are placed, respectively on the right $(+z)$ and left $(-z)$ side of the source then the resulting probability of joint detection of photons, taken experimentally to be proportional to the coincidence rate R, compared to the probability of joint detection with the optical elements removed, taken as proportional to the coincidence rate R_0 is given by

$$\begin{aligned} \frac{R}{R_0} &= \frac{\text{Trace}\left[B \otimes A\rho A^\dagger \otimes B^\dagger\right]}{\text{Trace}(\rho)} \\ &= \text{Trace}\left[B \otimes A\rho A^\dagger \otimes B^\dagger\right] \end{aligned} \tag{24}$$

since Trace $(\rho) = 1$. The quantities $A^\dagger$, $B^\dagger$ are the Hermitian conjugate operators and $\otimes$ denotes the direct product operation.

For example, in the case that A represents the action of a linear polariser orientated with its transmission axis at angle θ to the x-axis, then (Haji-Hassan [57]),

$$A = \frac{1}{2} \begin{bmatrix} (A_1+A_2) + (A_1-A_2)\cos 2\theta & (A_1-A_2)\sin 2\theta \\ (A_1-A_2)\sin 2\theta & (A_1+A_2) - (A_1-A_2)\cos 2\theta \end{bmatrix} \tag{25}$$

where A_1 and A_2 are the *transmission amplitudes* (in general complex) for light polarised parallel to and perpendicular to the polariser transmission axis respectively; assign $|A_1|^2 = M$, and $|A_2|^2 = m$. For two polarisers, one placed in each side of the source, straightforward application of Equation (24) then leads directly to Equation (20) with the geometrical factor $F(\delta)$ taken as unity.

On the other hand, the effect of a retardation plate whose fast axis makes an angle α with the x axis and which introduces a phase delay ϕ between the ordinary and extraordinary rays, may be represented by the operator

$$R_{\phi,\alpha} = \begin{bmatrix} \cos \alpha \\ \sin \alpha \end{bmatrix} [\cos \alpha, \sin \alpha] + e^{i\phi} \begin{bmatrix} -\sin \alpha \\ \cos \alpha \end{bmatrix} [-\sin \alpha, \cos \alpha] . \tag{26}$$

6 Measurement of polarisation correlation

6.1 Linear polarisation correlation experiment

With two polarisers placed as shown in Figure 7, measurements were made of the coincidence rate $R(\theta)$ for various angles between the transmission axes of the polarisers, and the rate R_0 with both sets of polariser plates removed (Perrie *et al.* [35]). The photomultipliers were linked to their corresponding polarisers and rotated with them so that any sensitivity to polarisation that might exist in the photomultiplier windows or cathodes could be ignored. In addition, the rotational invariance of the detection system was checked by making sure that the background singles rate was constant as the angle of the polariser transmission axis was varied in each arm independently. All coincidence rates were normalised to a typical Lyman-α count rate, and the resulting linear polarisation correlation measurement is shown in Figure 10.

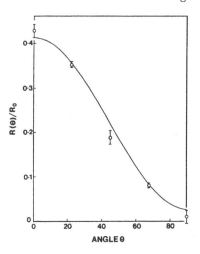

Figure 10. *Coincidence signal $R(\theta)$ in the linear polarisation correlation experiment as a function of the angle θ between the transmission axes of the two polarisers, relative to R_0, the coincidence signal with the polariser plates removed. The solid curve represents the η_{QM} prediction using the median values for M and m (Perrie et al. [35]).*

The error bars represent one standard deviation and vary from point to point because of the different total counting times allocated to different angles. In particular, the points at 22.5° and 67.5° resulted from measurements that each involved a total counting time of 240 h. The solid curve, which represents the quantum mechanical prediction derived from Equation (20) with $M = 0.908$, $m = 0.0299$, and $F(\delta) = 0.996$ corresponding to $\delta = 23°$, clearly agrees with the experimental results within the limits of error. Also, from the experimental points at 22.5° and 67.5°, it was found that $\eta = 0.268 \pm 0.011$, which violates the Freedman form of Bell's inequality by just less than two standard deviations, and agrees with the quantum mechanical prediction $\eta_{QM} = 0.272 \pm 0.008$ calculated from Equation (21).

6.2   Circular polarisation correlation experiment

Although many experiments have been carried out to measure the linear polarisation correlation of the two photons emitted in certain atomic decay processes, measurements of the circular polarisation correlation are rare being limited to those of Clauser [58] and Haji-Hassan [38]. In this latter experiment carried out at Stirling using a metastable atomic deuterium source, an achromatic quarter wave plate was placed between the source and linear pile-of-plates polariser on each side of the apparatus as shown in Figure 11.

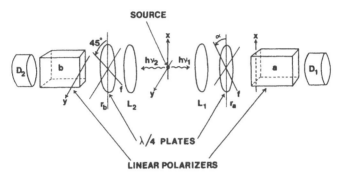

Figure 11. *Schematic diagram of the apparatus to measure circular polarisation correlation. D_1 and D_2 are photomultipliers. The transmission axis of polariser b is fixed in the y direction, that of polariser a in the x direction. Quarter-wave plate r_b is orientated with its fast axis (f) at 45° to the x axis while the fast axis of quarter-wave plate r_a is rotated through angle α relative to the x axis.*

The achromatic quarter-wave plates consisted of a combination of two double-plates one of crystal quartz and one of magnesium fluoride. Each plate of diameter 19.5mm was cut parallel to the optic axis and polished flat to an accuracy of $\lambda/10$. The retardation of the plates was nominally achromatic with a retardation which varied by 10% over the wavelength range from 180nm to 300nm provided the direction of the incident light did not deviate by more than 1° from normal incidence. The performance of the plates at wavelengths of 254nm and 283nm was checked in a subsidiary experiment.

In carrying out the experiments, as indicated in Figure 11, the linear polariser *b* was orientated with its transmission axis fixed in the *y* direction and that of the linear polariser *a* was fixed in the *x* direction. Thus, in the absence of the quarter-wave plates, the coincidence signal was a minimum (ideally zero). The quarter-wave plates r_a and r_b were inserted so that r_b in the left-hand arm had its fast axis orientated at an angle of 45° to the *x* axis to form, along with the linear polariser in that arm, an analyser for photons of right-handed helicity, while r_a in the right-hand arm was rotated through an angle α, relative to the *x* axis, from a position where the combination of linear polariser and quarter-wave plate acted as an analyser for photons of right-handed helicity (fast axes parallel, $\alpha = 45°$) to a position where the combination acted as an analyser for photons of left-handed helicity (fast axes perpendicular, $\alpha = -45°$). The experimental coincidence signal $R(\alpha)$ thus obtained is shown in Figure 12 for various angles, α, relative to R_0 the coincidence signal with the plates of the linear polarisers

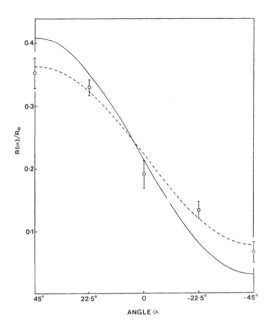

Figure 12. *Coincidence signal $R(\alpha)$ as a function of the angle α made by the fast axis of quarter-wave plate r_a with the x axis, relative to R_0 the coincidence rate with the plates of both linear polarisers removed. The error bars denote one standard deviation. The broken curve represents a least squares fit to the data of the form $P + Q\sin 2\alpha + R\sin^2 2\alpha$ with $P = 0.222$, $Q = 0.143$ and $R = -0.001$. The full curve shows the wavelength averaged quantum mechanical prediction. (Haji-Hassan et al. [38]).*

removed. The results are fitted well by a curve of the form $P + Q\sin 2a + R\sin^2 2a$ with $P = 0.222 \pm 0.029$, $Q = 0.143 \pm 0.014$, $R = -0.001 \pm 0.036$, shown as the broken curve in Figure 12.

The quantum mechanical prediction, shown as the full curve in Figure 12 was obtained by evaluating the trace of $AR_a \otimes BR_b\rho R_a^\dagger A^\dagger \otimes R_b^\dagger B^\dagger$ where A, B, R_a and R_b are the matrix operators representing the action of the linear polarisers and quarter-wave plates a, b, r_a and r_b respectively, $A^\dagger, B^\dagger, R_a^\dagger$ and $R_b^\dagger$ are their Hermitian conjugates and ρ is the density matrix describing the two-photon state. If M_A and m_A are, respectively, the transmission efficiencies for light polarised parallel to and perpendicular to the transmission axis of polariser a with similar definitions for M_B and m_B, and ϕ_a and ϕ_b are the retardations of the quarter-wave plates r_a and r_b, then it follows that

$$\left[\frac{R(\alpha)}{R_0}\right]_\lambda = \frac{1}{4}\Big\{(M_A + m_A)(M_B + m_A) - (M_A - m_A)(M_B - m_A)\cos\phi_b$$
$$+ (M_A - m_A)(M_B - m_B)\sin\phi_a\sin\phi_b\sin 2\alpha$$
$$+ (M_A - m_A)(M_B - m_B)\cos\phi_b(1 - \cos\phi_a)\sin^2 2\alpha\Big\} \qquad (27)$$

where the subscript λ on the left-hand side indicates that the expression is wavelength dependent. Averaging over wavelength taking account of the spectral distribution of the source and the absorption characteristics of the various lenses and windows in the optical system gives

$$[R(\alpha)/R_0]_{\text{mean}} = 0.215 + 0.189 \sin 2\alpha + 0.0061 \sin^2 2\alpha \qquad (28)$$

and this quantity was plotted as the full curve in Figure 5. Examining Figure 12 it is clear that, even taking account of all the factors described above, significant differences remain between theory and experiment.

Finally, it should be noted that the measurements above do not suffice to form a test of Bell's inequality. In previous tests, using linear polarisers alone, the transmission axis of one of the linear polarisers was rotated about the observation axis. In order to use the simple Freedman form of Bell's inequality it was necessary to assume that :

1. The source was rotationally symmetrical about the observation axis.

2. The coincidence signal depended only on the relative angle between the transmission axes of the polarisers on either side of the source.

In this experiment it was the fast axis of one of the quarter-wave plates which was rotated and the angle this fast axis makes with the x axis becomes the variable parameter corresponding to the orientation of the transmission axis of the linear polariser in previous measurements. However, the coincidence signal in this case is not expected to be dependent only on the relative angle between the fast axes of the two quarter-wave plates and the Freedman form of the inequality is not, therefore, applicable. It would, in principle, be possible to test a more general form of Bell's inequality which did not depend on assumption 2 above but measurements additional to those reported here would be required. In view of the difficulties encountered with the use of quarter-wave plates it is unlikely that such a test would result in a violation of Bell's inequality.

6.3 Test for enhancement using a half-wave plate

Almost all the experiments making use of either a positronium annihilation or atomic cascade source have produced results that have violated Bell's inequality and have agreed with the predictions of quantum mechanics. However, the positronium annihilation experiments are open to the criticism (Pipkin [39]; Clauser and Shimony [40]) that the validity of quantum mechanics itself must be assumed in order to analyse the experimental results, and some discussion (Aspect [45]; Marshall *et al.* [59]; Selleri [60]; Pascazio [61]) has taken place regarding the validity of the results from atomic cascade experiments where the effects of absorption and reemission could possibly be important. The metastable atomic deuterium experiment cannot be criticised on either of these grounds, and also the simultaneous nature of the decay ensured that the detection events for every pair were spacelike separated. Nevertheless, because of the low detection efficiency of photomultipliers in the ultraviolet and visible part of the spectrum and the finite solid angle subtended by the detection system at the source, the possibility remained that the results could be explained in local realistic terms if an additional assumption of the kind discussed in Section 3 regarding lack of bias or enhancement in the detection process was not made. Therefore, from a purely logical viewpoint, no existing experiment could be considered to have provided grounds for a complete rejection of local realistic theories. It followed that it was important to carry out experimental tests either in situations where a distinction could be drawn between the predictions of quantum mechanics and local realism without making any additional assumptions or in situations where the validity of the assumptions themselves could be investigated directly.

Of particular interest here is the work of Garuccio and Selleri [62] who described how all existing results involving single-photon physics and two-polariser-type experiments could be explained in local realistic terms if a detection vector $\boldsymbol{\lambda}$, in addition to a polarisation vector $\mathbf{I}$, was attributed to each photon. The detection vector was assumed to be unaffected by passage through linear polarisers, but the detection probability was postulated to depend on the angle between $\mathbf{I}$ and $\boldsymbol{\lambda}$. In effect, then, they denied the validity of the additional assumptions and assumed a particular mechanism leading to enhancement. The Clauser, Horne, Shimony, and Holt [52] form of the additional assumption and the above proposal of Garuccio and Selleri [62] involved the concept of photons emerging from polarisers and both were tested in a fairly simple way by Haji-Hassan *et al.* [36] in an extension to the previously described linear polarisation correlation measurements, in which a half-wave plate was inserted in one detection arm of the apparatus between the linear polariser and photomultiplier: see Figure 13.

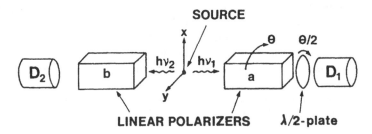

Figure 13. *Arrangement for two-photon polarisation correlation measurements with a half-wave plate placed between the linear polariser and photomultiplier in one of the detection arms. The transmission axis of the right-hand linear polariser is rotated through angle θ, the fast axis of the half-wave plate through angle $\theta/2$, relative to the x axis.*

Assuming that only the relative orientation of the polarisers' axes was important and that the polarisation state of each photon emerging from a polariser was determined by the setting of the polariser through which it had passed, the Clauser, Horne, Shimony, and Holt [52] form of the additional assumption could be expressed in the following form. If a pair of photons emerges from two polarisers, the probability of their joint detection is independent of the relative angle between the planes of polarisation of the two photons just prior to detection. The approach of Garuccio and Selleri [62] was based precisely on the denial of this assumption since, if their detection vector was the same for both members of a photon pair, then the detection probability for the pair had to depend on the angle between their respective planes of polarisation as they impinged on the cathodes of the photomultipliers. In the previous linear polarisation correlation measurements, of course, this angle changed as the polarisers were rotated relative to each other. However, by adding a half-wave plate to the system, as shown in Figure 12, the plane of polarisation of the photon incident on photomultiplier D, could be varied independently of the setting of polariser a. Initially, the transmission axes of both polarisers and the fast axis of the half-wave plate were arranged to be parallel. Then, when the polariser a was rotated through an angle θ, the fast axis of the half-wave plate was rotated through an angle $\phi = \theta/2$. In this way, it was arranged that, within the limits set by the imperfections of the nominally achromatic half-wave plate, which were similar to those of the quarter-wave plate described in Section 6.2, the orientation of the plane of polarisation of the photon incident on the photomultiplier did not change as the rotation of the polariser and half-wave plate took place. It followed that the relative angle between the planes of the polarisation did not change significantly as the measurement proceeded and that any enhancement effect that might have existed in previous experiments could not now distort the results.

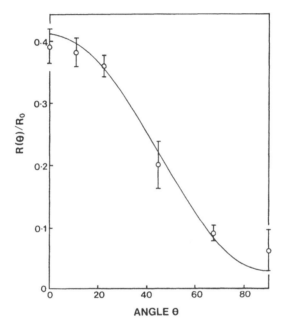

Figure 14. *Coincidence signal $R(\theta)$ as a function of the angle between the transmission axes of the linear polarisers, relative to R_0, the coincidence signal with the linear polariser plates removed but the half-wave plate still in place. The solid curve represents the quantum mechanical prediction, and is identical to that shown in Figure 10 (Haji-Hassan et al. [36]).*

Results of the experiment are shown in Figure 14 where $R(\theta)$ represents the coincidence rate with angle between the transmission axes of the polarisers and R_0 the rate with the plates of both linear polarisers removed. The quantum mechanical prediction derived from Equation (18), with the values of M, m, and $F(\delta)$ given in Section 6.1, is shown as a solid line and agrees with the experimental points within the limits of error.

In a further experiment, the transmission axes of the polarisers were fixed parallel to each other and the fast axis of the half-wave plate rotated through angle ϕ relative to the axes of the polarisers so that it could be argued that the relative angle between

the planes of polarisation of the photons impinging on the photomultipliers could be
varied continuously. There was, however, no indication within the limits of experimental
error, of any dependence of the signal on the orientation of the fast axis of the half-wave
plate. Finally, it was verified that the singles rate did not vary as the half-wave plate
was rotated.

These measurements using a half-wave plate indicate that, insofar as it is acceptable
to talk in terms of individual photons emerging from the polarisers, the relative angle
between the planes of polarisation of the two photons just prior to detection plays no
role in establishing the observed linear polarisation correlation and, hence, that the
additional assumption in the form proposed by Clauser, Horne, Shimony, and Holt
[52] is valid. There is, also, no evidence here to support the idea of enhanced photon
detection along the lines proposed by Garuccio and Selleri [62].

If it is assumed that the presence of the half-wave plate did, indeed, have no effect on
the results, then the measurements shown in Figure 14 can be used to test Freedman's
form of Bell's inequality, Equation (19). In this case $\eta = 0.271 \pm 0.021$, violating Bell's
inequality, although only by one standard deviation. However, combining this result
with that given for η in Section 6.1 gave $\eta = 0.269 \pm 0.009$ in violation of Bell's inequality
by just more than two standard deviations.

6.4   Test for enhancement using three polarisers

The experiments with the half-wave plate described in Section 6.3 did not rule out the
possibility that enhancement might occur as a result of the setting of the linear polarisers
themselves since, although the detection vector θ might not be affected by passage
through a linear polariser, it was possible to imagine that it was rotated in the same
way as the polarisation vector $\mathbf{I}$ by passage through a half-wave plate. Consequently,
as suggested by Garuccio and Selleri [62] it appeared that it would be interesting to
carry out experiments with a linear polariser in place of the half-wave plate, since, in
particular, in this situation a finite and measurable difference was predicted to exist
between quantum mechanics and their local realistic model. The introduction of a third
linear polariser also afforded an opportunity to test the quantum formalism in hitherto
unexplored and novel circumstances.

The experiment was performed by Haji-Hassan *et al.* [37] with an additional linear
polariser inserted as shown in Figure 15. The orientation of the polariser a was held
fixed while polariser b was rotated through an angle β in a clockwise sense and polariser
a' through an angle α' in the opposite sense. The objective of the experiment was then
to measure, for various angles β, the ratio $R(\beta, \alpha')/R(\beta, \infty)$, where $R(\beta, \alpha')$ was the
coincidence rate with all polarisers in place, and $R(\beta, \infty)$ the rate with the plates of
polariser a' removed. The corresponding quantum mechanical prediction is obtained by
evaluating the trace of expressions of the form $(A'A \otimes B\rho A^\dagger A'^\dagger B^\dagger)$ where A', A, and B
are matrices representing the action of polarisers a', a, and b, $A'^\dagger, A^\dagger$, and $B^\dagger$ are their
Hermitian conjugates, and ρ is the density matrix describing the two-photon state upon
emission. The result is

$$\frac{R(\beta\alpha')}{R\beta,\infty)} = \frac{1}{2}(M_{A'}+m_{A'}) + \frac{1}{2}(M_{A'}-m_{A'})\left(\frac{M_A P - m_A Q}{M_A P + m_A Q}\right)\cos 2\alpha' - \Delta(\beta, \alpha'),\ (29)$$

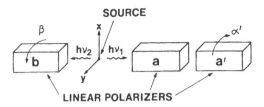

Figure 15. *Arrangement of apparatus for the three-polariser experiment. The orientation of polariser a is fixed with its transmission axis parallel to the x axis, while the transmission axes of polarisers b and a' are rotated, respectively, through angles β and α' relative to the x axis.*

where

$$P = \frac{1}{2}[(M_B + m_B) + (M_B - m_B)\cos 2\beta],$$

$$Q = \frac{1}{2}[(M_B + m_B) - (M_B - m_B)\cos 2\beta].$$

The term $\Delta(\beta, \alpha')$ results from interference between the wanted light (light polarised in a direction parallel to the transmission axis of the polariser) and the unwanted light (light polarised in a direction perpendicular to the transmission axis) passing through polariser a, and is given by

$$\Delta(\beta, \alpha') = \frac{(M_A m_A)^{1/2}(M_B - m_B)(M_{A'} - m_{A'})}{2(M_A P + m_A Q)} \sin 2\beta \sin 2\alpha' \cos \phi$$

where M_A and m_A are the transmission efficiencies (the moduli squared of the transmission amplitudes) for light polarised parallel and perpendicular to the transmission axis of polariser a, respectively, with similar definitions for M_B, m_B, $M_{A'}$, and $m_{A'}$. The angle ϕ represents the relative phase between the complex transmission amplitudes for wanted and unwanted light through polariser a, and, for light passing directly through the polariser, it would be expected that $\phi = 0°$, $\cos \phi = 1$. Of course, if the polariser were perfect, the interference term would not occur since in that case $m_A = 0$.

A complication arises with the use of imperfect pile-of-plates polarisers in that a portion of the transmitted light results from internal reflections from the plates of the polarisers. It is assumed that the contribution of internal reflections to the wanted component is negligibly small since they occur near to Brewster's angle, but a significant part of the unwanted component does arise from these reflections. However, because of the small deviations of the plate alignment from Brewster's angle, the lack of parallelism of the surfaces of the individual plates and the imperfect polish of the plate surfaces, it is unlikely that the component of the unwanted light resulting from these internal reflections will interfere with the light passing straight through the polariser. If this assumption is made then, in the expression for $\Delta(\beta, \alpha')$, the factor $(M_A m_A)^{1/2}$ must be modified to $(M_A h_A)^{1/2}$ where h_A represents the transmission efficiency of the unwanted component not resulting from internal reflections. The quantity h_A is wavelength dependent and cannot be readily measured so, in practice, a weighted mean value was calculated, taking account of the optical properties of the polariser plates

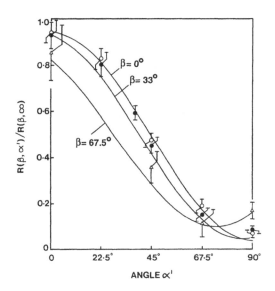

Figure 16. *Variation of the ratio $R(\beta,\alpha')/R(\beta,\infty)$ as a function of α' for $\beta = 0°$ (○), 33o (●), and 67.5° (△). The solid curves represent the quantum-mechanical predictions. (From Haji-Hassan et al. [37]).*

and the spectral distribution of the radiation from the two-photon source. This procedure resulted in the value $h_A = 0.0182$. The other transmission efficiencies were measured in a subsidiary experiment described previously. For polarisers a and b, $M_A = M_B = 0.908 \pm 0.013$, $m_A = m_B = 0.0299 \pm 0.0020$, whereas for polariser a', $M_{A'} = 0.938 \pm 0.010$, $m_{A'} = 0.040 \pm 0.002$.

It would be possible to use the above values for the transmission efficiencies directly in Equation (29) to obtain the quantum-mechanical predictions. However, strictly, the quantity $R(\beta,\alpha')/R(\beta,\infty)$ is wavelength-dependent and depends nonlinearly on wavelength dependent transmission efficiencies. To evaluate the importance of this fact, a wavelength-averaged value for $R(\beta,\alpha')/R(\beta,\infty)$ was computed taking into account the spectral distribution of the two-photon source, the quantum efficiency of the photomultipliers and the wavelength dependence of the transmission efficiencies of the polarisers. The result of these calculations are shown as the quantum-mechanical predictions in Figure 16 but, as can easily be verified, the curves shown do not differ significantly from those obtained by direct substitution of the above quoted values of the transmission efficiencies in Equation (29).

The experimental and theoretical results for $\beta = 0°, 33°$, and $67.5°$ are shown in Figure 16. Clearly, within the limits of experimental error, the results are in good agreement with the predictions, although there is a suggestion that the results for $\beta = 33°$ are systematically slightly too high. It is also worth noting that the quantum-mechanical prediction for $\beta = 0°$ is extremely close to the form $M_{A'} \cos 2\alpha' + m_{A'} \sin 2\alpha'$, and the experimental results for this case can be considered a test of Malus' law for the transmission of single photons through polariser a'.

Referring to Equation (29), it is clear that, according to quantum mechanics, the

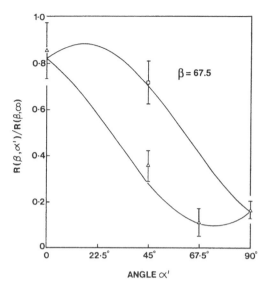

Figure 17. *The ratio $R(\beta, \alpha')/R(\beta, \infty)$ as a function of α' for $\beta = 67.5°$. The upper (lower) curve is lthe theoretical result for $\alpha' < 0 (\alpha' > 0)$. The experimental points are marked $\square$ for $\alpha' < 0$, $\triangle$ for $\alpha' > 0$.*

ratio $R(\beta, \alpha')/R(\beta, \infty)$ should not be symmetrical with respect to a change of sign of angle α', because of the presence of interference between the coherent wanted and unwanted components of radiation transmitted through polariser a. To investigate this prediction, an additional measurement for $\beta = 67.5°, \alpha' = -45°$ was carried out and the result is shown, in comparison with the results for positive α', in Figure 17 along with the appropriate quantum-mechanical prediction. Again the results clearly indicate the quantum-mechanical prediction, particularly with regard to the asymmetry with respect to angle α'. It is, perhaps, interesting to note how the use of imperfect polarisers reveals these interesting features and provides even more convincing verification of the quantum-mechanical formalism than might otherwise be obtained.

Finally, returning to the predictions of the class of local realistic theories proposed by Garuccio and Selleri [62], for the three-polariser experiment they showed that for any angle $\theta < 90°$, if it is arranged that the angles of the transmission axes of the three polarisers satisfy the relation $\beta = 3\theta$, $\alpha' + \beta = \theta$, then the ratio of the quantum-mechanical prediction to their local realistic prediction must always be greater than some minimum value, γ_L say, which depends on θ in a specified way. For the parameters of this experiment the important range of angles occurs for $58° < \theta < 80°$, since for these values of the lower limit $\gamma_L = 1.447$ occurs here for $\theta = 71°$, corresponding to the experimental angles $\beta = 33°$, $\alpha' = 38°$, so that the approach of Garuccio and Selleri sets an upper limit on $R(33°, 38°)/R(33°, \infty)$ of 0.413 obtained by dividing the quantum mechanical prediction for $\beta = 0°$, $\alpha' = 38°$ by 1.447. In contrast the actual experimental point has the value 0.585 ± 0.029, violating the prediction of the Garuccio-Selleri model by over 6 standard deviations. More recently Selleri [63], taking other factors into account, modified the prediction of the local realistic model to give $\gamma_L = 1.162$ with a corresponding upper limit on $R(33°, 38°)/R(33°, \infty)$ of 0.514 which, however, is still violated by the experimental result by almost 3 standard deviations. The three-polariser experiment, therefore, provides further strong evidence against the possibility of enhancement in the detection process and appears to rule out the class of local realistic theories proposed by Garuccio and Selleri.

7 Discussion

This article has described a series of experiments carried out to measure the polarisation correlation properties of the two photons emitted simultaneously in the spontaneous decay of metastable atomic deuterium. The measurements have, in general, confirmed the predictions of quantum mechanics for this true second order process and, in particular, the linear polarisation correlation results have violated Bell's inequality, thus rendering untenable any local realistic explanation of the observations, provided it is assumed there was no bias or enhancement in the detection process caused by the presence of the polarisers used in the experiment. The measurements of the circular polarisation correlation, although consistent with conservation of angular momentum for each photon pair, were not in a form suitable to test Bell's inequality directly and indicated that the imperfections of the nominally achromatic quarter-wave plates were such as to render unlikely a successful test along these lines. The experiments in which a half-wave plate or a third linear polariser was added to one detection arm of the apparatus produced results in agreement with the predictions of quantum mechanics. They gave strong support to the assumption that the presence of the polarisers did not introduce any bias or enhancement into the detection process and were able to eliminate a large class of local realistic theories in which enhancement was assumed as a central mechanism.

Finally, it should be noted that although recently a series of interesting experiments have been carried our (for example, Ou and Mandel [47]; Brendel *et al.* [31]) to test Bell's inequality and investigate nonlocal effects using the two photons produced in parametric down conversion, the work in atomic hydrogen remains a unique example of such work in which the two photons are emitted simultaneously from a single atom with a wide bandwidth and coherence length for single photons of approximately only 350nm, corresponding to just one and a half wavelengths at the centre wavelength of 243nm.

References

[1] W E Lamb and R C Retherford, *Phys Rev* , **72**, 241 (1947); *ibid*, **79**, 549 (1950)

[2] M Göppert-Mayer, *Ann Phys (NY)*, **9**, 273 (1931)

[3] G Breit and E Teller, *Astrophys J* , **91**, 215 (1940)

[4] L Spitzer and J L Greenstein, *Astrophys J* , **114**, 407 (1951)

[5] J Shapiro and G Breit, *Phys Rev* , **113**, 179 (1959)

[6] B A Zon and L P Rapaport, 1968, *JETP Lett* , **7**, 52 (1968)

[7] S Klarsfield, *Phys Lett* , **30A**, 382 (1969)

[8] W R Johnson, *Phys Rev Lett* , **29**, 1123 (1972)

[9] S P Goldman and G W F Drake, *Phys Rev A*, **24**, 183 (1981)

[10] F A Parpia and W R Johnson, *Phys Rev A*, **26**, 1142 (1982)

[11] J H Tung, X M Ye, G J Salamo and F T Chan, *Phys Rev A*, **30**, 1175 (1984)

[12] V Florescu, *Phys Rev A*, **30**, 2441 (1984)

[13] A Costescu, I Brandus and N Mezincescu, *J Phys B*, **18**, L11 (1985)

[14] G W F Drake, *Phys Rev A*, **34**, 2871 (1986)

[15] M Lipeles, R Novick and N Tolk, *Phys Rev Lett* , **15** 690, 815 (1965)

[16] C J Artura, N Tolk and R Novick , *Astrophys J* , **157**, L181 (1969)

[17] R W Schmeider and R Marrus, *Phys Rev Lett* , **25**, 1692 (1970)

[18] R Marrus and R W Schmeider, *Phys Rev A*, **5**, 1160 (2973)

[19] C L Cocke, B Curnette, J R MacDonald, J A Bednar, and R Marrus, *Phys Rev A*, **9**, 2242 (1974)

[20] M H Prior, *Phys Rev Lett* , **29**, 611 (1972)

[21] C A Kocher, J E Clendenin and R Novick, *Phys Rev Lett* , **29**, 615 (1972)

[22] E A Hinds, J E Clendinnin and R Novick, *Phys Rev A*, **17**, 670 (1978)

[23] H Gould and R Marrus, *Phys Rev A*, **28**, 2001 (1983)

[24] Y Bannett and I Freund, *Phys Rev Lett* , **49**, 539 (1982)

[25] P Braunlich, R Hall and P Lambropoulos, *Phys Rev A*, **5**, 1013 (1972)

[26] R Hippler, *Phys Rev Lett* , **66**, 2197 (1991)

[27] D C Burnham and D L Weinberg, *Phys Rev Lett* , **25**, 84 (1973)

[28] C K Hong, Z Y Ou and L Mandel, *Phys Rev Lett* , **59**, 2044 (1987)

[29] A M Steinberg, P G Kwiat and R Y Chiao, *Phys Rev Lett* , **68**, 2421 (1992)

[30] Z Y Ou, X Y Zou, L J Wang and L Mandel, *Phys Rev Lett* , **65**, 321 (1990)

[31] J Brendel, E Mohler and W Martienssen, *Europhys Lett* , **20**, 575 (1992)

[32] D O'Connell, K J Kollath, A J Duncan and H Kleinpoppen, *J Phys B*, **8**, L214 (1975)

[33] H Kruger and A Oed, *Phys Lett* , **54A**, 251 (1975)

[34] J S Bell, *Physics (NY)*, **1**, 195 (1964)

[35] W Perrie, A J Duncan, H J Beyer and H Kleinpoppen, *Phys Rev Lett* , **54**, 1790, 2647(E) (1985)

[36] T Haji-Hassan, A J Duncan, W Perrie, H J Beyer, and H Kleinpoppen, *Phys Lett* , **123A**, 110 (1987)

[37] T Haji-Hassan, A J Duncan, W Perrie, H Kleinpoppen and E Merzbacher, *Phys Rev Lett* , **62**, 237 (1989)

[38] T Haji-Hassan, A J Duncan, W Perrie, H Kleinpoppen and E Merzbacher, *J Phys B: At Mol Opt Phys* , **24**, 5035 (1991)

[39] F M Pipkin, in *Advances in Atomic and Molecular Physics*, edited by D R Bates and B Bederson (Academic, New York, 1978) Vol 14 pp 281-340

[40] J F Clauser and A Shimony, *Rep Prog Phys* , **41**, 1881 (1978)

[41] A J Duncan, in *Progress in Atomic Spectroscopy*, Part D, edited by H J Beyer and H Kleinpoppen (Plenum, New York, 1987) pp 477-505

[42] A J Duncan and H Kleinpoppen, in *Quantum Mechanics versus Local Realism*, edited by F Selleri, (Plenum, New York, 1988) pp 175-218

[43] A Aspect, P Grangier and G Roger, *Phys Rev Lett* , **47**, 460 (1981); *ibidem*, **49**, 91 (1982)

[44] A Aspect , J Dalibard and G Roger, *Phys Rev Lett* , **49**, 1804 (1982)

[45] A Aspect and P Grangier, Lett *Nuovo Cimento*, **43**, 345 (1985)

[46] C K Au, *Phys Rev A*, **14**, 531 (1976)

[47] Z Y Ou and L Mandel, *Phys Rev Lett* , **61**, 54 (1988); *ibidem*, **61**, 50 (1988)

[48] A Einstein, B Podolsky and N Rosen, *Phys Rev* , **47**, 777 (1935)

[49] D Bohm, *Quantum Theory* (Prentice-Hall, Englewood Cliffs, New Jersey, 1951)

[50] D Bohm, *Phys Rev* , **85**, 166 (1952); *ibidem*, **85**, 180 (1952)

[51] D Bohm and Y Aharonov, *Phys Rev* , **108**, 1070 (1957)

[52] J F Clauser, M A Horne, A Shimony and R A Holt, *Phys Rev Lett* , **23**, 880 (1969)

[53] J F Clauser and M A Horne, *Phys Rev D*, **10**, 526 (1974)

[54] S J Freedman, PhD Thesis, University of California, Berkeley, 1972

[55] M Bacal, A Truc, H J Doucet, H Lamain and M Chretien, *Nucl Instrum Methods*, **114**, 407 (1974)

[56] M Bacal and W Reichelt, *Rev Sci Instrum* , **20**, 769 (1974)

[57] T Haji-Hassan, PhD Thesis, University of Stirling, Stirling, 1987

[58] J F Clauser, *Nuovo Cimento B*, **33**, 740 (1976)

[59] T W Marshall, E Santos and F Selleri, *Lett Nuovo Cimento*, **38**, 41 (1983)

[60] F Selleri, *Nuovo Cimento*, **39**, 252 (1984)

[61] S Pascazio, *Nuovo Cimento D*, **5**, 23 (1985)

[62] A Garuccio and F Selleri, *Phys Lett* , **103A**, 99 (1984)

[63] F Selleri, private communication, 1988

Quantum Localisation

Shmuel Fishman

Israel Institute of Technology, Haifa

1   Introduction

The quantum behaviour of systems that are chaotic in the classical limit was studied extensively in recent years [1]–[5]. This field is sometimes called 'Quantum Chaos'. For bound systems with time dependent Hamiltonians the spectrum is discrete and the motion is quasiperiodic and therefore definitely cannot be considered chaotic. For such systems the 'fingerprints' of classical chaos in the quantum behaviour, that is not chaotic, have been investigated extensively. For systems with time dependent Hamiltonians the energy is not a good quantum number, and the phase space is in general not bounded. Consequently there is no such a general argument about the asymptotic nature of the quantum dynamics, ruling out its being chaotic. Classically for these systems diffusion in phase space is found. This diffusion is often suppressed by quantum interference effects, by a mechanism that is very similar to Anderson localisation, namely suppression of electronic diffusion in disordered solids at low temperatures.

The quantum suppression of classical diffusion in phase space (that is found for chaotic systems) will be the subject of these lectures. This *quantum localisation* in classically chaotic systems is sometimes called *dynamical localisation*. This is a very robust phenomenon that can be found for a variety of systems. It will be demonstrated that it is very insensitive to the fine details of the system considered. We will be interested in a general understanding of the global behaviour of a large class of systems, rather than in the fine details of particular systems. Chaotic systems are characterised by a high degree of complexity. Therefore theoretically one aims to understand the behaviour in terms of statistical concepts like Lyapunov exponents, diffusion coefficients, entropies and dimensions, rather than in terms of the behaviour of individual trajectories. This is similar to the methodology applied in the investigation of macroscopic systems. Very often one is interested in a statistical description of the systems of interest rather than in the investigation of the detailed behaviour of particular atoms or degrees of freedom. Some of the macroscopic phenomena are very universal, depending only on few properties of the system.

The behaviour of systems near second order phase transitions is highly universal and critical exponents depend usually only on the symmetry of the system and its dimensionality. They do not depend on details of its nature. Another problem, that is of direct relevance for problems in quantum chaos, is the electronic behaviour of disordered solids. At very low temperatures, where quantum interference is dominant, the behaviour of the solid depends on very few parameters. This enables one to predict the behaviour of a large variety of systems in terms of a scaling theory, and to describe them in terms of few parameters. Any of the macroscopic theories involves coarse graining, that enables one to eliminate fine details from the theory.

In order to demonstrate how methods that are used for the description of macroscopic systems can be applied to systems that are chaotic in the classical limit, the kicked rotor will be studied in detail. It is the standard system in the investigation of chaotic systems with time dependent Hamiltonians. In Section 2 its classical and quantum mechanical behaviour will be summarised. In Section 3 the localisation theory for electrons in solids will be briefly summarised. Localisation is a quantum coherent effect that results from interference. In Section 4 application of localisation theory to the investigation of the quantum behaviour of the kicked rotor will be presented in detail. A mathematical mapping of the kicked rotor model on the solid state localisation problem will be introduced. Quantum mechanical systems can exhibit also other forms of localisation. One of them is adiabatic localisation that will be demonstrated in Section 5 for a system that is closely related to the kicked rotor. The interplay between adiabatic localisation, that takes place also in classical mechanics and quantum localisation that results from interference will be demonstrated. In Section 6 the destruction of localisation by external noise will be discussed. Finally in Section 7 several experimental realisations will be presented. This includes a recent experiment on momentum transfer to atoms in a magneto-optic trap. This experiment is directly related to theoretical model systems. The important subject of driven hydrogen atoms and similar systems will only be mentioned briefly because of space and time constraints. In these lectures the discussion will be mostly for the kicked rotor but reference will be made to other systems as well. Although rigorous results will be mentioned, the general approach will be based on physical considerations, rather than on mathematical rigour.

2   The kicked rotor

The standard system that is used in the investigation of the dynamics of chaotic systems with time dependent Hamiltonians is the kicked rotor. It is defined by the Hamiltonian

$$H = H_0 + V \tag{1}$$

where

$$H_0 = \frac{p^2}{2I} \tag{2}$$

is the integrable part while

$$V = \overline{K} \cos\theta \sum_m \delta(t - mT) \tag{3}$$

is the nonlinear time dependent perturbation. It describes a rotor with moment of inertia I and angular momentum p that is periodically kicked. The angle that is conjugate to p is θ and T is the time between kicks. Quantum mechanically the angular momentum is quantised in integer multiples of $\hbar$. Therefore it is convenient to rewrite the Hamiltonian in the form

$$H = \frac{\hbar^2}{2I} n^2 + \overline{K} \cos \theta \sum_m \delta(t - mT) \,. \tag{4}$$

Transforming to dimensionless units by setting $t' = t/T$ and $H' = HT/\hbar$ gives (after dropping the primes)

$$H = \frac{1}{2} \tau n^2 + k \cos \theta \sum_m \delta(t - m) \tag{5}$$

where $\tau = \hbar T/I$, $k = \overline{K}/\hbar$ and n is the angular momentum conjugate to θ. In units where $T = 1$ and $I = 1$ one finds $\tau = \hbar$. Therefore in some examples τ will be replaced by $\hbar$. Sometimes the model is generalised to

$$H = \frac{1}{2} \tau n^2 + V(\theta) \sum_m \delta(t - m) \,. \tag{6}$$

This reduces to (5) if one takes

$$V(\theta) = k \cos \theta \,. \tag{7}$$

If one denotes by θ_m and n_m the angle and angular momentum immediately after the m-th kick, one finds that for the Hamiltonian (5) these satisfy the recursive map,

$$\begin{aligned}
\theta_{m+1} &= \theta_m + \tau n_m \\
n_{m+1} &= n_m + k \sin \theta_{m+1} \,.
\end{aligned} \tag{8}$$

Changing variables to $p = \tau n$ one obtains the standard map

$$\begin{aligned}
\theta_{m+1} &= \theta_m + p_m \\
p_{m+1} &= p_m + K \sin \theta_{m+1}
\end{aligned} \tag{9}$$

where $K = k\tau = \overline{K}T/I$. It is worthwhile to note that this map depends only on one parameter, K.

The behaviour of this map has been studied extensively [6]–[10]. Only the properties that are relevant for quantum localisation are briefly summarised. For small K the motion in phase space is bounded and chaotic in some regions. As K is increased more and more Kolmogorov-Arnold-Moser (KAM) trajectories are destroyed. There is a critical value of K, namely $K_c = 0.9716....$ where only the trajectories with the golden mean and the inverse golden mean winding numbers are left. At this value of K the motion in phase space is still confined.

For $K > K_c$ the last bounding trajectory is destroyed and global diffusion in phase space takes place. The last KAM trajectory and a typical chaotic trajectory are depicted in Figure 1. The critical behaviour near K_c has been the subject of many studies [8, 9].

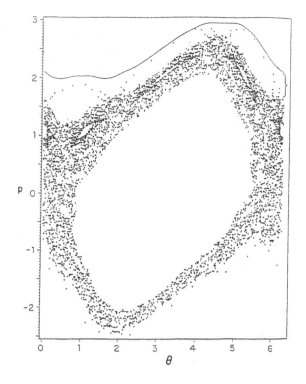

Figure 1. *A typical chaotic orbit for $K = 1.1$ (the calculation was terminated before it crossed the cantorus). The lower KAM torus for $K = K_c$ is displayed as well. The bracket indicates the strip of periodic cycles separating the main stochastic layer from the cantorus [Figure 1 of Reference 8(c)].*

It is easy to estimate the diffusion coefficient in the limit of large K. In this limit it is justified to assume that angular correlations can be ignored after one kick. Iterating the map t times one finds

$$
\begin{aligned}
\langle (p_t - p_0)^2 \rangle &= K^2 \sum_{m=1}^{t+1} \langle \sin^2 \theta_m \rangle + K^2 \sum_{m \neq m'} \langle \sin \theta_m \sin \theta_{m'} \rangle \\
&= \frac{K^2 t}{2}
\end{aligned}
\tag{10}
$$

where $\langle \cdots \rangle$ denotes an average with respect to the initial conditions. In this derivation ergodicity was assumed. The resulting diffusion coefficient is

$$
D_\infty = \frac{K^2}{2}.
\tag{11}
$$

Corrections resulting from angular correlations were calculated [6, 10] and found to be of the order $1/\sqrt{K}$ relatively to the leading term. The diffusive growth of momentum namely

$$
\langle p^2 \rangle = D(K) t
\tag{12}
$$

takes place for $K > K_c$ but $D(K)$ differs from D_∞. For some values of K, accelerator modes are important [6].

Unlike the classical system that depends only on one parameter K, the quantum system depends on both parameters k and τ. In order to understand the meaning of τ we note that all energy spacings of the unperturbed rotor that is defined by H_0, are multiples of $\hbar^2/2I$, the corresponding frequency being $\tilde{\omega} = \hbar/2I$ and the period being $\tilde{T} = 4\pi I/\hbar$. It is easy to see that

$$\frac{\tau}{4\pi} = \frac{T}{\tilde{T}} \tag{13}$$

i.e. the ratio between the period of the driving potential and the natural period of the rotor. Therefore rational values of τ/π correspond to quantum resonances [11]. For generic values of τ different behaviour was found. The Schrödinger equation

$$i\frac{\partial}{\partial t}\psi = H\psi \tag{14}$$

for the Hamiltonian (5) was first studied by Casati, Chirikov, Izrailev, Ford and Shepelyansky [12, 13]. The calculations were repeated by many investigators [14]–[18]. For a variety of systems it is found that diffusion is suppressed as shown in Figures 2 and 3.

Another problem where classical diffusion is suppressed by quantum interference is the electronic motion in disordered solids. This problem will be reviewed briefly in the following section and its relation to the kicked rotor will be presented in Section 4.

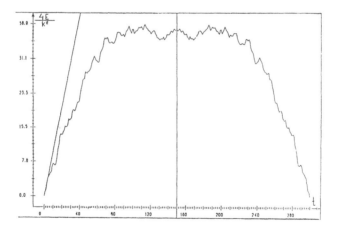

Figure 2. *The time dependence of the kinetic energy of the quantum kicked rotor for $k = 20$ and $K = 4$. The motion is reversed at $t = 150$ (vertical line) and small noise is added. The quantum evolution is completely reversible. The straight line corresponds to classical diffusion (Figure 4 of Reference [16]).*

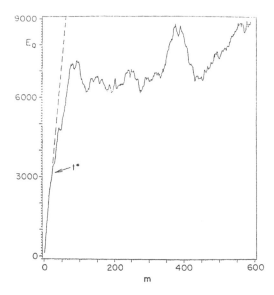

Figure 3. *The average momentum squared $E_Q = \langle p^2 \rangle$ for a kicked particle in a square well potential as a function of the number of kicks, m. The dashed line is the corresponding classical value (Figure 5 of Reference [17])*

3   Anderson localisation

We now deal with the investigation of electronic motion in disordered solids at low temperature. This is a vast field that was initiated by Anderson [19] and no attempt will be made to summarise it here. There are several good reviews [20]–[24]. Some of the mathematical work is reviewed in References [25]–[27]. Only results that are relevant for localisation in the kicked rotor problem are reviewed here.

3.1   One dimensional systems

There is reasonably good understanding of localisation in one dimensional systems [28]. The best way [29]–[31] to visualise this localisation is to consider a potential that consists of a sequence of bumps of bounded height with flat regions between them. The wave function in the n-th flat region can be written in the form

$$\psi_n(x) = A_n e^{ikx} + B_n e^{-ikx} . \tag{15}$$

The values of the expansion coefficients of the function in the various flat regions are related by the transfer matrix T_n namely

$$\begin{pmatrix} A_{n+1} \\ B_{n+1} \end{pmatrix} = T_n \begin{pmatrix} A_n \\ B_n \end{pmatrix} \tag{16}$$

where the transfer matrix

$$T_n = \begin{pmatrix} 1/t_n & -r_n/t_n \\ -r_n^*/t_n^* & 1/t_n^* \end{pmatrix} \tag{17}$$

depends on the potential in each bump. Time reversal symmetry of the Hamiltonian was assumed. The transmission and reflection coefficients of the n-th bump are $|t_n|^2$ and $|r_n|^2$ respectively. If electrons were classical particles with these transmission and reflection coefficients they would diffuse. But since in quantum mechanics amplitudes rather than probabilities should be followed, it is found that the average transmission coefficient through two bumps is smaller than the product of the transmission coefficients through these bumps. The transmission through the infinite chain vanishes. The difference between the random chain and the periodic chain is that for the periodic chain the conditions required for complete transmission through a bump are identical for all bumps and energies can be chosen so that they are satisfied. This is the content of Bloch's theorem. For random systems, for each bump there are different conditions for complete transmission. For the infinite chain, the transmission coefficient vanishes with probability one. This is correct even if the energy is higher than all the potential bumps.

Very often localisation is analysed in the framework of tight binding models of the form

$$\epsilon_n u_n + V(u_{n+1} + u_{n-1}) = E u_n \tag{18}$$

where ϵ_n is the diagonal energy, V is the hopping matrix element, E is the energy and u_n are the amplitudes of the wave function. It is assumed that the hopping is to nearest neighbours only, but most of the approach can be generalised also to problems with hopping to further neighbours. This problem can be formulated in the transfer matrix approach [28],

$$\mathbf{u}_{n+1} = T_n \, \mathbf{u}_n \tag{19}$$

where

$$T_n = \begin{pmatrix} E - \epsilon_n & -1 \\ 1 & 0 \end{pmatrix} \tag{20}$$

and

$$\mathbf{u}_n = \begin{pmatrix} u_n \\ u_{n-1} \end{pmatrix} . \tag{21}$$

The value $V = 1$ was taken for convenience. The values of the wave-function at an arbitrary point can be obtained by repeated application of the transfer matrix starting from an arbitrary initial point namely

$$\mathbf{u}_{N+1} = \tilde{T}_N \, \mathbf{u}_1 \tag{22}$$

where

$$\tilde{T}_N = T_N \cdot T_{N-1}...T_1 . \tag{23}$$

Furstenberg's theorem [32] assures the existence of the limit

$$\lim_{N \to \infty} \frac{1}{N} \ln \| u_N \| = \gamma > 0 \tag{24}$$

and the positivity of the constant γ, with probability one. For arbitrary boundary conditions it assures therefore exponential growth of the wave-functions. Moreover, it assures the self-averaging property of the growth rate γ, that is the Lyapunov exponent of the transfer matrix. Let us now take a chain of finite length and apply the transfer matrix from both ends with arbitrary initial conditions and arbitrary energy E. In general the wave-function will grow exponentially. One can, however, find initial conditions and energies so that the wave-function iterated from both ends match. This is the eigenvalue condition. This approach was introduced by Borland [33] and rigorously established by Delyon, Levy and Souillard [34].

The eigenstates of (18) are exponentially localised,

$$u_n = A_o v(n) e^{-|n-n_0|/\xi} \tag{25}$$

where ξ is the localisation length, n_0 is the maximum, $v(n)$ is a function that oscillates rapidly between values of unit magnitude, and A_o is the normalisation constant. The localisation length is

$$\xi = 1/\gamma. \tag{26}$$

All quantities in (25) depend on the energy E. The fact that all states are localised was proved by Goldsheid *et al.* [35] and by Kunz and Souillard [36].

This discussion can be generalised to strips and bars. The main difference is that the Hamiltonian matrix corresponding to (18) is not tridiagonal and the transfer matrix is not a 2×2 matrix. Oseledec's theorem [37] assures the existence of the eigenvalues of the product of transfer matrices, corresponding to $\tilde{T}_N$ of (23). The eigenvalues appear in pairs Λ and $1/\Lambda$. The largest length scale in the problem is the localisation length ξ. Therefore

$$\xi = 1/\ln \Lambda_0 \tag{27}$$

where Λ_0 is the eigenvalue of the transfer matrix that is closest to unity and is larger than unity [34]. Its numerical calculation requires, however, the calculation of all the other eigenvalues [6, 38, 39]. Localisation in strips was established rigorously as well [34]. For a more detailed discussion see [40, 41].

The tight binding model (18) that was studied most extensively is the Anderson model [19] that is defined by a square distribution of the diagonal matrix elements,

$$P(\epsilon_i) = \begin{cases} 1/w & |\epsilon_i| < (w/2) \\ 0 & \text{otherwise}. \end{cases} \tag{28}$$

The Lloyd model [42, 28] that turns out to be of particular importance for the kicked rotor is defined by the Lorentzian (or Cauchy) distribution

$$P(\epsilon_i) = \frac{\delta}{\pi \left(\epsilon_i^2 + \delta^2 \right)}. \tag{29}$$

For this model the localisation length is known exactly and is given by [28, 43]

$$\cosh \gamma = \frac{1}{4V} \left[\sqrt{(2V + E)^2 + \delta^2} + \sqrt{(2V - E)^2 + \delta^2} \right] \tag{30}$$

where $\gamma = 1/\xi$.

For models with hopping to nearest neighbours only, the localisation length is simply related to the density of states ρ by the Thouless formula [28, 43]

$$\gamma(E) = \int \rho(x) \ln |E - x| \, dx - \ln V . \tag{31}$$

3.2 Scaling and localisation in dimensions higher than one

The scaling theory for localisation [20, 22] enables one to describe properties of the system on large scales in terms of few parameters. The property of physical interest that is being calculated is the conductance on various scales (in natural units of $e^2/\hbar\pi$). Following ideas of Thouless [21, 44] one adds up blocks of size L, that are much larger than the microscopic scale. The coupling between the blocks is assumed to be determined by the ratio between the level spacing δW and δE the sensitivity to boundary conditions. δW is proportional to L^{-d} where d is the dimensionality of the system. δE is proportional to $\exp(-L/\xi)$ in the localised regime and to L^{-2} in the diffusive regime. Therefore in these extreme regimes $\delta E/\delta W$ is proportional to the conductance g. Consequently, the conductance on the scale L is the only parameter that is relevant for the description on larger scales. It contains all the relevant information about the microscopy of the system and the disorder. The scaling theory was formulated by Abrahams, Anderson, Licciardello and Ramakrishnan [45, 20, 22] and introduces the function

$$\beta = \frac{d \ln g}{d \ln L} . \tag{32}$$

Following the arguments that were presented above, we assume that β is a universal function depending only on g on the scale L. Abrahams *et al.* assume also that β is a monotonic function of $\ln g$. In the diffusive limit $\beta = d - 2$ while in the localised limit $\beta = \ln g$. In one dimension one finds indeed that on a sufficiently large scale localisation takes place, in agreement with known results. The theory predicts that this is correct in two dimensions as well. In three dimensions, on the other hand, there is a mobility edge and, depending on disorder and energy, the system is either a metal or an insulator on large scales.

To the first order in perturbation theory in $1/g$ the β function is

$$\beta = (d - 2) - a_o/g \tag{33}$$

where a_o is constant that depends only on the dimensionality d. The conductance on any given scale L can be obtained in principle from β by integration of (33). The boundary condition is the microscopic conductance g_0 on the scale ℓ_o, that is the elastic mean free path. This is the distance that the electron travels until its phase gets completely randomised due to scattering by impurities. In particular for $d = 2$ one finds the localisation length to be

$$\xi = \ell_o e^{g_0/a_o} = \ell_o e^{\pi g_0} = \ell_o e^{\pi k_F \ell_o/2} \tag{34}$$

where the momentum of an electron at the Fermi level is $\hbar k_F$. This value of the localisation length was obtained integrating (32) in the perturbative regime where (33) holds. The localisation length is the size of the system where perturbation theory breaks down.

In our discussion so far we referred to a conductance of a system on the scale L. However, the conductance is actually a random variable and we have considered a typical value of the conductance. The distribution of conductances in one dimension depends in general on two parameters and in the weak disorder limit it depends on one parameter [46]. In higher dimensions this was verified in the framework of the Migdal-Kadanoff approximation [46]. At the mobility edge the distribution is universal as was shown by B. Shapiro [47].

Existence of localisation for $d > 1$ was rigorously shown by Fröhlich, Spencer, Martinelli and Scoppola [48, 49].

3.3   Inhomogeneous random systems

So far it was assumed that the random system is homogeneous in space. It was assumed, for example, that the hopping terms, like V in (18) and the distributions of the diagonal energies (28) and (29), are translationally invariant. The behaviour is completely different if this invariance is not satisfied. For example, in one dimension, if ϵ_n in (18) is replaced by $\eta_n \epsilon_n$, where η_n falls off as $|n|^{-\alpha}$, one finds [50] that the eigenstates are extended for $\alpha > 1/2$ and localised for $0 \leq \alpha \leq 1/2$. For $\alpha = 1/2$ the behaviour depends on finer details.

3.4   Comments on the spectrum

The spectrum that one finds when the wave functions are localised in a random system is a dense point spectrum. The local density of states, with respect to some localised wave function (that is not an eigenstate) picks up contributions of very different sizes for various energies, since the eigenstates are localised at various locations in space. The integrated density of states is *discontinuous* at each energy in the spectrum. If the states are extended, each energy contributes an infinitesimal amount to the integrated density of states. Therefore it is continuous in this case. The integrated density of states may be a singular-continuous function, corresponding in one dimension, to states that are extremely small over large distances but assume large values at long distance [51]. In one dimension this type of spectrum tends to be very unstable [34].

4   Mapping the kicked-rotor to the Anderson model

In this section we show, following the work of the Maryland group (Prange, Grempel and Fishman [15]), that the kicked rotor that is defined by the Hamiltonian (5) or (6) can be mapped on a solid state problem of the form (18).

For a time dependent Hamiltonian that is periodic in time it is useful to define the evolution operator $\hat{U}$ that evolves the wave function by one step, namely

$$\hat{U}\psi(\theta, t) = \psi(\theta, t+1). \tag{35}$$

For the kicked systems of the form (6) the evolution operator takes a particularly simple form

$$\hat{U} = e^{-iH_0}\, e^{-iV(\theta)} \tag{36}$$

where the time t of (35) was chosen as the time immediately before a kick for sake of concreteness. The evolution of any function can be expanded in terms of quasi-energy states [52] $\psi_\omega(\theta, t)$ in the form

$$\psi(\theta, t) = \sum_\omega A_\omega \psi_\omega(\theta, t) \tag{37}$$

where A_ω are constants that are determined from the initial state $\psi(\theta, t = 0)$. The quasi-energy states are the eigenstates of the quasi-energy operator namely

$$\hat{U}\, \psi_\omega(\theta, t) = e^{-i\omega}\psi_\omega(\theta, t)\,. \tag{38}$$

The phase ω is called the quasi-energy, and it is real since $\hat{U}$ is unitary. The ψ_ω are just the Bloch-Floquet states in time. They can be written in the form [52]

$$\psi_\omega(\theta, t) = e^{-i\omega t}u_\omega(\theta, t) \tag{39}$$

where the periodicity of u_ω is identical to the one of the Hamiltonian namely

$$u_\omega(\theta, t) = u_\omega(\theta, t + 1)\,. \tag{40}$$

It can be shown that these states are orthogonal [52]. The expansion (37) assumes also the completeness of the basis of these states. It implies that the dynamics is completely determined by the nature of the states u_ω.

In order to investigate the properties of quasi-energy states, an equation for their projections on the angular momentum states, that are the eigenstates of H_0, will be derived. For this purpose we exploit the periodicity in time of u_ω (see Equation (40)) in order to map the problem on a static problem. Let us denote by u_ω^- and u_ω^+ the values of u_ω just before and after a kick respectively. The corresponding values of the wave function are $\psi_\omega^\pm = \exp(-i\omega t)\, u_\omega^\pm$. Between two consecutive kicks, say at times $t = m$ and $t = m + 1$, the evolution is

$$\psi_\omega^-(\theta, t = m + 1) = e^{-iH_0}\,\psi_+(\theta, t = m) \tag{41}$$

leading to

$$u_\omega^-(\theta, t = m + 1) = e^{i(\omega - H_0)}\, u_\omega^+(\theta, t = m)\,. \tag{42}$$

Projecting on a state of angular momentum n one finds

$$u_n^- = e^{i(\omega - \tau n^2/2)}u_n^+ \tag{43}$$

where $u_n^\pm = \langle n|u^\pm(\theta, t = m)\rangle$ and m is an integer (recall Equation (40)). Integrating over a kick one finds

$$u^+(\theta, m) = e^{-iV(\theta)}\, u^-(\theta, m)\,. \tag{44}$$

It turns out to be convenient to introduce the function

$$W(\theta) = -\tan(V(\theta)/2) \tag{45}$$

so that

$$e^{-iV(\theta)} = \frac{1 + iW(\theta)}{1 - iW(\theta)}. \tag{46}$$

With the help of the function $W(\theta)$, Equation (44) can be written as the set of equations

$$\begin{aligned}
u^{+}(\theta) &= \bar{u}(\theta)(1 + iW(\theta)) \\
u^{-}(\theta) &= \bar{u}(\theta)(1 - iW(\theta))
\end{aligned} \tag{47}$$

where

$$\bar{u}(\theta) = \frac{1}{2}\left[u^{+}(\theta) + u^{-}(\theta)\right]. \tag{48}$$

Projecting (47) on states of angular momentum (that is nothing but a Fourier transform) leads, after some manipulations to [15],

$$\tan\left[\frac{1}{2}\left(\omega - \frac{1}{2}\tau n^2\right)\right]\bar{u}_n + \sum_r W_{n-r}\bar{u}_r = 0 \tag{49}$$

where $\bar{u}_n$ and W_n are the projections of $\bar{u}(\theta)$ and $W(\theta)$ on angular momentum states.

This equation, as the Schrödinger equation with the Hamiltonian (6), cannot be solved exactly. Its advantage is that it is very similar to tight binding models in solid state physics of the form of (18). The diagonal matrix elements are

$$T_n = \tan\left[\frac{1}{2}\left(\omega - \frac{1}{2}\tau n^2\right)\right] \tag{50}$$

while the hopping matrix elements are W_{n-r}. The hopping elements are translationally invariant. Note that unlike Equation (18), hopping is not restricted to nearest neighbours

First consider the case $\tau = 4\pi r/q$, where r and q are integers [11]. In this case T_n is a periodic function of period q. On the basis of the analogy with solid-state problems, the quasi-energy states are extended in the momentum space. They form q bands. This usually results in ballistic transport, namely $\langle p^2 \rangle$ is proportional to t^2. This is indeed the case of the quantum resonance since r/q is the ratio between the natural frequency of the rotor and the driving frequency. It is instructive to note that the existence of extended states does not imply necessarily ballistic transport. If the bandwidth vanishes, as is the case for $q = 2$, one finds an infinite degeneracy. Therefore one can form from these states localised states, and the particle will remain localised near its initial value of n.

For generic values of τ, when τ/π is a generic irrational number, it was argued that the sequence $\{T_n\}$ of (50) is 'pseudorandom'. Pseudorandomness of a given sequence is not a precise term. It generally depends on the physical problem that is studied. Localisation for a variety of models with pseudorandom potentials was investigated

[15], [53]–[58]. It turns out that only very low degree of pseudorandomness, as obtained from various tests that are used in computer science [59], is required for localisation [53]. The reason why it was argued that $\{T_n\}$ is pseudorandom is that when n is changed by 1 the phase of the tangent appearing in (50) changes by $\tau n/2$—a large number. The tangent depends only on that phase mod π, that is a small fraction of a large number. If τ/π is irrational this phase is not simply related to π and therefore $\{T_n\}$ is assumed to be pseudorandom. Pseudorandomness of T_n implies localisation of the quasi-energy states, if W_n falls off sufficiently fast.

The simplest model that can be studied contains hopping to nearest neighbours only. Such a model is defined by choosing [15]

$$V(\theta) = -2\arctan(\kappa\cos\theta - E). \tag{51}$$

The tight binding equation for this potential resulting from (49) and (45) is

$$T_n u_n + \frac{1}{2}\kappa(u_{n+1} + u_{n-1}) = E\,u_n. \tag{52}$$

If the phase of the tangent in (50) is assumed to be random and uniformly distributed, $\{T_n\}$ satisfies the Cauchy or Lorentzian distribution

$$P(T_n) = \frac{1}{\pi\left(1 + T_n^2\right)}. \tag{53}$$

With this distribution, the model (52) is the Llyod model [28, 42] with $\delta = 1$, (see Equation (29)). For this model the localisation length is exactly known and is given by (30). Of particular interest for the correspondence to the kicked-rotor is the case $E = 0$. In this case

$$\cosh\gamma = \sqrt{1 + \kappa^{-2}}. \tag{54}$$

In Figure 4 this result is compared with results of numerical solutions of the Schrödinger equation for the Hamiltonian (6). The good agreement supports our assumption of the pseudorandomness of the model.

Exponential localisation of the quasi-energies results in suppression of diffusion in momentum space. Assume that the system is prepared in some initial state. Then only several of the A_ω in (37), corresponding to quasi-energies that are localised in the vicinity of the initial momentum, will have significant magnitude. Since the A_ω are time independent, only these quasi-energies contribute to $\psi(\theta,t)$. Since these are localised in the vicinity of the initial state, the motion will be confined to this region in angular momentum. This localisation is of pure *quantum mechanical* origin. It corresponds to *Anderson localisation* in disordered systems. In the context of suppression of diffusion in chaotic systems it is sometimes called *dynamical localisation*.

For the standard kicked rotor with the potential (7), the function $W(\theta)$ of (45) is singular for $k > \pi$. This is however a result of the specific definition of $\bar{u}(\theta)$ in (47). Shepelyansky proposed [60] replacing (47) with

$$\begin{aligned} u^+(\theta) &= e^{-iV(\theta)/2}\,\bar{u}(\theta) \\ u^-(\theta) &= e^{iV(\theta)/2}\,\bar{u}(\theta). \end{aligned} \tag{55}$$

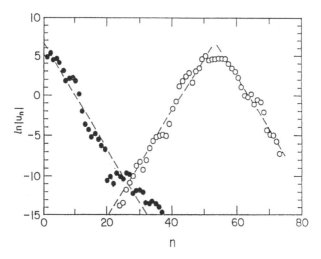

Figure 4. *Two quasi-energy eigenstates (solid and open circles) obtained from numerical solution of the Schrödinger equation for the Hamiltonian (6) with potential (51) and $E = 0$. The dashed lines correspond to the localisation length (54) that assumes true randomness (Figure 4 of Reference [15(b)])*

This satisfies (44) that is the only physical property that is required by the dynamics. The derivation can be continued along the lines leading to (49) for any potential $V(\theta)$. The result for the standard kicked rotor (5) is

$$\sum_r J_r(k/2) \sin\left[\frac{1}{2}\left(\omega - \frac{1}{2}\tau n^2\right) - \frac{\pi}{2}r\right] \bar{u}_{n+r} = 0 \tag{56}$$

where J_r are the Bessel functions of the first kind. This is an equation that is similar in nature to (49). The phases of the sine are pseudorandom for the same reason the phases of the tangent of (50) are such. The great advantage of (56) is that the hopping terms are all well defined. For $|r| > k/2$, the matrix elements are small and they decay factorially fast in r. The advantage of (49) is that it looks more like standard solid state models with diagonal disorder.

Localisation of the quasi-energy states could be expected from inspection of the evolution operator. In the basis of angular momentum states its matrix elements are

$$U_{nn'} = e^{-i\tau n^2/2}(-i)^{n-n'} J_{n-n'}(k). \tag{57}$$

The phases can be assumed to be pseudorandom for the same reasons that were presented for other similar phases. Application of transfer matrix arguments introduced for the localisation problem in Section 3.1 leads to localisation of the quasi-energy states in momentum space [61]. The matrix of $\hat{U}$ is unitary while the corresponding matrices in standard localisation theory are hermitian. This is one of the reasons why the problem was mapped on the standard tight binding problems (49) and (56).

Although localisation is found for a large variety of driving potentials, the localisation length is in general not identical to the one of the corresponding random model

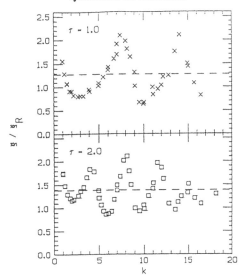

Figure 5. *The ratio between ξ, the localisation length of the kicked rotor with $\tau = 1$ and $\tau = 2$ (marked on the figure) and ξ_R, the localisation length of the corresponding random model (Figure 12 of Reference [61]).*

where the phase of the tangent in (49) and (50) is truly random [60, 61]. The ratio between these localisation lengths for the kicked rotor with the driving potential (7) is presented in Figure 5. Although they are of order unity, differences are systematic and require some understanding.

There is a simple relation between the localisation length and the classical diffusion coefficient in the semiclassical limit and for sufficiently large K (so that diffusion is not dominated by cantori). In this argument, that was introduced by Chirikov, Izrailev and Shepelyansky [13], one assumes that classical diffusion takes place on scales smaller than the localisation length ξ. After the wave-packet representing the system spreads over ξ momentum states, quantum suppression due to localisation takes place. It is assumed that this spreading goes on for a time t^* that is determined by

$$\tau^2 \, \xi^2 \approx D(K) \, t^* \tag{58}$$

where $D(K)$ is defined by (12) as the diffusion coefficient of classical momentum for the map (9). The argument leading to (58) assumes that classical diffusion takes place until $\langle p^2 \rangle$ reaches the value $\tau^2 \xi^2$. Note that the localisation length measures the number of angular momentum states over which the wave-functions spread, and τ is proportional to $\hbar$ (see Equations (5), (8) and (9)). It is reasonable to think that quantum effects become important and classical diffusion is suppressed after a time required to resolve the quasi-energy separations. Therefore

$$\Delta\omega \, t^* \approx 2\pi \tag{59}$$

where $\Delta\omega$ is the separation between quasi-energies corresponding to states that are localised in some region of space so that their overlap is appreciable. At any point there are ξ such states, therefore

$$\Delta\omega \approx 2\pi/\xi \tag{60}$$

and

$$t^* \approx \xi. \tag{61}$$

When substituted in (58) one finds that ξ equals $D(K)/\tau^2$ up to a constant of order unity. Because of the heuristic nature of this argument, it cannot predict the value of this constant. This relation was verified by Shepelyansky for several models [60], including (5) and (51–49). For all the models that he studied, that are kicked systems, he found that this constant takes the value $1/2$. Consequently,

$$\xi = \frac{D(K)}{2\tau^2} \tag{62}$$

This argument actually assumes scaling (see Section 3.2). It assumes that the localisation length is the only length scale in the problem and although it is defined in the regime where the wave-functions decay exponentially, it determines also the scale where diffusion stops. This argument *assumes* the existence of localisation and determines the localisation length *provided* there is localisation. A similar argument was presented in localisation theory for disordered solids by Allen [62]. The relation (62) can be derived also from the scaling theory for localisation [63]. This relation is very important for the numerical verification of localisation. Since the localisation length increases as $\hbar$ decreases, one may question [18] if it remains finite for all nonvanishing values of $\hbar$. Equation (62) is an asymptotic semiclassical relation. Its verification confirms localisation in the semiclassical regime that is most questionable, since classically diffusion takes place.

It is worthwhile to note that Equations (49), (52) and (56) are not standard eigenvalue equations. In these equations ω is part of an argument of a nonlinear function, and the quasi-energy state is a solution of the equation corresponding to a specific value of this parameter. The orthogonality of these states is assured by the fact that these are eigenstates of the evolution operator $\hat{U}$. The relation between ω and the eigenvalues of the tight-binding equation (49) was discussed in detail in Reference [15(b)].

Localisation of the quasi-energies was justified by the pseudorandomness of the phase in Equations (49), (52), (56) and (57). This argument should hold for generic values of τ. If τ/π is rational the quasi-energy states are extended and the spectrum is continuous [11]. If τ/π is some type of a Liuoville number (Liuoville numbers are irrational numbers that are very well approximated by rational numbers) it was shown by Casati and Guarneri that the spectrum is continuous [64]. In this case it is probably singular continuous.

For generic values of τ strong arguments were presented in favour of exponential localisation of quasi-energy states and consequently for point spectrum of quasi-energies (see Section 3.4). Mathematically the existence of pure point spectrum for generic values of τ should be considered a conjecture awaiting a rigorous proof.

In the arguments for localisation that were presented in this section we exploited the translational invariance of the hopping matrix elements, that enabled us to use the similarity with solid state models in order to argue for the existence of localisation. Quantum localisation holds, however, also for models where the hopping is not translationally invariant. Such a model that is very similar to the kicked rotor is the kicked particle in an infinite square well. Localisation very similar in nature to the one that was found for the kicked rotor was also found for this model [17] as shown in Figure 3.

The scaling theory for localisation that was presented in Section 3.2 predicts that in two dimensions the localisation length increases exponentially with the microscopic parameters (the microscopic conductance or the elastic mean free path (Equation (34))). This was tested for extensions of the kicked rotor to two dimensions in momentum space, and general agreement with localisation theory was found, although there are some details that should still be understood [65, 60]. For an extension that is effectively three dimensional, a transition between localised and extended states was found in agreement with the predictions of the scaling theory [66].

5   Adiabatic localisation

The purpose of this section is to demonstrate that quantum, Anderson localisation is not the only possible form of localisation that can be found in phase space. In other words, when one finds some type of localisation in phase space, it does not mean that it is Anderson localisation. A particular system may exhibit Anderson localisation combined with other types of localisation. An example of such localisation is the *adiabatic localisation* that is a trivial type of localisation that takes place in quantum as well as in classical mechanics.

To demonstrate this mechanism consider a rotor that is periodically driven by smooth pulses that are narrow in time [61, 67]. For this purpose we replace the periodic δ−function in the Hamiltonian (5) by a smooth function of the form

$$\Delta^{(\sigma)}(t) = \sum_m \frac{1}{\sqrt{2\pi}\sigma} e^{-(t-m)^2/2\sigma^2} \tag{63}$$

or

$$\Delta^{(M)}(t) = 1 + 2 \sum_{m=1}^{M} \cos(2\pi mt). \tag{64}$$

These functions, for one period, are depicted (shifted by 1/2) in Figure 6. In the limit $M \to \infty$ the function $\Delta^{(M)}(t)$ approaches the periodic δ-function by the Poisson summation formula. The specific form (64) will be used in what follows. The Hamiltonian for this smoothly driven rotor takes the form

$$H = \frac{1}{2}\tau n^2 + k \cos\theta \ \Delta^{(M)}(t) \tag{65}$$

For the purpose of the classical analysis it is useful to rewrite this Hamiltonian in the form

$$H = \frac{1}{2}\tau n^2 + k \sum_{m=-M}^{M} \cos(\theta - 2\pi mt). \tag{66}$$

To get its global behaviour, it is most useful to study it in the Chirikov resonance overlap picture [7, 68]. Define the phases

$$\varphi_m(t) = \theta - 2\pi mt. \tag{67}$$

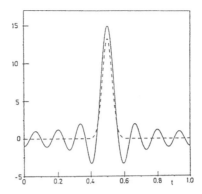

Figure 6. *The pulses (5.1) $\sigma = 0.03$ (dashed line) and (64) with $M = 7$ (solid line), for one period with time shifted by $1/2$ (Figure 1 of Reference [67]).*

The dominant contributions to the dynamics are from the resonances where the phase $\varphi_m(t)$ is stationary, namely $\dot\varphi_m(t) = 0$. This condition combined with the Hamilton equation for $\dot\theta$ leads to the resonance condition

$$\tau n_R^{(m)} = 2\pi m. \tag{68}$$

Around the m-th resonance the system is effectively a pendulum that is described by the Hamiltonian

$$H_R^{(m)} = \frac{1}{2}\tau(\Delta n)^2 + k\sin\varphi_m \tag{69}$$

where $\Delta n = n - n_R^{(m)}$. For

$$K = k\tau > \pi^2/4 \tag{70}$$

the resonances overlap. If the number of resonances is unlimited, as in the case for the kicked rotor (5), diffusion in phase space takes place whenever (70) is satisfied. This condition is not accurate numerically. We know that the correct condition is $K > K_{\mathrm{cr}} = 0.9716....$ [9], but the resonance overlap picture gives the correct global behaviour. For the Hamiltonian (65) or (66) that is the subject of this section, the largest resonance is M. Consequently the maximum value for resonating momentum is

$$n_R^{(M)} = \frac{2\pi M}{\tau}. \tag{71}$$

For larger values of n there are no more resonances and the momentum does not increase with time. The physical reason is that in the driving potential there are no high frequency components. Therefore the high angular momentum regime is adiabatic. If the resonances overlap, or, more accurately, if $K > K_{\mathrm{cr}}$, diffusion will take place for $|n| < n_R^{(M)}$. If the system is started initially at $n = 0$ it will initially diffuse and then saturate as shown in Figure 7.

  The quantum mechanical behaviour can be easily analysed in the framework of first order perturbation theory. The wave function can be expanded in the eigenstates of

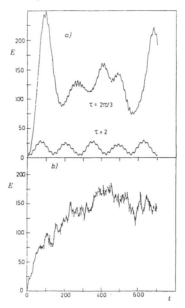

Figure 7. *The average energy of the smoothly driven rotor (65) with $M = 7$ and $k = 2$ as a function of the number of periods t. (a) Quantum mechanical calculations. (b) Classical calculation for $\tau = 2$. (Figure 4 of Reference [67]).*

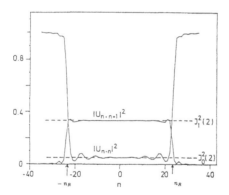

Figure 8. *Modulus square of diagonal and first off diagonal matrix elements of the evolution operator for the Hamiltonian (65) with $k = 2$, $\tau = 2$ and $M = 7$ (Figure 2 of Reference [67]).*

angular momentum, namely

$$|\psi(\theta, t)\rangle = \sum_n a_n(t) e^{-\tau n^2 t/2} |n\rangle . \tag{72}$$

In order to obtain the evolution operator, one iterates the equations for $a_n(t)$, that are obtained from the Schrödinger equation, over one period. One finds that the condition for the adiabatic approximation to hold is that $|n| > n_R^{(M)}$. For $|n| < n_R^{(M)}$ the driving couples strongly all angular momenta. In Figure 8 the evolution operator that is calcu-

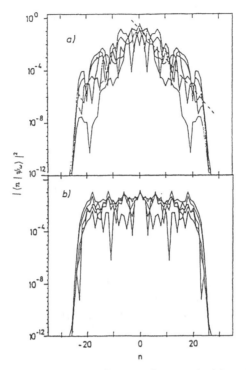

Figure 9. *Some quasi-energy states that are characterised by a large overlap with the rotor ground state* $|0\rangle$, *for* $k = 2$ *and (a)* $\tau = 2$, *(b)* $\tau = 2\pi/3$ *(Figure 3 of Reference [67]).*

lated by numerical integration of the Schrödinger equation with the Hamiltonian (65) is presented. In the regime $|n| < n_R^{(M)}$ the behaviour is identical to the one of the kicked rotor (compare with Equation (57)). One expects therefore that the system behaves as a finite solid of length $2n_R^{(M)}$. One finds indeed that for irrational τ/π the eigenstates are exponentially localised in the regime $|n| < n_R^{(M)}$ while for rational τ/π they extend over this regime as shown in Figure 9. The resulting variation of the energy with time is depicted in Figure 7. Localisation found within the range $|n| < n_R$ is Anderson localisation, while localisation that is found for $|n| > n_R$ is adiabatic, trivial localisation and it is found in the corresponding classical system as well. Localisation for the Hamiltonian of the form of (65) namely with smooth driving was shown rigorously by Bellissard [69] and by Howland [70].

Classical objects may enhance localisation by mechanisms that are different from Anderson localisation [71].

6 Effects of noise

The systems that were discussed so far were idealised in the sense that they are decoupled from the environment. The coupling to the environment, that is present in any realistic situation, introduces noise into the system. This noise destroys quantum coherence after some characteristic time t_c [72, 73, 74]. In this section the effect of noise is discussed. In particular quantum localisation is destroyed by noise. According to the argument that was presented in the end of Section 4, the time that it takes for localisation to be generated is $t^* \approx \xi$ (in units where $T = 1$ and $I = 1$ resulting in $\tau = \hbar$; thess units will be used in what follows in this section). The most interesting case is the one of weak noise that satisfies,

$$t^* \ll t_c . \tag{73}$$

In this case effects of localisation will be present. If the duration of the experiment is much shorter than t_c, localisation will be observed and the effects of noise can be neglected. If, on the other hand, the noise is sufficiently strong so that $t_c \le t^*$, but it is weak compared to the driving (7), the effects of quantum localisation are completely destroyed and classical diffusion is found for $K > K_{cr}$. The interesting regime is therefore the one of weak noise where (73) is satisfied and it will be discussed in some detail in what follows.

If the noise is white and satisfies (73) it induces a random walk on a lattice of localised states, with spacing $\hbar\xi$ and hopping time t_c. It leads to diffusion with the quantum diffusion coefficient [72, 74]

$$D_{qu} \approx \frac{(\hbar\xi)^2}{t_c} . \tag{74}$$

This result was obtained by a careful derivation in [72, 74]. It actually does not require the noise to be white and shows that Equation (74) holds also for many forms of coloured noise that are of physical interest [74]. From (74) and (58) one obtains a general expression for the diffusion rate in the semiclassical limit $\hbar \ll 1$ (in the units that were specified in the beginning of this section),

$$D_{qu} = \frac{t^*}{t_c} D(K) \tag{75}$$

that is much smaller than the classical diffusion coefficient $D(K)$. The effect of the noise on the diffusion coefficient is expressed in terms of one quantity, namely t_c, the time required for the destruction of coherence. If the variance of the noise is ν, the coherence is destroyed when there is an appreciable probability for noise induced transitions between levels. Consequently, $t_c = \hbar^2/\nu$ [72]. In the regime $K \gg 1$, where the classical diffusion coefficient can be approximated by (11) and in the semiclassical limit, the quantum diffusion coefficient resulting from the destruction of localisation by noise is,

$$D_{qu} = \frac{K^4 \nu}{16\hbar^4} . \tag{76}$$

Note that its dependence on K is very different from the one of the classical diffusion coefficient (see Equation (11)). This difference is a manifestation of the quantum localisation even after its destruction by noise.

A system that exhibits localisation like the kicked rotor is the kicked particle. It is described by the Hamiltonian (5) but the angle variable θ is replaced by a space variable x that varies from $-\infty$ to $+\infty$. The momentum is not quantised, but the transitions that are induced by the kicks couple only momenta that differ by integer multiples of $\hbar$. The matrix elements are identical to those of the kicked rotor and therefore localisation with the same localisation length is found. If initially the particle is prepared in a state p_0, it can reach states only within a ladder $p_0 + n\hbar$ that is determined by p_0, where n are integers. Within the ladder, quantum localisation takes place, and the dynamics is similar to that of the kicked rotor. Noise, however, induces transitions between ladders belonging to different values of p_0. Unlike the case of the kicked rotor the effect of noise on the kicked particle is *non-perturbative* in its nature, and therefore it leads to a much more effective destruction of localisation as shown by Cohen [74]. Due to noise induced diffusion the spreading in momentum is $\Delta p \approx \sqrt{\nu t}$. The resulting variation of the velocities of the particles leads to a spread in position

$$\delta x \approx \frac{2}{3}\sqrt{\nu}\, t^{3/2} . \tag{77}$$

For the description of a state with localisation length ξ in momentum, details on the scale $1/\xi$ are important. Consequently, localisation is destroyed when δx exceeds $1/\xi$ resulting in

$$t_c \approx \frac{(9/4)^{1/3}}{\xi^{2/3}\nu^{1/3}} . \tag{78}$$

Consequently the quantum diffusion coefficient is

$$D_{\text{qu}} = \frac{K^{4+2/3}\nu^{1/3}}{16\left(6^{2/3}\hbar^{2+4/3}\right)} . \tag{79}$$

For weak noise the quantum diffusion coefficient (79) of the kicked particle is much larger than the corresponding diffusion coefficient (76) of the kicked rotor due to the $\nu^{1/3}$ dependence on the variance of the noise. This result was verified numerically [74] and it is different from the perturbative result.

In the presence of noise, the behaviour of realistic systems is expected to be intermediate between the kicked rotor and the kicked particle [75]. Their proximity to either of these idealised systems is determined by the number of ladders of states such that the dynamics is restricted to be within each ladder. The noise induces transitions between these ladders. Investigation of the destruction of quantum localisation may be instructive for the understanding of quantum dynamics as well as the effect of the environment on this dynamics.

7 Experimental Realisations

Quantum, Anderson localisation was demonstrated for the kicked rotor in Section 4 (sometimes it is called dynamical localisation in this context). For a smoothly driven rotor quantum localisation and adiabatic localisation take place as shown in Section 5. These are model systems. The purpose of this section is to describe the relation of these models to realistic systems. The first three examples are direct realisations of the smoothly driven rotor, at least within some range of parameters. Only the second one was carried out experimentally while the other two are just theoretical suggestions at this stage. As was demonstrated in Section 5, for such systems classical diffusion may be suppressed by quantum localisation in regions of phase space where adiabatic localisation does not take place. The most direct and natural realisation of the smoothly driven rotor is the driven diatomic molecule, as described in Section 7.1. In the energy regime of the rotational spectrum, it can be considered as a rigid rotor and the angle variable is limited to $(0, 2\pi)$. In Section 7.2, quantum localisation in momentum of atoms in a magneto-optic trap will be described. This experiment was performed and results that are in agreement with theoretical predictions were found. This system is actually a driven particle with a Hamiltonian that is periodic in space. In absence of external noise it behaves like a driven rotor. Noise that breaks the translational symmetry in space is expected to distinguish between these two types of systems. In Section 7.3 a suggestion for the realisation of the rotor in the mode space of optical waveguides is presented. In this example localisation results from the interference of classical waves. Geometrical optics plays the role of classical mechanics. Finally in Section 7.4, driven systems where the density of states increases with the quantum number (contrary to the kicked rotor, where it decreases) are described. This class of systems includes the important example of the hydrogen atom that was extensively studied experimentally and theoretically. These systems will be described very briefly due to space limitations.

7.1 The driven diatomic molecule

In the regime of rotational energies a diatomic molecule can be considered as a three dimensional rigid rotor. If its dipole moment is d_m and it is driven periodically by an electric field of amplitude e_0, the driving can be modeled like in (65) with k equal to $d_m e_0 T / \hbar$. For the molecule, n should be replaced by the total angular momentum in three dimensions. Since the 'magnetic quantum number' is conserved, the only quantum number that changes is the total angular momentum j. The main difference between the Hamiltonian of the molecule and the model (65) is that unlike n, the quantum number j is positive. Otherwise it turns out [67] that the matrix elements of the evolution operator are similar to those corresponding to the Hamiltonian (65) that are presented in Figure 8. One expects therefore that the results of Section 5 hold also for the driven diatomic molecule [67]. The best candidates for this experiment are CsI and PbTe. The main experimental difficulty is to generate a pulse that is sufficiently narrow [M sufficiently large in (64)] so that there will be a sufficient number of angular momentum states where quantum (Anderson) localisation is observed before adiabatic localisation takes place.

7.2 Realisation of localisation in atomic momentum transfer

This is the first experimental realisation of a system that, in some range of parameters, is very similar to the quantum kicked rotor. In this experiment the momentum that is transferred from a modulated standing wave of a near-resonant laser to a sample of ultra cold atoms is measured. The experiment was performed by the group of Mark G. Raizen at the University of Texas at Austin [76] following a suggestion by Graham, Shlautmann and Zoller [77]. In this experiment sodium atoms are trapped and laser cooled in a magneto-optic trap. Then the trap is turned off and a modulated standing wave is turned on, resulting in momentum transfer to the trapped atoms. The mechanism of this momentum transfer will be described in detail in what follows. The momentum of the atoms in a specific direction corresponds to the angular momentum of the kicked rotor. Then the standing wave is turned off and the distribution of momentum of the atoms is measured.

In this experiment the driving is done by a standing polarised wave,

$$\mathbf{E}(x,t) = \hat{y}e_0 \cos\left[k_L(x - \Delta L\ \sin(\omega_m t))\right] e^{-i\omega_L t} + \text{c.c.} \tag{80}$$

The frequency ω_L is nearly resonant to transitions between the ground state $|g\rangle$ and the excited state $|e\rangle$. Therefore we assume in what follows that the atoms are two level systems and make use of the rotating wave approximation. If dipole transitions between these states are considered, the resulting Hamiltonian is

$$H = \frac{P^2}{2\mu} + \hbar\omega_0|e\rangle\langle e| - \frac{d_m e_0}{2}\left\{\cos\left[k_L\left(x - \Delta L\ \sin(\omega_m t)\right)\right] e^{i\omega_L t}\sigma_+ + \text{h.c.}\right\} \tag{81}$$

where P is the center of mass momentum of the atom, μ is the mass, $\hbar\omega_0$ is the energy difference between the ground state $|g\rangle$ and the excited state $|e\rangle$ while d_m is the dipole moment and $\sigma_\pm$ are the Pauli matrices. Assuming that the detuning $\delta_L = \omega_0 - \omega_L$ is large compared to the Rabi frequency $\Omega = 2d_m e_0/\hbar$ and to the kinetic energy of the atomic motion in the excited state, adiabatic elimination of the excited state is possible. The resulting Schrödinger equation for the motion of atoms in the ground state in the x direction is (see Reference [77] for details)

$$i\hbar\frac{\partial\psi_g}{\partial t} = -\frac{\hbar^2}{2\mu}\left(\frac{\partial^2\psi_g}{\partial x^2}\right) - \frac{\hbar\Omega_{\text{eff}}}{4}\cos^2[k_L(x - \Delta L\ \sin(\omega_m t))]\,\psi_g \tag{82}$$

where $\Omega_{\text{eff}} = \Omega^2/\delta_L$ is the effective Rabi frequency. In this derivation the spontaneous emission from the excited state was neglected. Therefore it is valid for measurement times that are sufficiently short. The effective Hamiltonian corresponding to (82), up to an energy shift, is

$$H = \frac{p_x^2}{2\mu} - \frac{\hbar\Omega_{\text{eff}}}{8}\cos[2k_L(x - \Delta L\ \sin(\omega_m t))]. \tag{83}$$

It is convenient to introduce dimensionless units via $t' = \omega_m t$ and the normalisations $p = 2k_L p_x/(\mu\omega_m)$, $\phi = 2k_L x$ and $H' = 4k_L^2 H/(\mu\omega_m^2)$. In these units the effective Hamiltonian takes the form (with the primes dropped),

$$H = \frac{p^2}{2} - K_a\cos(\phi - A\sin t) \tag{84}$$

with $A = 2k_L\Delta L$ and $K_a = \varepsilon_r\Omega_{\text{eff}}/\omega_m^2$ where $\varepsilon_r = \hbar k_L^2/(2\mu)$. The effective Planck's constant that is defined via the commutator of p and ϕ, is $\bar{k} = 8\varepsilon_r/\omega_m$.

We turn now to the classical analysis of the Hamiltonian (84) following Reference [77]. It will be shown that for a wide range of parameters it behaves like the kicked rotor (5). The Hamilton equations for the Hamiltonian (84) are

$$\dot{p} = K_a \sin \varphi(t) \tag{85}$$

and

$$\dot{\phi} = p \tag{86}$$

with

$$\varphi(t) = \phi - A \sin t. \tag{87}$$

The motion is dominated by the resonance points where $\dot{\varphi} = 0$ or

$$p = \dot{\phi} = A \cos t \tag{88}$$

If $K_a \ll A$, the crossing of the resonant points is typically fast and the resonances separate in time. Expanding (87) around the resonance point t_r one finds,

$$\varphi(t) \approx \varphi_r + \frac{1}{2} A \sin t_r \, (\delta t)^2 \tag{89}$$

where $\delta t = t - t_r$ and $\varphi_r = \varphi(t_r)$. Integrating (85) over the region of a resonance, with $\varphi(t)$ of (89) and taking into account the fact that this integral accumulates most of its contribution from a narrow region (of width $|A \sin t_r|^{-1/2}$) around the resonance, one finds that the momentum transferred at each resonance is

$$\Delta p_r = \sqrt{\frac{2\pi}{A|\sin t_r|}} \, K_a \sin \left(\varphi_r \pm \frac{\pi}{4} \right) \tag{90}$$

where the sign depends on the direction of the crossing of the resonance. For $|p| \ll A$ one finds from (88) that $\sin t_r \approx 1$ and the r.h.s. of (90) is independent of p. If also $K_a \ll |p|$, the momentum does not change much during a period and the variation of ϕ is

$$\Delta\varphi_{\text{period}} = \Delta\phi_{\text{period}} = 2\pi p \tag{91}$$

In this regime, without the driving, the pendulum (84) is approximately a rotor and the transitions at the resonances correspond to the kicks of the rotor. This analogy holds only for $|p| < A$, since according to (88) for larger values of the momentum there are no resonances and the driving is adiabatic. This limitation is similar to Equation (71). Therefore p is bounded, namely

$$|p| < p_{\text{max}} = A. \tag{92}$$

Actually p can be somewhat larger as a result of transitions where (88) is satisfied only approximately. In the regime where (92) is satisfied and K_a is sufficiently large, diffusion takes place. If the driving is sufficiently strong, so that the resonances are uncorrelated, the diffusion in momentum space will be similar to that found for a system with one resonance per period with the momentum transfer

$$\Delta p_{\text{period}} = 2\sqrt{\frac{\pi}{A}} K_a \sin \varphi_r \tag{93}$$

(for $|p| \ll A$). This results from the fact that under these assumptions $2\langle \Delta p_r^2 \rangle = \langle \Delta p_{\text{period}}^2 \rangle$ (during each period there are two resonant points). Equations (91) and (93) correspond to the standard map (9) with p replaced by $2\pi p$ and

$$K = 4\pi \sqrt{\frac{\pi}{A}} K_a . \tag{94}$$

For $K \gg 1$ the diffusion coefficient is approximated by (11), resulting in the diffusion coefficient

$$D_a = 8\pi^3 \frac{K_a^2}{A} \tag{95}$$

for the problem studied here.

The quantum dynamics of this system for $K_a < |p| < A$ is very similar to that of the kicked rotor in the semiclassical regime. Between the resonances each eigenstate of p accumulates a phase proportional to p^2. At resonances these states are coupled for a short time. In the restricted regime (92), this leads to quantum dynamics similar to that found for the kicked rotor. In this aspect this situation is similar to the one outlined in Section 5, resulting in quantum localisation if the localisation length is smaller than p_{max}. The localisation length can be estimated from the argument leading to (62), since this argument is very general and does not assume anything about the specific details of the kicked rotor. Taking into account the various factors of 2π introduced in the change of units relating (91) and (93) to (9) one finds,

$$\xi_p = \xi \bar{k} = \frac{D_a}{2(2\pi)^2 \bar{k}} \tag{96}$$

where ξ is the localisation length in terms of the number of momentum states while ξ_p is the localisation length in terms of momentum. Therefore,

$$\xi_p = \frac{\pi K_a^2}{A\bar{k}} = \frac{\pi \hbar k_L^2 \Omega_{\text{eff}}^2}{16\mu A \omega_m^3} . \tag{97}$$

In the experiment the diffusion in momentum is suppressed by the classical adiabatic mechanism if p_{max} of (92) is smaller than ξ_p, while quantum localisation takes place otherwise. The spread in momentum was measured as a function of A and these two regimes were explored. In the regime where the suppression of diffusion is classical, one finds

$$P_{\text{rms}} = \sqrt{\langle p^2 \rangle} = A/\sqrt{3} \tag{98}$$

where spreading in all the region (92) was assumed. For the case where suppression is due to quantum localisation the final wave function decays exponentially with an effective localisation length of $2\xi_p$ (see Reference [60]). Consequently in this regime

$$P_{\text{rms}} = \sqrt{\langle p^2 \rangle} = \sqrt{2}\, \xi_p . \tag{99}$$

The experimental results for P_{rms} are presented in Figure 10 in units of the number of momentum states, namely in terms of $p_x/(2\hbar k_L) = p/\bar{k}$. Both types of suppression of diffusion are clearly observed. Typical initial and final distributions (proportional to the fluorescence intensity that is actually measured) of atoms are presented in Figure 11.

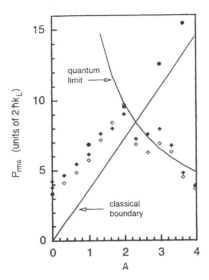

Figure 10. *Momentum spread as a function of A for $\Omega_{\text{eff}}/2\pi=30MHz$, $\delta_L/2\pi=5.4GHz$, $\omega_m/2\pi=1.3MHz$, $k=0.36$ and $\bar{k}=0.16$. Straight/curved solid lines denote classical boundary (98)/ quantum limit (99). Data for durations $10\mu s$ (empty diamonds) and $20\mu s$ (solid) shows that saturation was reached. Solid circles are a classical simulation and agree with the data up to a critical value of A. There are no adjustable parameters in this comparison between theory and experiment. (Figure 1 of [76]).*

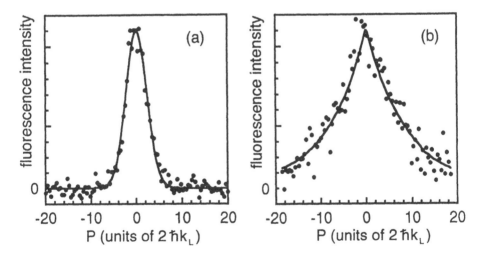

Figure 11. *Atomic momentum distribution. In (a) the distribution is for a magneto-optic trap—a typical initial condition for the measurements shown in Figure 10. The least squares fit is a Gaussian with standard deviation $4.6\hbar k_L$ representing an initial thermal distribution. In (b) the final distribution is for $A = 2.6$ and the parameters of Figure 10. Also shown is the least square fit to an exponential. (Figure 2 of [76]).*

So far it was assumed that ϕ is in the interval $(0, 2\pi)$ and the system was considered as a kicked rotor. Actually ϕ, that is proportional to x, is an extended coordinate. The evolution is however similar to that of the kicked rotor, since the evolution operator connects states with p (or more accurately the appropriate action variable) separated by integer multiples of $\bar{k}$. External noise may distinguish between these two types of systems, as explained in Section 6. If the noise does not break the translational symmetry in ϕ the resulting quantum diffusion coefficient is given by (76). On the contrary, if noise breaks this symmetry it is given by (79). Spontaneous emission from the excited state results in symmetry breaking noise [76]. So far the experiment was performed in a regime of parameters where observable effects of the noise are avoided. It turns out that it may be illuminating to study these effects.

7.3   Anderson localisation in the mode space of waveguides

The localisation that was found for the kicked rotor is a wave phenomenon and can be extended to classical waves. Of particular interest is the investigation of this phenomenon in dielectric optical waveguides [78, 79]. It has been proposed to investigate experimentally realisations of the kicked rotor and similar problems [80, 81]. Such realisations will take advantage of the advanced technology in the field of optical waveguides. In these realisations geometrical optics plays the role of classical mechanics in quantum chaos problems. It will be demonstrated that for some problems, the chaotic behaviour of rays resulting from the nonlinear dependence of the index of refraction on coordinates is suppressed by the wave nature of the electromagnetic radiation. For the sake of concreteness a waveguide in the slab geometry will be considered in some detail.

The slab dielectric waveguide consists of a core layer where the index of refraction takes the constant value n_{co} and a cladding where it takes the constant value n_{cl}. It is required that $n_{co} > n_{cl}$. It is convenient to introduce coordinates such that the index of refraction is

$$n(x) = \begin{cases} n_{co} & |x| < a \\ n_{cl} & |x| > a \, . \end{cases} \tag{100}$$

In what follows it will be assumed that the radiation propagates in the z-direction. For a medium where the index of refraction is piecewise constant, Maxwell's equations reduce to scalar equations. For the TE (transverse electric) modes the components E_x, E_z and H_y of the electric and magnetic field vanish while E_y satisfies

$$\nabla^2 E_y + k^2 n^2 E_y = 0 \, . \tag{101}$$

The wavenumber is $k = 2\pi/\lambda$ and λ is the wavelength of the electromagnetic radiation. The boundary conditions for the surfaces $x = \pm a$, where the index of refraction is discontinuous, require the continuity of E_y and its derivatives with respect to x and z. For TM (transverse magnetic) modes, H_y satisfies an equation that is similar in its nature to (101) but the boundary conditions are somewhat different. In what follows the discussion of the behaviour of TE modes will be presented. The behaviour of TM modes is very similar. Translational invariance in the z direction implies that the solutions of (101) are of the form

$$E_y(x, y, z) = e^{-i\beta_0 z} \psi(x) \tag{102}$$

where β_0 is the propagation constant. The function $\psi(x)$ satisfies

$$-\frac{\partial^2}{\partial x^2}\psi(x) + V(x)\psi(x) = \tilde{E}\psi(x) \tag{103}$$

where

$$\tilde{E} = k^2 n_{co}^2 - \beta_0^2 \tag{104}$$

and

$$V(x) = \begin{cases} V_0 & |x| > a \\ 0 & |x| < a \end{cases} \tag{105}$$

with

$$V_0 = k^2(n_{co}^2 - n_{cl}^2). \tag{106}$$

Equation (103) with the boundary conditions of the TE modes is just the time independent Schrödinger equation for a particle of mass $1/2$ in a square well of depth V_0. The guided modes correspond to the bound states of the particle in the well, *i.e.* $\tilde{E} < V_0$. Most waveguides operate in the limit of weak guidance where $(n_{co} - n_{cl}) \ll n_{co}$. In this limit one has for the guided modes

$$\beta_0 \approx k n_{co} - \frac{\tilde{E}}{2k n_{co}}. \tag{107}$$

The values of β_0 are determined by $\tilde{E}_\ell$ the eigenvalues of (103). For eigenvalues that are far from the cutoff, it can be assumed that the well is infinite, and the eigenvalues are

$$\tilde{E}_\ell = \frac{\pi^2 \ell^2}{4a^2}. \tag{108}$$

For these modes the electric field is,

$$E_{y,\ell} = e^{i\phi_\ell(z)}\psi_\ell(x) \tag{109}$$

where $\psi_\ell(x)$ are the eigenstates of the infinite well, while the phase is

$$\phi_\ell(z) = \beta_0 z = k n_{co} z + \frac{\pi^2 \ell^2}{8a^2 k n_{co}} z. \tag{110}$$

This behaviour is very similar to the motion of the free (unkicked) rotor and of the free particle in a potential well. Angular momentum of the rotor and the energy quantum number of the particle in the well correspond to the mode number ℓ, while time corresponds to the coordinate along the direction of propagation.

In order to simulate the effect of kicks one should introduce narrow regions where the index of refraction varies so that the modes are coupled within these narrow regions. In these regions the index of refraction is modified by $\Delta n^2(x, z)$ that satisfies,

$$k^2 \Delta n^2(x, z) \equiv \tilde{V}(x, z) = \overline{V}(x)\Delta(z). \tag{111}$$

The dependence on z is determined by the periodic function

$$\Delta(z) = \sum_r \delta_\sigma(z - rZ_0).$$ (112)

The functions δ_σ are narrow bumps of width σ. It will be assumed that $\sigma \ll Z_0$. Therefore $\Delta(z)$ is approximately a periodic δ-function. If the modulation (111) is added to the index of refraction, the potential $V(x)$ in (103) has to be replaced by $V(x) + \tilde{V}(x, z)$. The solutions are no more of the form (102) but $\psi(x)$ has to be replaced by $\psi(x, z)$ that is a function of both x and z. If, however, the width of the bumps is much larger than the wavelength λ and all the variations of the index of refraction within the bumps are negligible on the scale of λ, the function $\psi(x, z)$ varies slowly as a function of z when compared to the exponent in (102). Under these conditions, reflections from the bumps are exponentially small in (σ/λ) and $\partial^2\psi(x, z)/\partial z^2$ is negligible compared to $(i\beta_0)\partial\psi(x, z)/\partial z$. The ratio between the wave-function after and before the bump is [80, 81]

$$\frac{\psi_+}{\psi_-} = e^{-i\overline{V}(x)/2\beta_0}.$$ (113)

This corresponds to the kicks for the problems of the rotor and of the particle in the square well.

In order to establish a numerical correspondence between the transmission problem in the optical waveguide and the quantum chaos problems, the parameters $\hbar$ or τ and K should be expressed in terms of parameters of the optical waveguides. Comparing the free motion part of the evolution operator of a particle in the square well with (110) one finds that $(\pi^2\ell^2)Z_0/(8a^2kn_{co})$ corresponds to $E_\ell/\hbar = \hbar\ell^2/8$ (width of the well is 2π and mass of the particle unity) implying

$$\bar{\lambda} = \frac{Z_0\pi^2}{a^2kn_{co}} = \frac{Z_0\pi}{2a^2n_{co}}\lambda.$$ (114)

Note that for a fixed geometry, $\bar{\lambda}$ is proportional to λ and plays the role of Planck's constant $\hbar$ for this problem. The kick in the quantum chaos problems is related to (113). If one assumes that the potential in the bumps has the form

$$\overline{V}(x) = \epsilon \cos\left(\frac{\pi x}{a}\right)$$ (115)

then

$$\epsilon = k^2 \Delta n^2 \sigma$$ (116)

where Δn^2 is the typical difference between n^2 and n_{co}^2 in the bump. Comparing (113) with (36) and (7) it is found that

$$\frac{\epsilon}{2\beta_0} = k = \frac{K}{\bar{\lambda}}.$$ (117)

Using the approximation $\beta_0 \approx kn_{co}$ leads to

$$K = \frac{\pi^2 Z_0 \sigma \Delta n}{a^2 n_{co}}$$ (118)

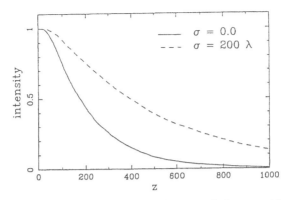

Figure 12. *Decrease of light intensity along the perturbed waveguide, according to the classical theory, for $n_{co} = 1.52$, $n_{cl} = 1.50$, $a = 100\lambda$, $\sigma = 200\lambda$, $\Delta n_{max} = 2.5 \times 10^{-3}$ and $\lambda = 10^4 \mathring{A}$. The stochastic parameter is $K = 8.005$. The initial intensity is normalised to 1 (Figure 2 of Reference [81]).*

where Δn is the typical difference between $n(x)$ and n_{co} inside the bump. For gaussian bumps used in the calculations presented in the following, $\Delta n = (\pi/2)^{1/2}\Delta_{max}$ where Δ_{max} is the maximal difference between $n(x)$ and n_{co} in the bump. Note that K does not depend on the wavelength. This is expected since it is a purely classical quantity in quantum chaos problems.

In the limit of geometrical optics the rays are deflected chaotically due to the nonlinear dependence on coordinates of the index of refraction in the bumps. This results in diffusive-like spreading over a variety of angles of incidence. Consequently the average angle of incidence and its variance grow as the rays propagate [81]. If the bumps are sufficiently narrow so that the propagation cannot be considered adiabatic (see conditions outlined in Section 5) the incidence angle of some rays will exceed the critical angle. Consequently the intensity of the guided radiation will decay, as depicted in Figure 12. The diffusive increase in the angle of incidence corresponds to the growth of the angular momentum of the kicked rotor or of the energy quantum number of the kicked particle in a well. In quantum chaos problems, this diffusive growth is suppressed by Anderson localisation. It is expected from the correspondence between these problems and the transmission in optical waveguides that localisation in the mode space takes place. This is confirmed by numerical calculations [81] that are presented in Figure 13. Localisation on modes corresponding to small angles of incidence is found. The intensity of these modes decreases exponentially with the mode number as clearly displayed in Figure 14. The intensity of the radiation is concentrated in low laying propagating modes and does not reach the radiating modes. Consequently the intensity transmitted by the waveguide does not decay, in contradiction with the results found in the geometrical optics limit. It is found that the results depend only on $\bar{\lambda}$ and K and are insensitive to many experimental details such as the width of the bumps, the depth of the well, the width of the waveguide (if the results are appropriately rescaled) and the precise functional dependence of the index of refraction in a bump on the coordinates.

Transmission through circular fibres with bumps was studied in detail theoretically

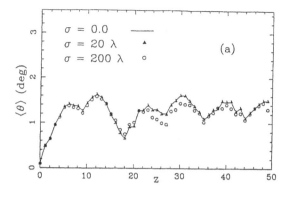

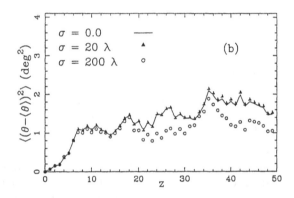

Figure 13. *Moments of the angle distribution according to the wave theory assuming an infinite potential well for* $n_{co} = 1.52$, $a = 100\lambda$, $\sigma = 200\lambda$, $Z_0 = 2 \times 10^4\lambda$, $\Delta n_{max} = 2.5 \cdot 10^{-3}$ *and* $\lambda = 10^4 \text{\AA}$. *The normalised parameters are* $K = 8.005$ *and* $\bar{\lambda} = 2.1$. *(a) Angle average. (b) Angle variance (Figure 4 of Reference [81]).*

[81]. For bumps with circular symmetry results that are similar to those of the slab geometry were obtained. If the bumps break the circular symmetry, an interesting crossover to a two dimensional localisation problem is found.

The most obvious way to introduce the bumps is by doping. This, however, requires the use of a new waveguide for each value of $\bar{\lambda}$ and K. It is also possible to apply heating or stress in order to vary the index of refraction. In this way the values of K and $\bar{\lambda}$ of (118) and (114) can be varied experimentally.

Randomness in the location and the shape of the bumps as well as other imperfections in the fibre act as noise in the localisation problem corresponding to the quantum chaos problem. This leads to diffusion in mode space. It results in decay of the intensity that is transmitted in the direction of propagation. This corresponds to Anderson localisation in real space that is expected for random wave guides [82]. If z_c is the characteristic length scale over which coherence of the wave is preserved (corresponding

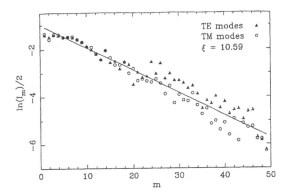

Figure 14. *Localisation in mode space. The intensity I_m of the m-th mode as a function of m for $n_{co} = 1.52$, $n_{cl} = 1.50$, $a = 100\lambda$, $Z_0 = 2 \cdot 10^4 \lambda$, $\sigma = 200\lambda$, $\Delta n_{max} = 2.5 \cdot 10^{-3}$ and $\lambda = 10^4 \AA$, leading to $K = 8.005$, $\bar{\lambda} = 2.1$. The solid line is the best fit corresponding to the effective localisation length $\xi = 10.59$ (Figure 5 of Reference [81]).*

to t_c of Section 6), localisation effects in mode space that have been discussed here, can be observed if $\xi \ll z_c$. The proposed experiments on waveguides may enable one to study experimentally the effects of noise on localisation in a controlled way.

7.4 Driven systems with an increasing density of states

The driven hydrogen atom, that belongs to the class of systems analysed here, was extensively studied theoretically [83, 84] and experimentally [85]. Effects of quantum localisation were found for driven hydrogen [85] and rubidium [86] atoms. Due to the increasing density of states found for the hydrogen atom and similar systems, the arguments for localisation in energy that were used for the kicked rotor (see Section 4) are not directly applicable to these systems. The reason is that in this case, because of the increasing density of states, the sequence $E_n \bmod (2\pi)$ (where $\{E_n\}$ is the spectrum) cannot be considered pseudorandom. For these systems however, states that are separated by approximately an integer multiple of the photon energy of the driving field, are much more important for the dynamics than other states. Heuristic theories have been developed for the dynamics in this restricted space. In particular it was proposed that locally the system behaves like a kicked rotor with parameters that depend on energy. On the basis of this analogy the regime where quantum localisation takes place was predicted. Some parts of this stimulating theory require further justification and understanding, in my opinion.

Another model that is somewhat simpler for theoretical analysis is of a particle bouncing on a fixed wall under the action of a constant field and an oscillating field. The analysis that was applied to the hydrogen atom was extended to this model [87]. It was also studied in the framework of an adiabatic perturbation theory combined with a semiclassical approximation [88]. Eigenstates that decay exponentially in one part of the phase space and are extended in another part were found [88]. Therefore the behaviour of this system, as well as of other systems with an increasing density of states, can be very different from the one of the kicked rotor and the other systems that were discussed in these lectures, and should be the subject of further studies.

Acknowledgements

During the years I worked in this field, I enjoyed collaborations and discussions with many colleagues. I would like to acknowledge, in particular, the stimulating and continuous collaboration with Richard E Prange. The approach to the problem that was described in these lectures was initiated and developed in collaboration with him, and with Daniel R Grempel, whom I deeply thank. Many results that were presented in these lectures reflect collaboration with Uzy Smilansky, Reinhold Blümel, Dima L Shepelyansky, O Agam, N Brenner, A Cohen, D Cohen, I Dana, E Doron, M Feingold, M Griniasty and D Iliescu, whom it is my pleasure to thank. The work was continuously supported by the US-Israel Bi-national Science Foundation (BSF) and by the Fund for the Promotion of Research at the Technion.

* Member of the Minerva Center for Nonlinear Physics of Complex Systems.

References

[1] F Haake, *Quantum Signatures of Chaos* (Springer, New York, 1991)

[2] M C Gutzwiller, *Chaos in Classical and Quantum Mechanics* (Springer, New York, 1990)

[3] K Nakamura, *Quantum Chaos, a New Paradigm of Nonlinear Dynamics* (Cambridge, Cambridge, 1993)

[4] *Chaos and Quantum Physics*, Proceedings of the Les-Houches Summer School, Session LII, 1989, edited by M J Giannoni, A Voros and J Zinn-Justin (North Holland, Amsterdam, 1991)

[5] *Quantum Chaos*, Proceedings of the International School of Physics 'Enrico Fermi', Varenna, July 1991, edited by G Casati, I Guarneri and U Smilansky (North-Holland, New York, 1993)

[6] A J Lichtenberg and M A Lieberman, *Regular and Stochastic Motion* (Springer, New York, 1983)

[7] B V Chirikov, *Phys Rep* , **52**, 263 (1979)

[8] (a) D Ben-Simon and L P Kadanoff, *Physica D*, **13**, 82 (1984); (b) R S MacKay, J D Meiss and I C Percival, *Phys Rev Lett* , **52**, 697 (1984); *Physica D*, **13**, 55 (1984); (c) I Dana and S Fishman, *Physica D*, **17**, 63 (1985) and references therein

[9] J M Greene, *J Math Phys* , **20**, 1183 (1981)

[10] A B Rechester and R B White, *Phys Rev Lett* , **44**, 1586 (1980); A B Rechester, M N Rosenbluth and R B White, *Phys Rev A*, **23**, 2664 (1981)

[11] F M Izrailev and D L Shepelyansky, *Teor Mat Fiz* , **43**, 417 (1980) [*Theor Math Phys* , **43**, 553 (1980)]

[12] G Casati, B V Chirikov, F M Izrailev and J Ford in *Stochastic Behaviour in Classical and Quantum Hamiltonian Systems*, Vol 93 of Lecture Notes in Physics, edited by G Casati and J Ford (Springer, Berlin 1979), p 334

[13] B V Chirikov, F M Izrailev and D L Shepelyansky, *Sov Sci Rev Sec* , **C2**, 209 (1981)

[14] T Hogg and B A Huberman *Phys Rev Lett* , **48**, 711 (1982); *Phys Rev A*, **28**, 22 (1983)

[15] (a) S Fishman, D R Grempel and R E Prange, *Phys Rev Lett* , **49**, 509 (1982); (b) D R Grempel, R E Prange and S Fishman, *Phys Rev A*, **29**, 1639 (1984); (c) R E Prange, D R Grempel and S Fishman, in *Chaotic Behaviour in Quantum Systems*, Proceedings of the Como Conference on Quantum Chaos, edited by G Casati (Plenum Press, New York, 1984), p 205

[16] D L Shepelyansky, *Physica D*, **8**, 208 (1983)

[17] A Cohen and S Fishman, *Int J of Mod Phys B*, **2**, 103 (1988)

[18] B Dorrizi, B Gramaticos and Y Pomeau, *J Stat Phys* , **37**, 93 (1984)

[19] P W Anderson, *Phys Rev* , **109**, 1492 (1958)

[20] P A Lee and T V Ramakrishnan, *Rev Mod Phys* , **57**, 287 (1985)

[21] D J Thouless, in *Ill-Condensed Matter*, Proceedings of the Les Houches Summer School, edited by R Balian, R Maynard G Toulouse (North-Holland, Amsterdam, 1979), p 5

[22] D J Thouless, in *Critical phenomena, random systems, gauge theories*, Proceedings of the Les-Houches Summer School, edited by K Osterwalder and R Stora (North Holland, Amsterdam, 1986), p 681

[23] I M Lifshits, S A Gredeskul and L A Pastur, *Introduction to the theory of disordered systems* (Wiley, New York, 1988)

[24] B Kramer and A MacKinnon, *Rep Prog Phys* , **56**, 1469 (1993)

[25] R Carmona, in *École d'Été de Probabilités de Saint-Flour XIV - 1984*, edited by P L Hennequin, Lecture notes in Mathematics, Vol 1180 (Springer, Berlin 1986), p 2

[26] H L Cycon, R G Froese, W Kirsch and B Simon, *Schrödinger Operators, with Applications to Quantum Mechanics and Global Geometry* (Springer, Berlin 1987) I thank Y Avron and J Bellissard for drawing my attention to this book and for discussions about some of its contents

[27] T C Spencer, in *Critical phenomena, random systems, gauge theories*, Proceedings of the Les-Houches Summer-School, edited by K Osterwalder and R Stora (North-Holland, 1986), p 895

[28] For a review of many one dimensional models see K Ishii, *Prog Theor Phys Suppl* , **53**, 77 (1973)

[29] N F Mott and W D Twose, *Adv Phys* , **10**, 107 (1960)

[30] R Landauer, *Phil Mag*, **21**, 863 (1970)

[31] P W Anderson, D J Thouless, E Abrahams and D S Fisher, *Phys Rev B*, **22**, 3519 (1980)

[32] H Furstenberg, *Trans Am Math Soc* , **108**, 377 (1963)

[33] R E Borland, *Proc Roy Soc A*, **274**, 529 (1963)

[34] F Delyon, Y Levy and B Souillard, *J Stat Phys* , **41**, 375 (1985); *Phys Rev Lett* , **55**, 618 (1985) and references therein

[35] I Goldsheid, S Molcanov and L A Pastur, *Funct Anal Appl* , **11** 1 (1977); L A Pastur, *Comm Math Phys* , **75**, 179 (1980)

[36] H Kunz and B Souillard, *Comm Math Phys* , **78**, 201 (1980); F Delyon, H Kunz and B Souillard, *J Phys A*, **16**, 25 (1983)

[37] V I Oseledec, *Tran Moscow Math Soc* , **19**, 197 (1968)

[38] I Shimada and T Nagashima, *Prog Theor Phys* , **61**, 1605 (1979)

[39] J L Pichard and G André, *Europhys Lett* , **2**, 477 (1986)

[40] A D Stone, P A Mello, K A Muttalib and J L Pichard, in *Mesoscopic Phenomena in Solids*, edited by P A Lee and R A Webb (North-Holland, Amsterdam, 1991), p 369

[41] A Crisanti, G Paladin and A Vulpiani, *Products of Random Matrices in Statistical Physics* (Springer, Berlin, 1993)

[42] P Lloyd, *J Phys C*, **2**, 1717 (1969)

[43] D J Thouless, *J Phys C*, **5**, 77 (1972)

[44] D J Thouless, *Phys Rev Lett* , **39**, 1167 (1977)

[45] E Abrahams, P W Anderson, D C Licciardello and T V Ramakrishnan, *Phys Rev Lett* , **42**, 673 (1979)

[46] A Cohen, Y Roth and B Shapiro, *Phys Rev B*, **38**, 12125 (1988) and references therein

[47] B Shapiro, *Phys Rev Lett* , **65**, 1510 (1990) and references therein

[48] J Fröhlich and T C Spencer, *Comm Math Phys* , **88**, 151 (1983)

[49] J Fröhlich, F Martinelli, E Scoppola and T C Spencer, *Comm Math Phys* , **101**, 21 (1985)

[50] F Delyon, B Simon and B Souillard, *Ann Inst Henri Poincare*, **42**, 283 (1985) and *Phys Rev Lett* , **52**, 2187 (1984)

[51] Such states are presented by R E Prange, D R Grempel and S Fishman, *Phys Rev B*, **29**, 6500 (1984)

[52] Ya B Zeldovich, *Soviet Phys JETP*, **24**, 1006 (1967)

[53] N Brenner and S Fishman, *Nonlinearity*, **4**, 211 (1992)

[54] L S Levitov, *Europhys Lett* , **7**, 343 (1988)

[55] M Griniasty and S Fishman, *Phys Rev Lett* , **60**, 1334 (1988)

[56] D J Thouless, *Phys Rev Lett* , **61**, 2141 (1988)

[57] S Das-Sarma, S He and X C Xie, *Phys Rev Lett* , **61**, 2144 (1988); *Phys Rev B*, **41**, 5544 (1990)

[58] G Casati, I Guarneri and F M Izrailev, *Phys Lett* , **124A**, 263 (1987)

[59] D Knuth, *Seminumerical Algorithms* (Eddison-Wesley, Hoovavu, 1981)

[60] D L Shepelyansky, *Phys Rev Lett* , **56**, 677 (1986); *Physica D* **28**, 103 (1987)

[61] R Blümel, S Fishman, M Griniasty and U Smilansky, in *Quantum Chaos and Statistical Nuclear Physics*, Proceedings of the 2nd International Conference on Quantum Chaos, Curnevaca, Mexico, edited by T H Seligman and H Nishioka, (Springer, Heidelberg, 1986), p 212

[62] P B Allen, *J Phys C*, **13**, L667 (1980) I thank B Shapiro for a critical discussion of this work

[63] S Fishman, R E Prange and M Griniasty, *Phys Rev A* **39**, 1628 (1989) There are several misprints in this paper The RHS of Equation (3 25) has to be multiplied by 2

[64] G Casati and I Guarneri, *Comm Math Phys* , **95**, 121 (1984)

[65] E Doron and S Fishman, *Phys Rev Lett* , **60**, 867 (1988)

[66] G Casati, I Guarneri and D L Shepelyansky, *Phys Rev Lett* , **62**, 345 (1989)

[67] R Blümel, S Fishman and U Smilansky, *J Chem Phys* , **84**, 2604 (1986)

[68] G M Zaslavsky and B V Chirikov, *Sov Phys Uspekhi*, **14**, 549 (1972)

[69] J Bellissard, in *Trends and Developments in the Eighties*, edited by S Albeverio and P Blanchard, (World Scientific, Singapore, 1985), p 1

[70] J S Howland, *Ann Inst Henri Poincare*, **49**, 309 (1989); **49**, 325 (1989)

[71] R S MacKay and J D Meiss, *Phys Rev A* **37**, 4702 (1988)

[72] E Ott, T M Antonsen and J D Hanson, *Phys Rev Lett* , **53**, 2187 (1984)

[73] T Dittrich and R Graham, *Z Phys B*, **62**, 515 (1986); *Europhys Lett* , **4**, 263 (1987); *Europhys Lett* , **7**, 287 (1988); *Ann Phys* , **200**, 363 (1990)

[74] D Cohen, *Phys Rev A*, **43**, 639 (1991); *Phys Rev A*, **44**, 2292 (1991); *Phys Rev Lett* , **67**, 1945 (1991)

[75] S Fishman and D L Shepelyansky, *Europhys Lett* , **16**, 643 (1991)

[76] F L Moore, J C Robinson, C Bharucha, P E Williams and M G Raizen, University of Texas preprint It is my great pleasure to thank Mark G Raizen for bringing this work to my attention and for illuminating discussions

[77] R Graham, M Schlautmann and P Zoller, *Phys Rev A*, **45**, R19 (1992); See also R Graham, M Schlautmann and D L Shepelyansky, *Phys Rev Lett* , **67**, 255 (1991) and other references therein

[78] A W Snyder and S D Love, *Optical Waveguide Theory* (Chapman and Hall, London, 1983)

[79] D Marcuse, *Theory of Dielectric Optical Waveguides* (Academic, New York, 1974)

[80] R E Prange and S Fishman, *Phys Rev Lett* , **63**, 704 (1989)

[81] O Agam, S Fishman and R E Prange, *Phys Rev A*, **45**, 6773 (1992)

[82] U Sivan and A Saar, *Europhys Lett* , **5**, 139 (1988)

[83] G Casati, I Guarneri and D L Shepelyansky, *IEEE J of Quantum El* , **24**, 1420 (1988) and references therein

[84] R Blümel and U Smilansky, *Z Phys D*, **6**, 83 (1987) and references therein

[85] E J Galvez, B E Sauer, L Moorman, P M Koch and D Richards, *Phys Rev Lett* , **61**, 2011 (1988); J E Bayfield, G Casati, I Guarneri and D W Sokol, *Phys Rev Lett* , **63**, 364 (1989) and references therein

[86] R Blümel, R Graham, L Sirko, U Smilansky, H Walther and K Yamada, *Phys Rev Lett* **62**, 341 (1989); R Blümel, A Buchleitner, R Graham, L Sirko, U Smilansky and H Walther, *Phys Rev A*, **44**, 4521 (1991) and references therein

[87] F Benvenuto, G Casati, I Guarneri, D L Shepelyansky, *Z Phys B*, **24**, 159 (1991)

[88] N Brenner and S Fishman, Technion preprint

Studies of fundamental interactions using molecules

E A Hinds

Yale University, USA

1 Introduction

This course is about a series of experiments performed at Yale University using molecular beams to study fundamental questions in physics. The central topic is the search for information about time-reversal (T) symmetry in elementary particle interactions. The experimental approach is to observe the electric dipole interaction between a molecule and an external electric field and to search for a part of the energy that changes sign when the electric field is reversed. In this way it is possible, as we shall see, to learn about T symmetry in the nucleus, in the electrons and in the interaction between the two. In addition, we discuss the Aharonov-Casher geometric phase , which also concerns the dipole interactions of a neutral particle with external fields, although in that case we are interested in normal (T-conserving) electromagnetic interactions.

In Section 2, I provide some background information about the discrete symmetries C (charge), P (space), and T (time) and discuss how they are connected with the electric dipole moment (EDM) of an atomic system. This leads to a discussion of Schiff's theorem and of mechanisms which allow an atom to have an EDM. In this section I focus on the finite nuclear size mechanism or volume effect, leaving relativistic effects for Section 5.

Section 3 begins by making the distinction between nuclear-spin-dependent and electron-spin-dependent T-violations. This is followed by an account of our molecular beam experiment on TlF, which is an example of the former since it looks for T-violation associated with the Tl nuclear spin.

In Section 4, I recall the main points about the Aharonov-Bohm and Aharonov-Casher phases and describe the various effects one might hope to see. There follows an account of our high-precision test of the Aharonov-Casher phase, which was done using the TlF molecular beam.

Returning to Schiff's theorem and relativistic effects, I show in Section 5 how the electron EDM can give rise to an atomic EDM. This leads to a discussion about the use paramagnetic molecules to determine the electron EDM and an evaluation of which are the most promising candidates. Finally, I describe the status of our own work on the YbF molecule and the prospects for using it to measure the electron EDM.

2   Discrete symmetries, Schiff's theorem and the volume effect

2.1   Discrete symmetries

There are three interrelated reflection symmetries to consider: space inversion (parity, P), interchange of particles and anti-particles (charge conjugation, C), and time reversal (T). Until the early 1950's, it was thought to be axiomatic that all of these were symmetry operation, *i.e.* that the properties of physical systems were invariant under the action of any of them. This point of view was challenged by Purcell and Ramsey [1] who proposed looking for a permanent electric dipole moment (EDM) of the neutron, and by Lee and Yang [2] who discussed in some detail the possibility that weak interactions may not have parity symmetry. In 1957 a famous experiment by Wu *et al.* [3] showed that weak interactions do indeed violate P symmetry very strongly (in fact, maximally). The method they used was to polarise ^{60}Co nuclei in a cryostat by means of nuclear demagnetisation and to search for a correlation $\langle \boldsymbol{\sigma} \cdot \mathbf{p} \rangle$ between the nuclear spin $\boldsymbol{\sigma}$ and the momentum $\mathbf{p}$ of the electron emitted in β–decay. They found the strong correlation shown schematically on the left of Figure 1, in which the electrons prefer the left-handed final state $\langle \boldsymbol{\sigma} \cdot \mathbf{p} \rangle < 0$ to the right handed version $\langle \boldsymbol{\sigma} \cdot \mathbf{p} \rangle > 0$. Since this property $(\langle \boldsymbol{\sigma} \cdot \mathbf{p} \rangle < 0)$ is not invariant under space inversion, P is violated somewhere in the system. The left-handedness of weak interaction is now a very well established phenomenon, tested by numerous experiments in particle, nuclear, and atomic physics (Commins and Bucksbaum [4]).

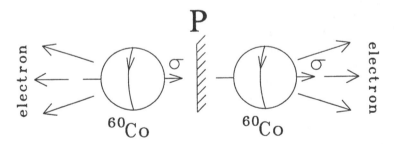

Figure 1. *Beta-decay of the ^{60}Co nucleus exhibits a corkscrew sense, thereby violating P symmetry.*

Soon after the discovery of P-violation, Landau [5] proposed that symmetry might not be entirely lost; the combined action CP of charge conjugation and space inversion could still be a symmetry operation. To illustrate the point, I imagine (somewhat unrealistically) repeating the cobalt experiment with anti-cobalt. With CP symmetry the rates must be exactly equal and the angular distributions exactly opposite. In that case, the world under CP inversion is indistinguishable from the real world, as illustrated in Figure 2; P-violation and C-violation conspire to produce CP symmetry.

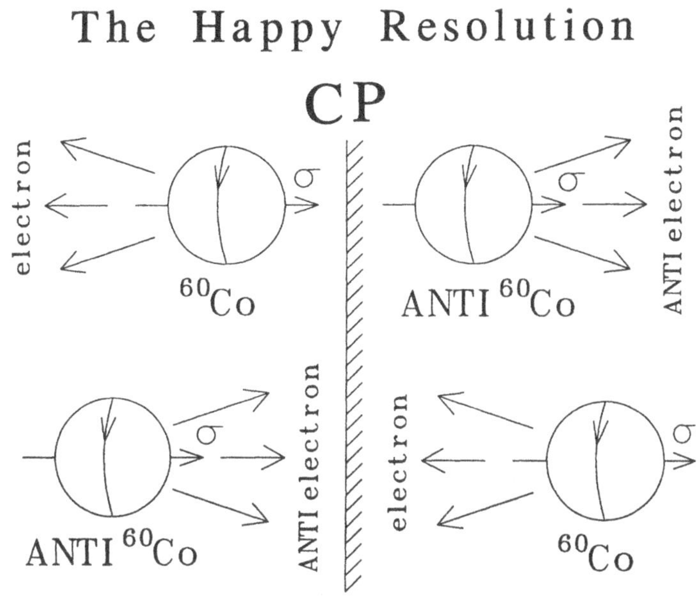

The Happy Resolution

CP

Figure 2. *If ^{60}Co and anti-^{60}Co nuclei decay with opposite handedness, the world is the same after a CP transformation.*

This happy resolution of broken P-symmetry did not last long. In 1964 Christenson *et al.* [6] discovered that the long-lived neutral kaon, K_L^0, can decay occasionally into two pions as well as the more usual three. Since the 2π and 3π final states have opposite CP symmetry, one is forced to conclude that K_L^0 is not an eigenstate of CP and/or the decay process itself can change the CP symmetry of the system. In either case CP symmetry is violated. Our current understanding is that K_L^0 is close to being an equal antisymmetric superposition $(K^0 - \overline{K^0})/\sqrt{2}$ of the strong eigenstate K^0 and its charge conjugate $\overline{K^0}$. If the two coefficients were exactly equal to $\pm 1/\sqrt{2}$, this state could not decay into 2π because the amplitudes for $K^0 \to 2\pi$ and $\overline{K^0} \to 2\pi$ would cancel. The small decay branch ($\sim 2 \times 10^{-3}$) into 2π indicates that this cancellation is not perfect, and it is thought that indeed the K_L^0 spends a little more of its time being a K^0 than it does being a $\overline{K^0}$. This CP non-symmetric state is illustrated in Figure 3.

If nature is described by a relativistic local field theory, then symmetry is still not entirely lost because the even more complicated transformation CPT is necessarily a symmetry operation (the CPT theorem due to Pauli [7]). In that case T symmetry must be violated in just the right way to make up for CP-violation. The role of T-violation

The CP-odd Kaon

CP

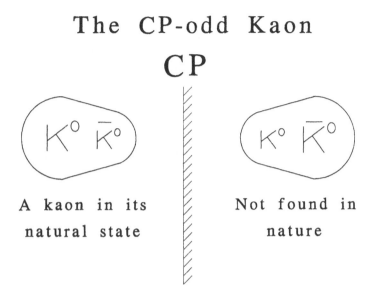

A kaon in its	Not found in
natural state	nature

Figure 3. *The long-lived kaon K_L^0 is a superposition of K^0 and $\overline{K^0}$, with a slightly larger amplitude for K^0. This violates CP symmetry.*

can be seen in the following simple model. CPT invariance tells us that K^0 and $\overline{K^0}$ have the same mass, width and internal degrees of freedom. Nevertheless the dynamic equilibrium of $K^0 \Longleftrightarrow \overline{K^0}$ produces the unbalanced state in Figure 3. This implies that the rate $K^0 \to \overline{K^0}$ is faster than the reverse process $\overline{K^0} \to K^0$, in violation of T symmetry. In the rest of this course we will assume the validity of the CPT theorem and will refer to T violation and CP-violation interchangeably.

We come back now to the beginning of this section and the proposal by Purcell and Ramsey that the neutron may have an EDM. Figure 4 shows that the existence of a neutron EDM would indicate a violation of P-symmetry since the EDM must reverse relative to the angular momentum σ under space inversion P. A similar argument, also illustrated in Figure 4, shows that the EDM violates T-symmetry as well. When Smith *et al.* [8] measured the neutron EDM, they obtained a result consistent with zero, and after 37 years, although the precision has been improved enormously, the result is still consistent with zero. Since we know that P-symmetry is violated by the weak interaction , this indicates a very high degree of T-symmetry in the neutron. This seems to be generally true. Apart from the kaon system, no instance of T-symmetry violation has been found anywhere despite a large number of searches in atomic, nuclear, and particle physics. In this course I will discuss new possibilities for detecting T-violation by molecular spectroscopy on a diatomic molecule. The basic idea (as with the neutron and with atoms) is to look for an EDM due to some P- and T-violating interaction within the molecule. Whereas the normal dipole moment lies among the internuclear axis and vanishes on average (except in the co-rotating frame), this T-violating EDM is along the angular momentum and is nonzero, even in zero electric field.

P and T reflections of an Electric Dipole Moment

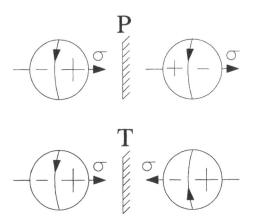

Figure 4. *An elementary system having angular momentum* σ *and a permanent electric dipole moment. This violated both P and T symmetry.*

2.2 Schiff's theorem

Let us suppose that one of the particles making up our molecule (*e.g.* the electron or proton) has an intrinsic EDM. This should induce an EDM of the molecule, which could be detected by looking at the interaction with an external field $\mathbf{E}_{\text{ext}}$. Of course the field also induces a dipole moment, giving rise to the normal Stark shift, but the two can be distinguished simply by reversing the sign of $\mathbf{E}_{\text{ext}}$. This changes the sign of the T-violating interaction but not the usual Stark effect.

Let us approximate an atom or molecule as a nonrelativistic collection of point charges interacting only through Coulomb forces. To begin, we will take the EDMs of all the charged particles to be zero. In an external electric field $\mathbf{E}_{\text{ext}}$, the Hamiltonian can be written as a sum of kinetic and potential terms $H^0 = T + V$ where V includes the external potential. Now let particle i at position $\mathbf{r}_i$ have a small EDM $\mathbf{d}_i$. This interacts with the total electric field $\mathbf{E}(\mathbf{r}_i)$ at $\mathbf{r}_i$ according to

$$H^1 = -\mathbf{d}_i\cdot\mathbf{E}(\mathbf{r}_i) = \mathbf{d}_i\cdot\left[\boldsymbol{\nabla}_i, V/q_i\right]. \tag{1}$$

The kinetic term T in our nonrelativistic Hamiltonian H^0 commutes with $\boldsymbol{\nabla}_i$, so we may equally well write

$$H^1 = \left[\frac{\mathbf{d}_i\cdot\boldsymbol{\nabla}_i}{q_i}, H^0\right]. \tag{2}$$

The first-order contribution of H^1 to the energy of the system is $\langle\psi^0|H^1|\psi^0\rangle$, where $|\psi^0\rangle$, the unperturbed state, is an eigenfunction of H^0. Because H^1 can be written as a commutator with H^0 (Equation (2)), this diagonal matrix element vanishes. We conclude therefore that in an applied electric field $\mathbf{E}_{\text{ext}}$, there is no interaction energy to first order in d_i; the atom or molecule has no permanent EDM even if the constituents do!

This is generally called Schiff's theorem, although Schiff's famous paper [9] is mainly devoted to showing how first-order effects can occur in spite of the theorem. Schiff proposed two mechanisms.

1. The volume effect: particles are not points, but extended objects with distributions ρ_c of charge and ρ_d of electric dipole moment.

2. The relativistic effect: when the problem is treated relativistically, H^1 can no longer be written as a commutator with H^0 .

In the rest of this section I will discuss the case of a nonrelativistic atom in which the nucleus has finite size and EDM, giving rise to a first-order energy shift in an external field through the volume effect.

2.3   Volume effect

Let the nucleus have charge Q_N and an electric dipole moment $\mathbf{D_N}$. The expectation value of $\mathbf{D_N}$ must lie along the nuclear angular momentum direction $\boldsymbol{\sigma}$, allowing us to define $\langle \mathbf{D_N} \rangle = D_N\boldsymbol{\sigma}$. Let the charge in an elementary volume d^3x of the nucleus be $Q_N\rho_c(\mathbf{x})d^3x$ and let the EDM of the element be $D_N\boldsymbol{\sigma}\rho_d(\mathbf{x})d^3x$. (We are assuming for the sake of simplicity that the spin of each volume element averages along the axis of total spin.) This defines the distributions ρ_c and ρ_d of charge and EDM. For a neutral atom or molecule, there is no average force on the nucleus, even in the presence of an uniform applied electric field $\mathbf{E_{ext}}$, so

$$\langle \mathbf{F_N} \rangle = Q_N \int d^3x \rho_c \langle \psi^0 | \mathbf{E}(\mathbf{x}) | \psi^0 \rangle = 0 \,. \tag{3}$$

Here $\mathbf{E}(\mathbf{x})$ is the total electric field at the nuclear volume element d^3x due to both $\mathbf{E_{ext}}$ and the electrons. The electric dipole interaction energy is

$$\langle H^1 \rangle = -D_N\boldsymbol{\sigma} \cdot \int d^3x \rho_d \langle \psi^0 | \mathbf{E}(\mathbf{x}) | \psi^0 \rangle. \tag{4}$$

Since $\langle \mathbf{F_N} \rangle = 0$, we are free to add $\boldsymbol{\sigma} \cdot \langle \mathbf{F_N} \rangle D_N/Q_N$ to Equation (4), with the result

$$\langle H^1 \rangle = -D_N\boldsymbol{\sigma} \cdot \int d^3x (\rho_d - \rho_c) \langle \psi^0 | \mathbf{E}(\mathbf{x}) | \psi^0 \rangle \,. \tag{5}$$

When $\rho_d = \rho_c$, as in the case of a point nucleus, we recover Schiff's theorem ($\langle H^1 \rangle = 0$). Since it follows from the vanishing of $\langle \mathbf{F_N} \rangle$, we see that Schiff's theorem is related to the shielding of the external field by the electrons. In fact each particle is shielded from $\mathbf{E_{ext}}$ by the polarisation of the others. However, if ρ_d differs from ρ_c, $\mathbf{E_{ext}}$ is not entirely hidden from the EDM and the residual interaction is given by Equation (5).

2.3.1 The field E(x)

The evaluation of Equation (5), requires an explicit form for $\mathbf{E}(\mathbf{x})$:

$$\mathbf{E}(\mathbf{x}) = \mathbf{E}_{\text{ext}} - \frac{e}{4\pi\epsilon_0} \sum_i \nabla r_i \frac{1}{|\mathbf{x} - \mathbf{r}_i|}. \tag{6}$$

In order to keep the notation as simple as possible, I will use cgs atomic units here (set $e/4\pi\epsilon_0 = 1$) and drop the explicit summation over electrons. Now we make a multipole expansion of $|\mathbf{x} - \mathbf{r}|^{-1}$, keeping only the monopole term in $\mathbf{x}$.

$$
\begin{aligned}
\mathbf{E}(\mathbf{x}) &= \mathbf{E}_{\text{ext}} + \frac{\hat{\mathbf{r}}}{r^2}\theta(r - x) + \text{higher multipoles in } \mathbf{x} \\
&= \mathbf{E}_{\text{ext}} + \frac{\hat{\mathbf{r}}}{r^2}[1 - \theta(x - r)] + \text{higher multipoles in } \mathbf{x},
\end{aligned} \tag{7}
$$

where θ is the Heaviside step function.

In this section, we consider only the case of a spherical nucleus. With spherical ρ_c and ρ_d the higher multipoles of Equation (7) contribute nothing to the integral over d^3x in Equation (5) and they can be omitted. Also the first two terms of Equation (7) contribute nothing because $\mathbf{E}_{\text{ext}}$ and $\hat{\mathbf{r}}/r^2$ are independent of $\mathbf{x}$ and $\int d^3x(\rho_d - \rho_c) = 0$. Thus we can make the effective replacement

$$\mathbf{E}(\mathbf{x}) \Rightarrow \frac{\hat{\mathbf{r}}}{r^2}\theta(x - r) \tag{8}$$

which acts only on electrons *inside* the nucleus, and yields

$$\langle\psi^0|\mathbf{E}(\mathbf{x})|\psi^0\rangle \Rightarrow -\int_0^x r^2 dr \int_0^{4\pi} d\Omega \psi^{0*} \frac{\hat{\mathbf{r}}}{r^2}\psi^0. \tag{9}$$

Since $\hat{\mathbf{r}}$ is an odd parity operator, this must vanish unless ψ^0 is a state of mixed parity, as it is when in the external field $\mathbf{E}_{\text{ext}}$. In non-zero external field, we can expand ψ^0 in hydrogenic partial waves around the nucleus

$$\psi^0(r) = a_s Z^{1/2}Y_0^0 + a_p Z^{3/2}rY_0^1 + \dots \tag{10}$$

where the first two terms are the S and P waves and terms of higher angular momentum l involve factors r^l. We have factored out explicitly the dependence on nuclear charge Z from the amplitudes a_s and a_p. Since the nuclear radius is of the order of 10^{-4} (we are still using atomic units), the contributions of higher partial waves to $\langle\mathbf{E}(\mathbf{x})\rangle$ are strongly suppressed and we will assume they are negligible in comparison with the $s-p$ contribution. Then

$$\langle\psi^0|\mathbf{E}(\mathbf{x})|\psi^0\rangle \Rightarrow -x^2 Z^2 \frac{a_s a_p}{\sqrt{3}}\hat{\boldsymbol{\lambda}}, \tag{11}$$

where $\hat{\boldsymbol{\lambda}}$ is the axis of the electronic $s-p$ mixing. Now that we have an effective expression for the field, we can return to the main theme; evaluation of Equation (5), the volume effect.

2.3.2   The Schiff moment

Inserting Equation (11) into Equation (5) we obtain $\langle H^1 \rangle$ as a product of a nuclear factor and an electronic factor.

$$\langle H^1 \rangle = \left[\frac{1}{6} D_N \int d^3x (\rho_d - \rho_c) x^2 \boldsymbol{\sigma} \right] \cdot \left[6Z \frac{a_s s_p}{\sqrt{3}} \hat{\boldsymbol{\lambda}} \cdot \right] \tag{12}$$

The nuclear part is known as the Schiff moment $Q\boldsymbol{\sigma}$

$$Q = \frac{1}{6} D_N \int d^3x (\rho_d - \rho_c) x^2 . \tag{13}$$

2.3.3   The Schiff interaction

Consider the derivative $[\nabla \nabla^2 V]_0$ of the potential due to the electron evaluated at the origin. This is related to the electron density through Poisson's equation. Still in cgs atomic units,

$$\nabla \nabla^2 V = -4\pi \nabla \rho = 4\pi \nabla |\psi^0|^2 \tag{14}$$

and using Equation (10) we obtain

$$[\nabla \nabla^2 V]_0 = 6Z^2 \frac{a_s a_p}{\sqrt{3}} \hat{\boldsymbol{\lambda}} . \tag{15}$$

Collecting together Equations (12), (13), (15) we find

$$\langle H^1 \rangle = Q\boldsymbol{\sigma} \cdot [\nabla \nabla^2 V]_0 . \tag{16}$$

This interaction is what remains as a result of the finite nuclear size when the electric dipole interaction is suppressed by Schiff's theorem. We call it the Schiff interaction.

An aside

One can recognise the Schiff interaction as the first non-zero term in a Taylor expansion. Loosely speaking we expand $\langle \psi^0 | \mathbf{E}(x) | \psi^0 \rangle$ in Equation (5) around the center of charge as

$$\langle \mathbf{E}(x) \rangle \sim E(0) + xE' + x^2 E'' + ... \tag{17}$$

in which $E(0) = 0$ because the force must be zero. The first-derivative term E' contributes nothing to the integral over the nucleus (Equation (5)) because it has odd parity. Hence the leading effect is due to $x^2 E''$ which yields in Equation (5)

$$\langle H^1 \rangle = -D\boldsymbol{\sigma} \cdot \langle (\rho_d - \rho_c) x^2 E'' \rangle . \tag{18}$$

Since $E'' \sim -V'''$, this has all the essential features of the more rigorous Equation (16).

2.3.4 The Schiff moment is actually more general

In deriving the Schiff moment Q (Equation (13)), we assumed a spherical distribution $\rho_d\boldsymbol{\sigma}$ of electric dipole moment due, for example, to an intrinsic EDM of the proton or neutron. But even when the nucleons have no EDM, it is still possible for the nucleus to acquire a dipole distortion if the forces between nucleons violate P and T symmetry. If the dipolar part of the charge distribution is $\delta(\mathbf{x})$, we can define the distributed EDM as $\mathbf{d} = \mathbf{x}\delta(\mathbf{x})$, and the total EDM as $D_N\boldsymbol{\sigma} = d^3x\mathbf{d}$. It can be shown (from the V''' term in a Taylor expansion of $V(\mathbf{x})$ (Sushkov *et al.* [10]) that once again there is a Schiff interaction of the form given in Equation (16), with the Schiff moment being given now by

$$Q = -\frac{1}{6}\int d^3x\left[\frac{3}{5}\mathbf{d} - D_N\rho_c\boldsymbol{\sigma}\right]x^2. \tag{19}$$

2.3.5 Conclusions about the volume effect

The volume effect leads to an effective interaction given by Equation (16) which is the product of two parts. The first is the Schiff moment $Q\boldsymbol{\sigma}$ and it is of order $D_N r_N^2$ where r_N is the nuclear radius (see Equations (13) and (19)). Q is sensitive to the EDM of the nucleons (due, for example, to P and T-odd interactions between the quarks) and to a dipole distortion of the nuclear charge (due for example to P and T-odd interactions between nucleons). The second part is the electronic factor $[\boldsymbol{\nabla}\nabla^2 V]_0$, which is proportional to the gradient of electron density at the nucleus. This can be related to the $l = 0$ and $l = 1$ parts of the electronic wave-function by Equation (15). In systems where neither the s nor the p part of the wave-function is large, the Schiff interaction is suppressed since higher partial waves have a factor $(r_N a_0)^l$ at the origin.

2.3.6 Schiff interaction in an atom

For an atom in an s state, the polarisation induced by the electric field $\mathbf{E}_{\text{ext}}$ is due to an $s-p$ admixture given to first order in $\mathbf{E}_{\text{ext}}$ by

$$a_s a_p\hat{\boldsymbol{\lambda}} = \sigma_n\frac{\langle s|ez|np\rangle}{W_{np} - W_s}\mathbf{E}_{\text{ext}}. \tag{20}$$

For a very rough numerical estimate, we take $\langle s|ez|np\rangle \sim 1$ a.u., and $W_{np} - W_s \sim 1$ a.u., and hence $a_s a_p\hat{\boldsymbol{\lambda}} \sim \mathbf{E}_{\text{ext}}$ in atomic units (1 a.u. of electric field is 5×10^9 V/cm). Equation (15) then gives $[\boldsymbol{\nabla}\nabla^2 V]_0 \sim 4Z^2\mathbf{E}_{\text{ext}}$. With $Q \sim D_N r_N^2$ it follows from Equation (16) that the Schiff interaction is

$$\langle H^1\rangle \sim D_N\boldsymbol{\sigma}\cdot 4Z^2(r_N/a_0)^2\mathbf{E}_{\text{ext}} \tag{21}$$

where we have written the Bohr radius explicitly in order to return to normal units. Since $r_N/a_0 \sim 10^{-4}$ we find that in a heavy atom ($Z=80$), the Schiff shielding effectively suppresses the applied electric field by a factor of order 3×10^{-4}. Heavy atoms are better than light ones because of the factor Z^2, which is due to the enhanced electron density near a highly charged nucleus. A more careful calculation shows that there is an additional relativistic enhancement factor K_r, which can be as large as 10 for heavy atoms (see Khriplovich [11]), but that is beyond the scope of this section.

2.3.7 Why a molecule is better

A molecule can be much more sensitive than an atom because $a_s a_p$ can be much larger. The s and p atomic orbitals of a polar molecule are strongly mixed since the atoms are highly polarised along the internuclear axis $\hat{\boldsymbol{\lambda}}$. Without any external electric field $\mathbf{E}_{ext}$, the molecular rotation averages $[\nabla\nabla^2 V]_0$ to zero, but it is easy to apply enough field to polarise $\hat{\boldsymbol{\lambda}}$ completely in a heavy molecule. Taking $a_s a_p \sim 0.1$ as a typical value, Equation (15) gives $[\nabla\nabla^2 V]_0 \sim 0.4 Z^2 \hat{\boldsymbol{\lambda}}$ a.u. independent of $\mathbf{E}_{ext}$ once it is sufficiently strong. The same quantity in an atom is $4Z^2 \mathbf{E}_{ext}$, which means that the Schiff interaction of a molecule is larger by a factor $(0.1 \mathbf{E}_{ext})$ a.u. or

$$\frac{\langle H^1 \rangle_{\text{molecule}}}{\langle H^1 \rangle_{\text{atom}}} \sim \frac{5 \times 10^8}{E_{ext}} \tag{22}$$

where E_{ext} is the field in V/cm applied to the atom. For a typical field of 5kV/cm this is a factor of 10^5, quite an improvement in sensitivity!

3 The TlF experiment on nuclear spin dependent T-violation

3.1 Possible origins of T-violation

In Section 2 we discussed the discrete symmetries C, P, and T, and saw that the existence of an EDM in an elementary system such as a neutron, atom, or molecule would imply the violation of T-symmetry in nature. One might crudely divide the possible origins of the T-violations into the following categories:

1. nucleon-nucleon interactions or intrinsic nucleon EDM;

2. electron-electron interactions or intrinsic electron EDM;

3. electron-nucleon interactions.

The first of them is characterised by the Schiff moment , as we saw in Section 2, and corresponds to an interaction of the form $\hat{\boldsymbol{\sigma}}_N \cdot \hat{\boldsymbol{\lambda}}$ where $\hat{\boldsymbol{\sigma}}_N$ is along the nuclear spin, and $\hat{\boldsymbol{\lambda}}$ is a polar axis associated with the (shielded) applied electric field. The second gives rise to a $\hat{\boldsymbol{\sigma}}_e \cdot \hat{\boldsymbol{\lambda}}$ interaction, where $\hat{\boldsymbol{\sigma}}_e$ is along the electron spin, as we will discuss more fully in Section 5. Interactions of the third type can give rise both to a nuclear spin dependent interaction $\hat{\boldsymbol{\sigma}}_N \cdot \hat{\boldsymbol{\lambda}}$ and to an electron spin interaction $\hat{\boldsymbol{\sigma}}_e \cdot \hat{\boldsymbol{\lambda}}$.

The simplest suitable electron-nucleon interaction we can postulate is the neutral-current weak interaction

$$H_{P,T} = i\frac{G_F}{\sqrt{2}}(C_T \bar{n}\sigma^{\mu\nu}n\bar{e}\gamma_5\sigma_{\mu\nu}e + C_s \bar{n}ne\gamma_5 e). \tag{23}$$

Here we are using an effective 4-fermion interaction (Commins and Bucksbaum [4]), which is suitable only in our low-energy atomic context. The scale of the interaction is characterised by G_F the Fermi coupling constant, which has the value 2.2×10^{14} a.u.. The first term, known as the tensor-pseudotensor interaction (since $\sigma^{\mu\nu}$ is the Lorentz-invariant tensor constructed from Dirac matrices), is characterised by a constant C_T and reduces in the non-relativistic limit to the effective interaction $\hat{\sigma}_N \cdot \hat{\lambda}$. The second term, the scalar-pseudoscalar interaction, corresponds to $\hat{\sigma}_e \cdot \hat{\lambda}$ and is characterised by C_s. More details of weak interaction theory can be found in Commins and Bucksbaum [4]). The point I am making here is that C_T describes the strength of the T-violating weak interaction that gives a $\hat{\sigma}_N \cdot \hat{\lambda}$ effective coupling. $C_T = 1$ means it is as strong as the normal weak interaction. Similarly C_s characterises the weak interaction that gives a $\hat{\sigma}_e \cdot \hat{\lambda}$ coupling.

Some experiments search for T-violation associated with nuclear spin, while others look for effects connected with electron spin. The former determine the nuclear Schiff moment Q and the weak interaction constant C_T. From Q one obtains information about the nucleon-nucleon interactions and the intrinsic nucleon EDMs, which can, in turn, be related to the constituent quarks and gluons. The latter give information about C_s and the electron EDM.

3.1.1   Current experiments

At the forefront of sensitivity to the nuclear spin dependent EDM there are three experiments in close competition, involving the neutron (Golub and Lamoreaux [12]), the ^{199}Hg atom (Jacobs *et al.* [13]), and the molecule ^{205}TlF (Cho *et al.* [14]). In the rest of this section we will describe the TlF experiment and its results. The most sensitive of the experiments to detect an electron spin dependent interaction involves the atoms Tl (Commins *et al.* [15]) and Cs (Murthy [16]). We are planning a new electron-spin experiment using the paramagnetic molecule YbF, which will be discussed in Section 5.

3.2   The TlF experiment

3.2.1   Principle of the experiment

The experiment (Cho *et al.* [14]) involved the spin-polarised TlF molecule, placed in an electric field $\mathbf{E}_{\text{ext}}$. Nuclear magnetic resonance (NMR) was performed on the Tl nucleus and we looked for a linear Stark effect by searching for a shift of the NMR frequency when $\mathbf{E}_{\text{ext}}$ was reversed. The interaction of the Tl nuclear spin $(1/2)\hbar\boldsymbol{\sigma}$ with the rest of the molecule can be described by the effective Hamiltonian

$$H = -\mu_{\text{Tl}}\hat{\boldsymbol{\sigma}} \cdot \mathbf{B}_0 - dh\hat{\boldsymbol{\sigma}} \cdot \hat{\boldsymbol{\lambda}}. \tag{24}$$

The first term is the usual (T-conserving) hyperfine interaction of the nuclear magnetic dipole moment $\mu_{\text{Tl}}\hat{\boldsymbol{\sigma}}$ with the internal magnetic field $\mathbf{B}_0$ of the molecule. The second term describes the P and T-violating electric dipole interaction that we are interested in. Here $\hat{\boldsymbol{\lambda}}$ is a unit vector pointing from the Tl nucleus to the F nucleus, d is a measure of T-violation in TlF and h is Planck's constant. In free space, such an interaction would tip the internuclear axis, giving it a small projection along $\boldsymbol{\sigma}$ hence producing a small permanent EDM. We prefer, however, to detect this interaction by applying the strong field $\mathbf{E}_{\text{ext}}$ which substantially polarises $\hat{\boldsymbol{\lambda}}$, and to look for an energy of the form

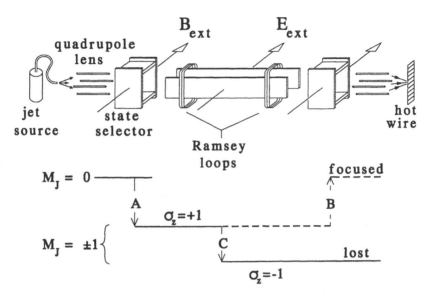

Figure 5. *Schematic diagram of TlF apparatus. State selector transition A places $M_J = 0$ molecules into a single magnetic sublevel of the $M_J = \pm 1$ manifold. Transition B is the inverse of A. The main resonance, labelled C, is the Tl nuclear spin flip.*

$\sigma \cdot \mathbf{E}_{\text{ext}}$ as discussed in Section 2. This energy appears as a shift of the NMR frequency when $\mathbf{E}_{\text{ext}}$ is reversed.

A schematic view of the experiment is shown in Figure 5. A beam of molecules was produced by a supersonic jet source. The temperature of the molecules was sufficiently low that they were all in the electronic ground state and most of them were in the vibrational ground state ($^1\Sigma, v = 0$). However, a large number of rotational and hyperfine states were occupied. Our measurement was performed using only one particular magnetic hyperfine sublevel of the first excited rotational state $J=1$.

Within $J=1$ there are twelve magnetic sublevels corresponding to the three projections of $|J = 1\rangle$ and the two projections of each spin-1/2 nucleus. In order to select a particular one of these, we used the combination of an electrostatic quadrupole lens together with a so-called state selector. First, the lens focussed those molecules having ($J = 1$, $M_J = 0$) and deflected the $M_J = \pm 1$ molecules out of the beam. Next, the four nuclear spin states were resolved in the state selector by a 27 G magnetic field, B_{ext}, and an oscillating field drove a transition (schematically shown as A in Figure 5) from one of these four states to one of the eight sublevels in the manifold ($J=1$, $M_J = \pm 1$). These selected molecules were the ones on which our measurement was made. The Tl nuclear spin transition (labeled C in Figure 5) was induced in them using separated oscillating magnetic fields to produce a narrow Ramsey resonance line. Finally, a second combination of state selector (transition B) and electric quadrupole rendered the NMR transition observable: it focussed the beam onto a hot wire detector when $\hat{\sigma}$ was unchanged but defocused it when $\hat{\sigma}$ had been flipped.

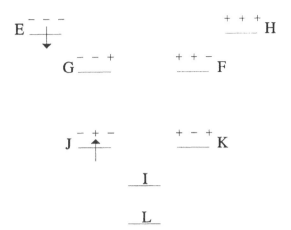

Figure 6. *The eight hyperfine sublevels having $M_J = \pm 1$ in high electric field. States are levelled by M_J, M_{Tl}, M_F and by a letter. Thus state J has $M_J = -1$, $M_{Tl} = +1/2$, $M_F = -1/2$. States I and L are symmetric and antisymmetric superpositions of $(+--)$ and $(-++)$. Arrows show the Tl spin flip relating states J and E.*

3.2.2 The Ramsey resonance

Figure 6 shows the eight sublevels of the manifold $(J=1,\ M_J \neq 1)$ in high electric field $\mathbf{E}_{ext}$, together with their magnetic quantum numbers $M_J = \pm 1$, $M_{Tl} = \pm 1/2$, $M_F = \pm 1/2$. The NMR transition that we excited was one of the Tl nuclear spin flips; either $J \to E$ or $K \to H$ depending upon the state chosen by the state selectors. Since levels J and K are degenerate, as are E and H (they differ only in the signs of the magnetic quantum numbers) the two transitions are resonant at the same frequency $f_0 = 119.57\text{kHz}$ at $E_c = 29.5\text{kV/cm}$.

The transition was driven by a pair of separated oscillating magnetic fields of frequency f whose relative phase was switched under computer control between $\pm\pi/2$. The difference signal, plotted as a function of frequency, produced the antisymmetric Ramsey lineshape [17] shown in Figure 7. Near the center of the line this difference signal is well characterised by

$$S = I_0 \sin\left[2\pi T(f - f_0)\right], \tag{25}$$

where I_0 is the peak number of molecules per second in the resonance signal and T is the time of flight between the two separated oscillating fields.

3.2.3 Reversals

Now we turn to the three main reversals employed in our experiment. The first of these was a reversal of the sign of the electric field $\mathbf{E}_{ext}$. This reversed the polarisation of the molecule $\langle \hat{\lambda} \rangle$ but did not affect any of the angular momenta and, in accordance with Equation (24), produced a shift of $-4d|\langle \boldsymbol{\sigma} \cdot \hat{\lambda} \rangle|$ in the resonance frequency. We looked for the corresponding change in the signal S, given by Equation (25). Note that electric

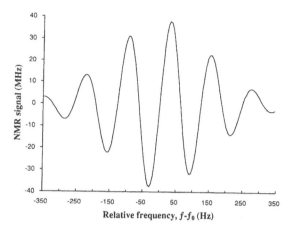

Figure 7. *Antisymmetric lineshape of Tl NMR signal produced by the separated oscillating field method. We plot the difference in the signal strength as the relative phase between the separated oscillating fields is switched between $+\pi/2$ and $-\pi/2$. The curve crosses zero at the resonance frequency $f_0 = 119.57\,kHz$, and has a linewidth of 130Hz. The slope of the line at resonance determined the sensitivity of the measurement.*

field reversal is similar to performing a parity transformation. In order to approximate this as closely as possible in our experiment, we reversed all the electric fields in the apparatus, *i.e.* state selector and quadrupole fields as well as $\mathbf{E}_{ext}$. Ideally one might also have hoped to reverse the beam velocity, but this was not done.

Our second main reversal was to change the sign of the magnetic fields in the state selectors. Since the state selector resonance populated a specific magnetic sublevel (either J or K) relative to the direction of the field, this modulation reversed the signs of all the chosen angular momenta relative to an axis fixed in the laboratory. In the language of Equation (24), this corresponded to a reversal of the Tl spin $\hat{\boldsymbol{\sigma}}$ and of the internal magnetic field $\mathbf{B}_0$. Consequently it produced the same frequency shift as reversal of $\mathbf{E}_{ext}$, but by an experimentally independent method. Note that this reversal of the angular momenta and magnetic fields was very similar to a time reversal transformation although, again, we did not reverse the beam velocity.

The third primary modulation also allowed us to reverse the magnetisation state of the molecule but this time without changing the state selector magnetic fields. This was accomplished by changing the frequency of the RF fields in the state selectors.

3.2.4   Results and their implications

The shift of the resonance frequency that was properly synchronous with all these reversals gave the experimental result

$$d = -0.13 \pm 0.22\,\mathrm{mHz} \tag{26}$$

for the T-violating coupling constant in Equation (24). In order to extract a Schiff moment for the Tl nucleus from this result, it is necessary to calculate the electronic integral $\langle \nabla \nabla^2 V \rangle$, as we discussed in Section 2. This calculation has been done by Coveney and Sandars [18] and when combined with our measurement of d gives

$$Q_{Tl} = (2.3 \pm 3.9) \times 10^{-10} \, \text{e.fm}^3 . \tag{27}$$

The experiment on atomic ^{199}Hg (Jacobs *et al.* [13]) measures an upper limit on the Schiff moment of that nucleus,

$$|Q_{Hg}| \leq 0.3 \times 10^{-10} \, \text{e.fm}^3 \tag{28}$$

while the most recent neutron EDM measurement yields

$$d_n = (-3 \pm 5) \times 10^{-13} \, \text{e.fm} . \tag{29}$$

These are small limits indeed: for example, the dipole distortion of the neutron is less than 10^{-12} of its diameter.

The experimental result in Equation (26) can also be used to determine the weak electron-nucleon coupling constant C_T discussed in Section 3.1:

$$C_T = (-2 \pm 3) \times 10^{-7}, \tag{30}$$

and an even stronger limit comes from the experiment on ^{199}Hg

$$|C_T| < (0.2) \times 10^{-7} . \tag{31}$$

It is interesting to compare this atomic physics limit with the only known T-violating weak interaction effect, namely the CP-violating K_L^0 decay, where the effective weak interaction 2×10^{-3}. These measurements of C_T show that the T-violating electron-nucleon interaction is five orders of magnitude smaller than that.

When the experimental limits given above are used to limit possible T-violating elementary particle interactions, the exact sensitivity of each experiment to a given hypothetical effect depends upon the details of the effect. However, the general trend is that the neutron gives the strongest limit, followed closely by Hg, while the TlF constraint is typically a factor of ten weaker than the neutron. It is nevertheless important to make measurements in several different systems because the nature and extent of T-violation is completely uncertain at present and such studies provide an excellent prospect for learning something new about physics beyond the standard model.

4 Aharonov-Bohm and Aharonov-Casher phases

We leave the subject of T-violation for a moment to discuss a recent experiment on the Aharonov-Casher phase performed by Sangster *et al.* [28, 29] using the TlF molecular beam.

4.1 Geometric phases

4.1.1 The Aharonov-Bohm effect

When a particle propagates along some path P_1, its wave-function acquires a phase factor $\exp(iS/\hbar)$ where S is the action along that path. If the particle is charged and there is an electromagnetic potential $(\phi, \mathbf{A})$ along the path then

$$S_{P_1} = q \int_{P_1} (-\phi dt + \mathbf{A} \cdot d\mathbf{x}) \tag{32}$$

is the part of the action associated with the electromagnetic interaction. In many cases, this phase factor is of no physical interest, but if there are two significantly different paths P_1 and P_2 between the same initial and final state, then the final state exhibits interference fringes depending on the phase difference $(S_{P_2} - S_{P_1})/\hbar$ between the two paths. This was first pointed out by Aharonov and Bohm [19].

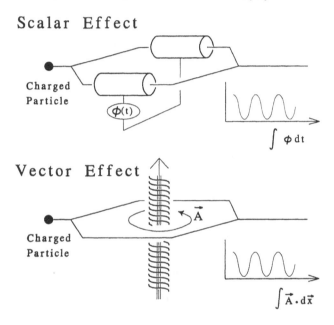

Figure 8. *Schemes for realising the scalar and vector Aharonov-Bohm effects.*

Figure 8 shows two schemes for realising the Aharonov-Bohm (AB) effect. In the first, the particle passes through a beam splitter and into two long conducting cylinders. While it is inside the cylinders, a potential difference $\phi(t)$ is turned on and then turned off. The particle exits and the two paths are brought together on a second beam splitter. According to Equation (32) the phase difference between the two paths has an electromagnetic part $-(q/\hbar) \int \phi(t) dt$ which can be varied to produce interference fringes in the final state probability. Figure 8 also shows a magnetic version in which the two paths encircle a long solenoid, and the phase difference $(q/\hbar) \oint \mathbf{A} \cdot d\mathbf{x}$ controls the final state. These are known as the scalar and vector AB effects. The vector effect has been studied experimentally (Tonomura *et al.* [20]) but the scalar one has not.

The interesting point about these experiments is that the trajectory of the particle leaving the final beam splitter can be altered by varying ϕ or A, even though there is no force acting on it at any time. A further interesting aspect of the vector effect is that the path integral $\oint \mathbf{A} \cdot d\mathbf{x}$ is just equal to the enclosed flux and therefore does not depend on any details of the path. Since any path of this geometry gives the same result, this is sometimes called a geometric phase. Finally, we remark that the division between scalar and vector contributions is rather arbitrary because each transforms into the other according to the velocity of the observer. It is worth noting that the phase given by Equation (32) can be rewritten in terms of the 4-vectors $(A^0, \mathbf{A})$ and $(x^0, \mathbf{x})$ as $-(q/\hbar) \int A^\mu dx_\mu$. In this form it is evident that the phase of the interference pattern is invariant under Lorentz transformations, as indeed it must be.

4.1.2 The Aharonov-Casher effect

Neutral particles do not exhibit the Aharonov-Bohm effect because their charge is zero. However the magnetic dipole moment $\boldsymbol{\mu}$ couples to electromagnetic fields, resulting in phase shifts and interference effects that are analogous. These were first discussed by Aharonov and Casher [21]. The phase associated with this coupling can be written as

$$\Phi = \frac{1}{\hbar} \int \left(-\boldsymbol{\mu} \cdot \mathbf{B} dt + \frac{1}{c^2} (\boldsymbol{\mu} \times \mathbf{E}) \cdot d\mathbf{x} \right). \tag{33}$$

Figure 9 shows two schemes for realising the Aharonov-Casher (AC) effect. The first probes the scalar part by applying a magnetic field to one arm of the interferometer so that the final state of the particle is determined by the phase shift $-(1/\hbar) \int \boldsymbol{\mu} \cdot \mathbf{B} dt$. This analog of the scalar AB effect has been studied by Allman *et al.* [22]. The second scheme demonstrates the vector effect, using a line charge to produce an electric field and hence a phase difference $(1/\hbar c^2) \oint (\boldsymbol{\mu} \times \mathbf{E}) \cdot d\mathbf{x}$. If $\boldsymbol{\mu}$ is always perpendicular to the plane of the path (or, more precisely, the component of $\boldsymbol{\mu}$ along $\mathbf{E} \times d\mathbf{x}$ is constant over the path) this integral is just proportional to the strength of the line charge and independent of the exact path. In this respect it seems to be geometric , like the AB phase. An important difference, however, is that the flux enclosed by a loop is well-defined, whereas the enclosed line charge is defined only if the charge per unit length is uniform. Thus the vector AC phase is geometric only when we restrict ourselves to a uniform line charge and a magnetic moment perpendicular to the path.

As we have already mentioned, part of the appeal of the AB effect is that one can arrange to have no forces exerted on the particle. This can also be accomplished in the AC case. For the scalar effect, one only needs to turn the solenoid on and off while the particle is inside, where the field is uniform. The question of forces in the vector case is much more subtle, but after substantial discussion in the literature it seems that there is no force in this case either (Aharonov *et al.* [23]). The primary difference between the AB and the AC effects is that the former involves an electromagnetic potential acting on a charge, while in the latter it is a field which couples to a dipole moment. From one point of view, this seems to make the Aharonov-Bohm effect more interesting since we are used to thinking that fields act on particles while potentials are just mathematical abstractions. However, the AC case is no less mysterious since there are still no forces acting on the particle and yet its final trajectory can be altered by changing the fields.

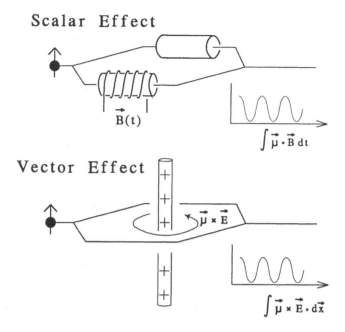

Figure 9. *Schemes for realising the scalar and vector Aharonov-Casher effects.*

4.2 Measurement of AC phase

4.2.1 Principle of experiment

In the rest of this section, I discuss the experimental verification of the vector AC phase. It was first tested in a neutron interferometer (Kaiser *et al.* [24]; Cimmino *et al.* [25]), where the measured phase shift was 2.11 ± 0.34mrad, compared with the predicted value of 1.52mrad. Although the observed phase was nearly two standard deviations above the theoretical value, the experiment did seem to confirm the existence of the effect. However, there was no experimental verification of the two most notable features, velocity-independence and proportionality to electric field. Subsequently, there were suggestions for observing the AC effect in similar interferometers using atoms instead of neutrons (Kasevich and Chu [26]; Keith *et al.* [27]), but the first results using atoms and the first tests of the velocity-dependence and field-dependence were obtained by a different technique in our laboratory (Sangster *et al.* [28, 29]) using the TlF beam.

The scheme shown in Figure 10(a) involves two coherent beams with the same magnetic moment traveling on different paths around a charged wire. The beam on path **a** acquires an AC phase shift

$$\Phi_a = \frac{1}{\hbar c^2} \int (\mu \times \mathbf{E}(\mathbf{r}_a)) \cdot d\mathbf{r}_a . \tag{34}$$

A similar expression applies to path **b**, and the net accumulated phase difference is $\Delta\Phi = \Phi_a - \Phi_b$.

It is not necessary for the paths **a** and **b** to enclose a line of charge in order to observe the AC effect. Sangster *et al.* [28] have pointed out another possible configuration,

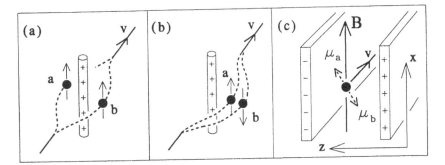

Figure 10. *Experimental configurations for observing the AC effect. (a) Geometry of the original measurement using a neutron interferometer, in which the two interfering states encircle a charge and have the same magnetic moments. (b) Alternative geometry in which the interfering states travel on the same path, but with different magnetic moments. (c) Geometry of the experiment by Sangster et al. [28, 29] using a uniform electric field, and a superposition of magnetic moments μ_a and μ_b. The polarising magnetic field B_x is not shown here.*

shown in Figure 10(b), where the two coherent beams have *different* magnetic moments μ_a and μ_b and are not spatially separated; they pass through the *same* electric field. In this arrangement the AC phase shift between the two arms of the interferometer is given by

$$\Delta\Phi_{AC} = \frac{1}{\hbar c^2} \int (\boldsymbol{\mu}_a - \boldsymbol{\mu}_b) \times \mathbf{E} \cdot d\mathbf{r}. \tag{35}$$

For simplicity we assume as shown in Figure 10(c) that the beam travels a distance L on a path along the y axis and that $\mathbf{E}$ lies in the z direction, then the x component of the magnetic moment is the only relevant one, and

$$\Delta\Phi_{AC} = \frac{1}{\hbar c^2} ([\mu_x]_a - [\mu_x]_b) \, EL. \tag{36}$$

The loop corresponding to the two paths in the interferometer does not now enclose any line charge and therefore appears different from the situation considered by Aharonov and Casher. In order to obtain a non-vanishing path integral we have taken advantage of the fact that μ in Equation (34) need not be fixed with respect to $\mathbf{E} \times d\mathbf{r}$. It is worthwhile to note, however, that path **a** of Figure 10(b) can be continuously deformed at constant $\Delta\Phi_{AC}$ to recover the geometry of Figure 10(a) by passing over the end of the line charge or, if it is infinitely long, through an infinitesimal cut in the line. This is possible because the AC phase is not geometric , except when the direction of μ is constrained as discussed in Section 4.1.2.

It is by no means necessary to use neutrons, any neutral particle with a magnetic moment should exhibit the AC effect; all that is required is a convenient way of preparing the magnetic moment in a coherent superposition of two states with different values of μ_x and detecting the accumulated AC phase difference. We have chosen to use Ramsey's method of separated oscillatory fields (Ramsey [17]), in which the first field prepares a coherent superposition of two spin states ($[\mu_x]_a$ and $[\mu_x]_b$) and the second probes the phase that has evolved between them. Thus the experiment involves magnetic resonance in the presence of an electric field. When the electric field polarity is reversed, the sign of the AC phase changes and this appears as a phase shift of the Ramsey resonance line.

We observed the AC phase using the fluorine nuclei in our TlF molecular beam in a strong (10–30kV/cm) external electric field **E**. As in the experiment of Section 3, the molecules were in the electronic and vibrational ground states $^1\Sigma^+, v = 0$, and in the first excited rotational state $J=1$. The rotational states are strongly mixed by the applied electric field, so J is not a good quantum number, but it serves adequately to identify which rotational state we use. Within the $J=1$ manifold there are twelve hyperfine sublevels corresponding to the magnetic quantum numbers of the rotation ($M_J = 0, \pm 1$), Tl nuclear spin ($M_{Tl} = \pm 1/2$), and F nuclear spin ($M_F = \pm 1/2$). In a strong electric field, these separate into four $M_J = 0$ and eight $M_J = \pm 1$ states, the latter being shown in Figure 6. The transitions studied in this experiment, were $F - H$ and $I - K$, which correspond closely to simple flips of the fluorine nuclear spin.

Using the state selector described in Section 3, we first prepared the molecules in the initial state a ($a = F$ or I). The beam then passed through the first of two Ramsey loops in which an RF magnetic field near-resonantly excited a coherent superposition of the states a and b ($b = H$ or K) with roughly equal amplitudes. This loop was effectively the beam splitter of our interferometer, providing the required coherent superposition of magnetic moments μ_a and μ_b. The molecules traveled in this state for a distance L before reaching the second RF loop which played the role of the re-combining beam splitter. The rest of the apparatus then determined what fraction P of the molecules made the transition from a to b. Close to resonance, the Ramsey fringe pattern has the usual form

$$P = \frac{1}{2}\left[1 + \cos\left((\omega - \omega_0)\frac{L}{v} + \delta + \Delta\Phi\right)\right] \tag{37}$$

where ω is the RF frequency, ω_0 is the resonance frequency, v is the beam velocity, δ is the phase difference between the two RF fields, and $\Delta\Phi$ is any additional phase shift between the two states a and b in the interferometer, such as the AC phase.

Since the molecule in external electric field is cylindrically symmetric around the field direction z, the expectation value of the transverse magnetic moment μ_x is zero in any of these states. It follows from Equation (36) that the AC effect is completely suppressed. While this is a great advantage in the search for an EDM, where the AC effect is a potential source of systematic error, it is obviously an obstacle to be overcome in the present context. In order to study the AC phase, we must rotate the magnetic symmetry axis so that the magnetic moment of the molecule can have a nonzero projection μ_x. This was done by applying a uniform magnetic field **B** along the x axis, as shown in Figure 10(c).

To summarise, the experiment involved a radio-frequency transition between two hyperfine sublevels a and b of the TlF molecule, which are separated in energy by $\hbar\omega_0$. A magnetic field $\mathbf{B}_x$ induced transverse magnetic moments $[\mu_x]_a$ and $[\mu_x]_b$, and a strong electric field $\mathbf{E}_x$ produced an AC phase shift between the two levels (in addition to the usual $\omega_0 t$ due to the energy difference between the levels). When either applied field was reversed, the AC phase changed sign, allowing us with the help of Equation (37) to deduce $\Delta\Phi$ from the measured changes in the transition probability P. This experimental phase shift could be compared with a theoretical prediction (based on

Equation (36) together with the calculated value of $([\mu_x]_a - [\mu_x]_b)$ in order to test the validity of the theory. Details about the calculation of the transverse moments $[\mu_x]_a$ and $[\mu_x]_b$ are beyond the scope of these lectures but can be found in Sangster *et al.* [29].

4.2.2 Experimental results

The AC phase shift $\Delta\Phi_{AC}$ was picked out by the fact that $\mathbf{\mu} \times \mathbf{E}$ changes sign when either E_z or B_x is reversed. This allowed us to use a form of phase-sensitive detection in which we looked for a phase shift of the Ramsey fringes in synchronism with reversals of E_z and B_x. At the central zero-crossing of the Ramsey pattern, a small phase shift appeared as a proportional change in the number of molecules hitting the detector. In order to convert this change in count rate to an equivalent frequency shift, we made an on-line measurement of the derivative of the resonance signal with respect to frequency at the zero-crossing. Since the frequency interval between zero-crossings corresponds to a phase shift of π, the equivalent frequency shift could finally be converted to a measured AC phase shift.

The molecular beam was focused by two electrostatic quadrupole lenses, whose focal lengths depend upon the strength of the quadrupole field and on the velocity of the molecules. Thus our resonance signal was derived from a narrow slice ($\sim$20%) of the full Maxwell Boltzmann distribution, which we were free to choose by adjusting the voltages on the quadrupole lenses. The velocity was measured by the fringe spacing of the Ramsey pattern, which goes through zero each time the quantity $(\omega - \omega_0)l/v$ increases by π, as can be seen from Equation (37).

We set the magnetic field B_x to a value (approximately 1.3 G) such that the resonance of the $I - K$ transition increased from 22.1677(3)kHz to 22.70(1)kHz. Knowing this frequency shift, we were able to calculate $[\mu_x]_a$ and $[\mu_x]_b$, as described by Sangster *et al.* [29]. The potential difference across the electric field plates was set to 20.1(1)kV/cm. Knowing that the spacing between the RF coils was L=2.066(5)m, we used Equation (36) to predict a value for the AC phase of 2.47(2)mrad, shown as the solid line in Figure 11. In the same figure, we also show the phases measured at seven different velocities ranging from 188m/s to 366m/sec. The weighted mean of the experimental points is 2.42(5)mrad in excellent agreement with the theoretical expectation. We see no evidence for any deviation from the predicted velocity-independence.

Next, the velocity was fixed at 254m/sec while the electric field was varied from 5-20kV/cm, as shown in Figure 12. The theoretical prediction, shown once again by a solid line, is not quite linear in E_z due to a small Stark shift of the resonance. The experimental points show the AC phase measured at four different values of the electric field, approximately 5, 10, 15, and 20kV/cm. Again, there is no evidence of any discrepancy between theory and experiment.

The most stringent check of the theory is obtained by dividing the value of each measured phase by the corresponding predicted value. A weighted average over all the points we measured using the 2-3 transition gives

$$\frac{\Delta\Phi_{exp}}{\Delta\Phi_{th}} = 0.98(2). \tag{38}$$

An earlier experiment using the $F-H$ transition also confirmed the theory but with

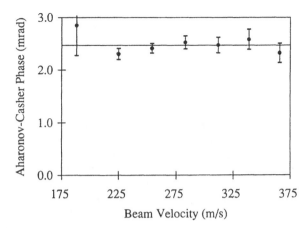

Figure 11. *AC phase versus beam velocity. The experimental points are in good agreement with the theoretical expectation. There are no free parameters.*

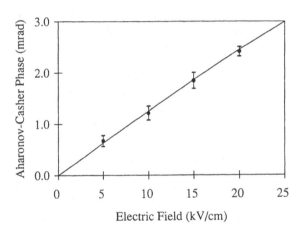

Figure 12. *AC phase versus electric field strength. The experimental points are in good agreement with the theoretical expectation. There are no free parameters.*

less precision (Sangster *et al.* [28]).

4.2.3 Conclusions

In this section we have studied the AC phase, which is an analogue found in neutral, spin 1/2 particles of the AB phase for charged particles. We have shown that although it may share some of the geometric features of the AB phase, this phase can be observed without requiring that the particles encircle a line charge. Finally, we have outlined an experiment to measure the phase shifts actually exhibited by molecules moving in an electric field, which shows that they are accurately described by the AC effect at the 2% level.

5 Searching for the electron EDM

5.1 Schiff's theorem and the electron EDM

We return now to the main theme—the search for T-violation in atomic and molecular systems. In Sections 2 and 3 we focussed on T-violating effects proportional to nuclear spin (characterised by the Schiff moment Q and the weak tensor coupling C_T) and we discussed the finite nuclear size as a mechanism for avoiding Schiff's theorem. The volume effect does not work for electrons because they are point particles but as you may recall from Section 2, relativistic effects provide another mechanism. If an atom is sufficiently heavy, the electrons can be quite relativistic and if they have an EDM it can lead to significant atomic EDM effects.

We take as our zeroth order Hamiltonian the Dirac equation

$$H^0 = \sum_{electrons} (\beta mc^2 + \boldsymbol{\alpha}\cdot\mathbf{c}\mathbf{p} + V), \tag{39}$$

where V includes the uniform applied field $\mathbf{E}_{ext}$ as well as the atomic Coulomb interaction. Now we would like to let the electron have an EDM d and to treat the electric dipole interaction as a perturbation, but what is the correct relativistic form of the interaction? The *magnetic* dipole interaction for a spin-1/2 particle of magnetic moment μ can be introduced into the Dirac equation by a term (Bjorken and Drell [30]) $H_{M1} = (\mu/2)\sigma_{\mu\nu}F^{\mu\nu}$, where $\sigma^{\mu\nu}$ is the tensor $(i/2)[\gamma^\mu,\gamma^\nu]$ and $F^{\mu\nu}$ is the electromagnetic field. Following Salpeter [31] we take it that the *electric* dipole interaction is just the odd-parity counterpart $\gamma_5 H_{M1}$. That is, we take

$$H_{E1} = \frac{d}{2}\gamma_5\sigma_{\mu\nu}F^{\mu\nu}, \tag{40}$$

where d is the supposed EDM of the electron. Converting from covariant notation to Dirac notation, the interaction becomes

$$H_{E1} = -d\beta(\boldsymbol{\sigma}\cdot\mathbf{E} + ic\boldsymbol{\alpha}\cdot\mathbf{B}). \tag{41}$$

Here β, $\boldsymbol{\alpha}$ and $\boldsymbol{\sigma}$ are the standard Dirac operators. To be concrete, let us choose the representation

$$\beta = \begin{bmatrix} 1 & 0 \\ 0 & -1 \end{bmatrix}, \quad \boldsymbol{\sigma} = \begin{bmatrix} \hat{\sigma} & 0 \\ 0 & \hat{\sigma} \end{bmatrix}, \quad \boldsymbol{\alpha} = \begin{bmatrix} 0 & \hat{\sigma} \\ \hat{\sigma} & 0 \end{bmatrix}, \tag{42}$$

$\hat{\sigma}$ being the Pauli matrix. The familiar classical expression $-\mathbf{d}\cdot(\mathbf{E} + \mathbf{v} \times \mathbf{B})$ is the non-relativistic limit of this interaction. In atoms and molecules with an unpaired electron, the electric term in Equation (41) is much larger then the magnetic one, so we take as our perturbation

$$H^1 = -d\beta\boldsymbol{\sigma}\cdot\mathbf{E}. \tag{43}$$

Notice the important difference between this and the non-relativistic operator, namely, the factor β. In a heavy atom we should really sum over electrons, but to keep the notation simple I will not write a $\sum$ sign. The EDM interaction energy of the atom is therefore

$$\langle H^1 \rangle = \langle \psi^0| - d\beta\boldsymbol{\sigma}\cdot\mathbf{E}|\psi^0\rangle. \tag{44}$$

The total field $\mathbf{E}$ on the electron can be written as $-[\nabla, V/q]$ where V is the potential in Equation (39). From this it follows (Sandars [32]) that

$$\boldsymbol{\sigma}\cdot\mathbf{E} = -[\boldsymbol{\sigma}\cdot\nabla, H^0/q] \tag{45}$$

where H^0 is the relativistic Hamiltonian of Equation (39). But this commutator vanishes, which means that $\langle\psi^0|d\boldsymbol{\sigma}\cdot\mathbf{E}|\psi^0\rangle$, *i.e.* the nonrelativistic interaction vanishes in accordance with Schiff's theorem. Finally, since this matrix element is zero, we can add it to the right-hand side of Equation (44) to obtain

$$\langle H^1\rangle = \langle\psi^0|(1-\beta)d\boldsymbol{\sigma}\cdot\mathbf{E}|\psi^0\rangle \tag{46}$$

which can be written explicitly as

$$\langle H^1\rangle = \left\langle\psi^0\left|\begin{array}{cc} 0 & 0 \\ 0 & 2d\hat{\boldsymbol{\sigma}}\cdot\mathbf{E} \end{array}\right|\psi^0\right\rangle . \tag{47}$$

This form of the interaction energy obviously satisfies Schiff's theorem since it only couples to the small components of the wave-function and therefore vanishes in the nonrelativistic limit.

Since the integral in Equation (46) is dominated by the region very close to the nucleus, where the electron is moving fast, the total field $\mathbf{E}$ on the electron can be reasonably approximated by $Ze\hat{\mathbf{r}}/r^2$, the Coulomb field near the nucleus. Writing the two small components of $|\psi^0\rangle$ as $|g^0\rangle$, we obtain the approximation

$$\langle H^1\rangle = \left\langle g^0\left|2d\hat{\boldsymbol{\sigma}}\cdot\frac{Ze\hat{\mathbf{r}}}{r^2}\right|g^0\right\rangle . \tag{48}$$

This vanishes unless g^0 is a state of mixed parity (because $\hat{\mathbf{r}}$ is an odd-parity operator), which means that the interaction requires the external field $\mathbf{E}_{\text{ext}}$. Let us expand the small component of the wave-function in angular momentum eigenstates, just as we did for the non-relativistic wave-function in Section 2. The small components of $p_{1/2}$ and $s_{1/2}$ states have orbital angular momenta $l = 0$ and 1 respectively, so the leading terms are

$$g^0 = a_p g_{1/2} + a_s g_{1/2} + \dots . \tag{49}$$

Once again, the higher angular momentum functions are suppressed at the origin and comes largely from these two leading terms. As a result, one finds in atomic units (the details are in Khriplovich's book [11])

$$\langle H^1\rangle \approx 8a_s a_p (Z\alpha)^2 Z d\,\hat{\boldsymbol{\sigma}}\cdot\hat{\boldsymbol{\lambda}} . \tag{50}$$

Here the factor $(Z\alpha)^2$ is due to the size of the small components of the wave-function, Z is due to the Coulomb field, and $\hat{\boldsymbol{\lambda}}$ is the axis of the $s-p$ mixing. For an atom we estimate once again that $a_s a_p \hat{\boldsymbol{\lambda}} \sim \mathbf{E}_{\text{ext}}$ (Equation (20) and the associated text). Then putting $Z=70$ into Equation (50) we find that

$$\langle H^1\rangle_{\text{Atom}} \sim 100d\,\hat{\boldsymbol{\sigma}}\cdot E_{\text{ext}} , \tag{51}$$

which means that in spite of Schiff's theorem, the EDM interaction in a heavy atom is 100 times larger than that of the free electron! This marvelous enhancement factor was first noticed by Sandars [32].

The situation in a polar molecule is even more favourable since we can have $a_s a_p \hat{\lambda} \sim 0.1 E_{\text{ext}}$ along the external field. In that case

$$\langle H^1 \rangle_{\text{Molecule}} \sim 10 \, d \, \hat{\boldsymbol{\sigma}} \cdot \hat{\boldsymbol{\lambda}}, \tag{52}$$

which is larger by a factor of order $5 \times 10^8 / E_{\text{ext}}$ than the atomic interaction. This additional enhancement of a molecule relative to an atom is exactly the same as the factor we encountered in Section 2 in the context of the Schiff interaction, namely the large mixing of s and p orbitals in a polar molecule. Our conclusion is that if the electron has an EDM, the effect in a heavy polar molecule should be similar to that of the free electron in a field of 10 a.u. $= 5 \times 10^{10} \text{V/cm}$. This is a huge effective electric field.

5.2 Suitable molecules

Our discussion of suitable molecules is restricted to diatomics because the spectra for polyatomic systems are so much more complex. Candidates of interest should be polar (for large $a_s a_p$), heavy (for large $Z^3 \alpha^2$) and paramagnetic (since the interaction depends on $\hat{\boldsymbol{\sigma}}$). We also need to be able to polarise the molecule fully along the external field $\mathbf{E}_{\text{ext}}$, so the rotational states should not be too widely split. Taking the molecular dipole to be $\mu_e \sim e a_0$ and the external field to be less than 50kV/cm, this places an upper limit on the rotational constant B of approximately 1.0cm^{-1}, which excludes the hydrides. The heaviest candidates, such as NpO or RaF, for example, are not very appealing because they are radioactive and cannot easily be handled in large amounts. There are a number of heavy but stable oxides and sulfides, such as ThN, LuO, and BiS but these tend to have very low vapour pressure and would involve working with extremely hot sources.

We do not discuss all the possibilities here, but go instead to the most promising group, the fluorides. Table 1 lists some of the heavier ones, together with some relevant information. From the standpoint of intrinsic sensitivity, HgF seems to be the best, with an effective electric field of almost 10^{11}V/cm on the electron spin. However, there does not seem to be a convenient way to detect the molecule with high efficiency because the ionisation energy of Hg is very high and the first optical transition to a bound state is in the UV at 256nm. The next most sensitive molecule is YbF, which can be

	Hot Wire	$A\text{--}X$(nm)	E_{int}(GV/cm)
PbF	No	443	28
HgF	No	256*	96
YbF	OK	552	30
DyF	OK	?	?
BaF	Good	859	9

Table 1. *Some properties of heavy paramagnetic fluorides. First column: detectability with hot wire. Second column: wavelength of first resonance line (* indicates $C\text{--}X$ transition). Third column: effective electric field on the electron EDM.*

detected both by a hot wire and by laser induced fluorescence. This makes it preferable
to PbF which has a similar effective electric field on the electron EDM. Finally, BaF is
an interesting case because it is very convenient to work with and the effective electric
field, though not the largest, is still very large. These considerations have led us to try
making an electron EDM measurement using YbF, and the rest of this section describes
the current status of that attempt.

5.3   Progress with YbF

Yb is a transition element in the lanthanide group, but because the ground state configuration is $4f^{14}6s^2$, it is in many respects similar to an alkaline earth. One of the two
$6s$ electrons bonds ionically with the F atom, leaving a simple valence electron whose
spectrum is very similar to that of an alkali. The ground state X is in a $^2\Sigma$ configuration, while the first excited state A is a $^2\Pi$. We find that a good way to produce a
molecular beam of YbF is to heat YbF_3 and Al to $\sim 1200°C$ in a zirconium crucible
with a small hole in the lid. The resulting beam has been observed by means of an oxy-
genated rhenium hot wire and by laser induced fluorescence (LIF) on the $A^2\Pi_{1/2}$–$X^2\Sigma$
transition. In order to perform an electron EDM measurement, we need to know the
resonance frequency for electron spin-flip transitions within the molecule, which means
understanding the magnetic structure of the ground state.

For the most abundant isotope, ^{174}YbF, which has no Yb nuclear spin, the magnetic
interaction Hamiltonian has the following structure

$$H = \gamma \mathbf{s} \cdot \mathbf{N} + b \, \mathbf{s} \cdot \mathbf{I} + c s_z I_z \qquad (53)$$

where $\mathbf{s}$ is the electron spin, $\mathbf{N}$ is the rotational angular momentum, $\mathbf{I}$ is the F nuclear
spin and the subscript z indicates projection along the internuclear axis. Since $s=1/2$
and $I=1/2$, there are four levels for each value of N, with relative energies determined
by the constants γ, b and c. In the rotational spectrum, each transition of the P-
branch ($\Delta N = -1$) is split into 4 lines by this structure, allowing us to determine
the constants from the LIF spectra. Figure 13 shows the $P(70)$ and $P(13)$ lines of the
$A^2\Pi_{1/2}$–$X^2\Sigma$ transition measured by Sauer et al. [33]. Although the laser is propagating
at right angles to the molecular beam, there is some residual Doppler width (due to
the divergence of the molecule beam) which is 15MHz in the upper trace and 32MHz
in the lower. From a large number of such scans we were able to build up the map of
ground-state energy levels versus the rotational quantum number N shown in Figure 14.

Analysis of Figure 14 leads to the surprising conclusion that γ is very far from
constant. In fact it is positive at low N, goes to zero at $N \sim 59$ and becomes negative
at larger N. We find that the variation of γ is well described by the empirical formula

$$\gamma = \gamma_0 + \gamma_1 N(N+1) \qquad (54)$$

in which $\gamma_0 = 13.31(8)$MHz and $\gamma_1 = -3.801(15)$kHz. A simple-minded estimate of γ_0
(as described by Sauer et al. [33]) gives $\gamma_0 \sim 270$MHz, which is 20 times larger, whereas,
the 'centrifugal stretching' correction γ_1 is expected to be positive and at least 10 times
smaller. This surprise has yet to be explained quantitatively, but it seems likely that it
is related to admixtures of the low-lying $4f^{13}$ configurations into the X and A

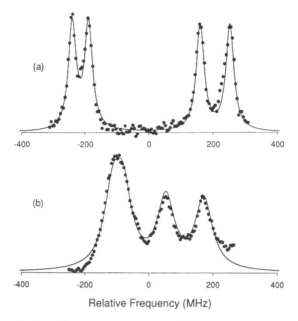

Figure 13. *Laser-induced fluorescence spectra of ^{174}YbF on the $A-X$ transition. Upper trace: the P-branch lines for $N=70$. Lower trace: the P-branch lines for $N=13$. In each case, the points are experimental data and the line is the sum of four Lorentzians.*

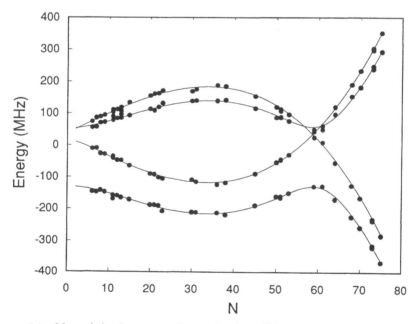

Figure 14. *Map of the four ground state levels of ^{174}YbF versus rotation N. Points are experimental. Lines are the best fit to the eigenvalues of Equation (53) with γ of the form $\gamma = \gamma_0 + \gamma_1 N(N+1)$.*

states. Values were also obtained for the fluorine hyperfine constants: b=142(2)MHz, c=84(8)MHz.

As a result of this spectroscopy, we can now read off the ground state electron spin resonance frequencies as a function of N using Figure 14. Typically the frequency required is 100–200MHz. The basic experimental setup for an electron EDM experiment is illustrated in Figure 15. The pump laser beam depopulates one of the four ground state levels by exciting $A-X$ transitions. The two separated RF fields then drive an electron spin resonance which repopulates that level. Finally the probe laser, tuned to the same transition as the pump, excites laser-induced fluorescence in proportion to the RF resonance probability. The resonance will be studied in a strong external field $\mathbf{E}_{ext}$, which can be reversed in order to look for a frequency shift due to the electron EDM. We think it should be possible to detect shifts at the level of 1mHz, which means that an electron EDM as small as 10^{-28} e.cm could be measured by this technique.

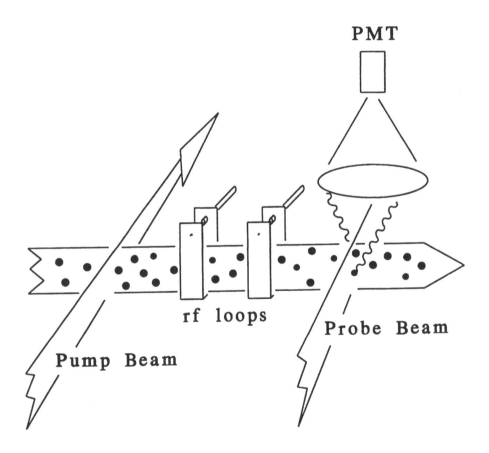

Figure 15. *General scheme for an experiment to measure the electron EDM using a YbF molecular beam.*

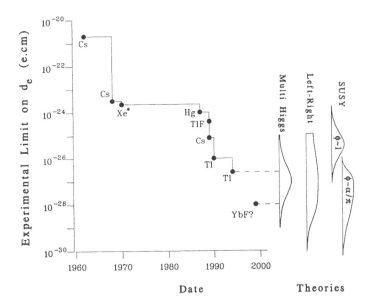

Figure 16. *A history of electron* EDM *measurements showing the experimental upper limit versus time. We also show possible values of d_e according to some extensions of the standard model.*

5.4 Connections with particle physics

Figure 16 shows a history of electron EDM measurements. Each point represents a new upper limit on the electron EDM and shows the atomic system used to measure it. The lack of activity in the '70s and early '80s corresponds to a period of intense interest in P-violation without T-violation. During this time attention was focussed on testing the electroweak theory of Weinberg, Salam and Glashow (WSG) which predicted P-violation in atomic physics due to the neutral Z^0 particle. By 1986, the WSG model had become part of the standard model of elementary particle physics and parity violation was a well-established effect in atomic physics. There was then a resurgence of interest in T-violation because it provides a sensitive way to probe new interactions at much higher energy than the Z^0 mass, corresponding to effects in atoms that are much weaker than G_F. Since we do not know what physics lies at these energy scales, the EDM measurements provide important constraints on theory, even though they have so far been only upper limits. Figure 16 shows the range of electron EDM values expected in a few of the most popular extensions to the standard model and one sees that the measurements are entering the range where these theories start to be constrained. Theories predicting a larger electron EDM are not shown since they are, of course, already ruled out by the experiments. A detailed discussion about theories of elementary particle physics beyond the standard model is outside the scope of this article, but a number of excellent reviews have been written on this subject, the most recent of which is by Barr [34]. This provides a clear explanation of the proposed theories and shows how they are tested by the parameters Q, C_T, d_e, C_S which we have discussed.

References

[1] E.M. Purcell and N.F. Ramsey, *Phys Rev.*, **78**, 807 (1950).

[2] D.T. Lee and C.N. Yang , *Phys. Rev.*, **104**, 254 (1956).

[3] C.S. Wu, E. Ambler, R.W. Hayward, D.D. Hoppes and R.P. Hudson, *Phys. Rev.*, **105**, 1413 (1957).

[4] E.D. Commins and P.H. Bucksbaum, *Weak Interactions of Leptons and Quarks* (Cambridge University Press, Cambridge, 1983).

[5] L. Landau, *Nucl. Phys.*, **3**, 127 (1957).

[6] J.H. Christenson, J.W. Cronin, V.L. Fitch and R.J.H. Turlay, *Phys. Rev. Lett.*, **13**, 138 (1964).

[7] W. Pauli, *Niels Bohr and the Development of Physics*, (Pergamon Press, New York, 1955).

[8] J.H. Smith, E.M. Purcell and N.F. Ramsey, *Phys. Rev.*, **108**, 120 (1957)

[9] L.I. Schiff, *Phys. Rev.*, **132**, 2194 (1963).

[10] O.P. Sushkov, V.V. Flambaum and I.B. Khriplovich, *JETP*, **60**, 873 (1984).

[11] I.B. Khriplovich, *Parity Nonconservation in Atomic Phenomena* (Gordon and Breach, Hoovavu, 1991).

[12] R. Golub and S.K. Lamoreaux, *Phys. Reports*, **237**, 1 (1994).

[13] J.P. Jacobs, W.M. Klipstein, S.K. Lamoreaux, B.R. Heckel, E.N. Fortson, *Phys. Rev. Lett.*, **71**, 3782 (1993).

[14] D. Cho, K. Sangster and E.A. Hinds, *Phys. Rev. A*, **44**, 2783 (1991).

[15] E.D. Commins, S.B. Ross, D. DeMille and B.C. Regan, submitted to *Phys. Rev. A* (1995).

[16] S.A. Murthy, D. Krause Jr., Z.L. Li, L.R. Hunter, *Phys. Rev. Lett.*, **63**, 965 (1989).

[17] N.F. Ramsey, *Molecular Beams*, (Oxford University Press, Oxford, 1956).

[18] P.V. Coveney and P.G.H. Sandars, *J. Phys. B: Atom. Mol. Phys.*, **16**, 3727 (1983).

[19] Y. Aharonov and D. Bohm, *Phys. Rev.*, **115**, 485 (1959). The relationship between the Aharonov Bohm and Aharonov-Casher effects has been discussed by Aharonov and Casher [21] and by C.R. Hagen, *Phys. Rev. Lett.*, **64**, 2347 (1990).

[20] A. Tonomura, T. Matsuda, R. Suzuki, A. Fukuhara, N. Osakabe, H. Umezaki, J. Endo, K. Shinagawa, Y. Sugita and H. Fujiwara, *Phys. Rev. Lett.*, **48**, 1443 (1982).

[21] Y. Aharonov, A. Casher, *Phys. Rev. Lett.*, **53**, 319 (1984). See also J. Anandan, *Phys. Rev. Lett.*, **48**, 1660 (1982).

[22] B.E. Allman, A. Cimmino, A.G. Klein, G.I. Opat, H. Kaiser and S.A. Werner, *Phys. Rev. A*, **48**, 1799 (1993).

[23] Y. Aharonov, P. Pearle and L. Vaidman, *Phys. Rev. A*, **37**, 4052 (1988).

[24] H. Kaiser, S.A. Werner, R. Clothier, M. Arif, A.G. Klein, G.I. Opat and A. Cimmino, in it Atomic Physics 12, edited by J. Zorn and R. Lewis (AIP, New York, 1991) p. 247. An earlier result of the same experiment was published by A. Cimmino *et al.* [25].

[25] A. Cimmino, G.I. Opat, A.G. Klein, H. Kaiser, S.A. Werner, M. Arif and R. Clothier, *Phys. Rev. Lett.*, **63**, 380 (1989).

[26] M. Kasevich and S. Chu, *Phys. Rev. Lett.*, **67**, 181 (1991).

[27] D.W. Keith, C.R. Ekstrom, Q.A. Turchette and D.E. Pritchard, *Phys. Rev. Lett.*, **66**, 2693, (1991).

[28] K. Sangster, E.A. Hinds, S.M. Barnett and E. Riis, *Phys. Rev. Lett.*, **71**, 3641 (1993).

[29] K. Sangster, E.A. Hinds, S.M. Barnett, E. Riis and A.G. Sinclair, *Phys. Rev. A*, **51**, 1776 (1995).

[30] J. D. Bjorken and S. D. Drell, *Relativistic Quantum Mechanics* (McGraw-Hill, New York, 1964).

[31] E.E. Salpeter, *Phys. Rev.*, **112**, 1642 (1958).

[32] P.G.H. Sandars, *Phys. Lett.*, **14**, 194 (1965).

[33] B.E. Sauer, Wang Jun and E.A. Hinds, *Phys. Rev. Lett.*, **74**, 1554 (1995).

[34] S.M. Barr, *Int. J. Mod. Phys. A*, **8**, 209 (1993).

Quantum fluctuations and applications of squeezed light

H J Kimble and G -L Oppo*

California Institute of Technology, Pasadena, USA

* SUSSP44 Editor,
University of Strathclyde, Glasgow, UK

1 Introduction

This chapter is a short summary of recent developments concerning quantum noise. A far longer and more detailed review has recently appeared in the literature [1] and we will often refer the reader to the material and the calculations presented there. Our main subjects are the theory and the experiments on preparation of squeezed states of light and possible applications of squeezed light, such as interferometry and spectroscopy. The chapter concludes with a discussion of dual-beam correlations where we consider the Einstein-Podolsky-Rosen paradox, back-action evading measurements and quantum non-demolition measurements.

2 Squeezing in optical parametric oscillators

A schematic diagram of a degenerate optical parametric oscillator (OPO) is shown in Figure 1. The Hamiltonian describing this OPO configuration where two field modes (a, b) of frequencies (ω_a, ω_b) interact is

$$H = H_S + H_R^a + H_{SR}^a + H_R^b + H_{SR}^b, \tag{1}$$

where the system Hamiltonian H_S is

$$H_S = \hbar\omega_a aa^\dagger + \hbar\omega_b bb^\dagger + \frac{i\hbar\chi}{2}\left[(a^\dagger)^2 b - a^2 b^\dagger\right], \tag{2}$$

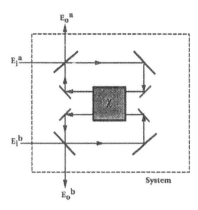

$E_o{}^a$
$E_i{}^a$
$E_i{}^b$
$E_o{}^b$
System

Figure 1. *Schematic diagram of the degenerate optical parametric oscillator.*

with χ being proportional to the second order susceptibility. The reservoir Hamiltonians H_R^a, H_R^b for the (a, b) modes are

$$H_R^a = \sum_j \hbar\omega_j r_{aj}^\dagger r_{aj}, \qquad\qquad H_R^b = \sum_j \hbar\omega_j r_{bj}^\dagger r_{bj}, \qquad (3)$$

and the interaction Hamiltonians of each reservoir with the respective mode are given by

$$H_{SR}^a = \hbar(a^\dagger \Gamma_a + a\Gamma_a^\dagger), \qquad\qquad H_{SR}^b = \hbar(b^\dagger \Gamma_b + b\Gamma_b^\dagger), \qquad (4)$$

with

$$\Gamma_a = \sum_j \kappa_j^a r_{aj}, \qquad\qquad \Gamma_b = \sum_j \kappa_j^b r_{bj}, \qquad (5)$$

and κ_j being the coupling coefficient between the system mode and reservoir mode r_j [1]. For simplicity, we consider the resonant case with $\omega_b = 2\omega_a = 2\omega_L$, where $2\omega_L$ is the frequency of the pump field, and denote the fluctuating input fields with E_i^a and E_i^b under the assumptions

$$\langle E_i^b(t) \rangle = \varepsilon e^{-2i\omega_L t}, \qquad\qquad \langle E_i^a(t) \rangle = 0, \qquad (6)$$

with ε real. Following the derivation of the Heisenberg equations presented in [1], we obtain

$$\dot{\tilde{a}}(t) = -\gamma_1 \tilde{a} + \chi \tilde{a}^\dagger \tilde{b} + \sqrt{2\gamma_1} \tilde{E}_i^a(t) \qquad (7)$$

$$\dot{\tilde{b}}(t) = -\gamma_2 \tilde{b} - \frac{\chi}{2}\tilde{a}^2 + \sqrt{2\gamma_2} \tilde{E}_i^b(t), \qquad (8)$$

where the γ_i are the decay rates in the cavity and where

$$\tilde{a}(t) = a(t)e^{i\omega_L t}, \qquad\qquad \tilde{b}(t) = b(t)e^{2i\omega_L t}, \qquad (9)$$

$$\tilde{E}_i^a(t) = E_i^a(t)e^{i\omega_L t}, \qquad\qquad \tilde{E}_i^b(t) = E_i^b(t)e^{2i\omega_L t}. \qquad (10)$$

If we now invoke the semiclassical approximation where

$$\langle \tilde{a} \rangle = \alpha, \qquad\qquad \langle \tilde{b} \rangle = \beta, \qquad\qquad (11)$$

and

$$\langle \tilde{a}^\dagger \tilde{b} \rangle = \alpha^* \beta, \qquad\qquad \langle \tilde{a}^2 \rangle = \alpha^2, \qquad\qquad (12)$$

then we derive the following coupled equations

$$\dot{\alpha}(t) = -\gamma_1 \alpha(t) + \chi \alpha^* \beta \qquad\qquad (13)$$
$$\dot{\beta}(t) = -\gamma_2 \beta(t) - \frac{\chi}{2} \alpha^2 + \sqrt{2\gamma_2}\, \varepsilon. \qquad\qquad (14)$$

The steady-state solutions of Equations 13 and 14 reveal two regimes of interest, namely the sub-threshold regime specified by

$$\varepsilon < \varepsilon_{th}, \qquad \alpha_s = 0, \qquad \beta_s = \sqrt{\frac{2}{\gamma_2}}\, \varepsilon, \qquad\qquad (15)$$

and the above threshold regime specified by

$$\varepsilon > \varepsilon_{th}, \qquad \alpha_s = \pm\sqrt{\frac{2\gamma_1\gamma_2}{\chi^2}\left(\frac{\varepsilon}{\varepsilon_{th}} - 1\right)}, \qquad \beta_s = \frac{\gamma_1}{\chi}, \qquad\qquad (16)$$

where

$$\varepsilon_{th} = \sqrt{\frac{\gamma_1^2\gamma_2^2}{2\chi^2}}. \qquad\qquad (17)$$

A quantum state is said to be squeezed when the fluctuations of one observable have been reduced below that of the vacuum state, *i.e.* the 'shot-noise' level of the coherent state. In the case where the observable which shows sub-shot noise fluctuations is a quadrature component of the electromagnetic field, one talks of *quadrature squeezing*. In order to quantify squeezing, one determines whether the normally ordered spectrum of fluctuations [1] is negative, the maximum of squeezing corresponding to the value -1. For example, Collett and Gardiner [2], [3] have shown that quadrature squeezing may occur in the sub-threshold regime of the OPO and that the normally ordered squeezing spectrum has the following form

$$S_\pm(\Omega) = \frac{\pm 4\chi \dfrac{\beta_s}{\gamma_1}}{\left(\dfrac{\Omega}{\gamma_1}\right)^2 + \left(1 \mp \dfrac{\chi\beta_s}{\gamma_1}\right)^2}, \qquad\qquad (18)$$

where Ω is the spectral frequency and where $\pm$ corresponds to the two quadratures of the output field. At threshold $\beta_s = \gamma_1/\chi$ so that near $\Omega = 0$ we have

$$S_+(\Omega) \to \infty \qquad\qquad S_-(\Omega) \to -1.$$

The divergence of S_+ near $\Omega = 0$ is not surprising as fluctuations are requested to diverge close to threshold due to the break-down of the linear analysis. $S_-(\Omega) \to -1$, instead, shows the possibility of generating large degrees of squeezing in the OPO output field. Quoting from [1], "*although Ponce de Leon never found his 'fountain of youth', we have (now) found the 'fountain of squeezing'; emanating from the OPO in which to bath our experiments with everlasting quantum quietness!*"

3   Generation of squeezed light

A schematic arrangement to generate squeezed light using a $\chi^{(2)}$ crystal is shown in Figure 2. More detailed diagrams of experimental arrangements may be found in the relevant literature [4, 5, 6].

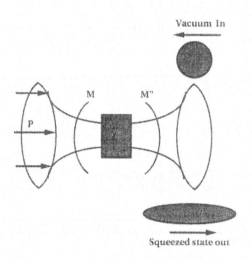

Figure 2. *Arrangement for the generation of squeezed light using a $\chi^{(2)}$ crystal.*

Squeezed states are detected using the technique known as balanced homodyne detection [7]. By varying the phase θ of a local oscillator, the photocount statistics reveal the existence of squeezed light [6]. What is measured in real experiments is the effective noise level

$$R(\Omega, \theta) = 1 + \xi S(\Omega, \theta) \tag{19}$$

where 1 is the shot noise level, ξ is the overall efficiency which was 0.934 in References [5, 6], and $S(\Omega, \theta)$ is the squeezing spectrum. For the maximum squeezing one obtains $R_- = 1 - \xi$ which was as low as 0.066. In a careful comparison between experiments and theory performed in [5], the predicted value of $R_- = 0.22$ has been contrasted with the observed value of $R_- = 0.25$ directly measured in the spectral density of photocurrent fluctuations. The main limitation of the experiments appears to be the blue-light induced infrared absorption.

In spite of the agreement between theory and experiments, it is important to note that real devices are subject to effects not considered by theoretical treatments of squeezed light generation. These experimental effects include: a) thermally excited noise, b) transverse effects in propagation c) Raman scattering and d) light induced absorption.

Table 1 charts some history of experimental squeezing results up to 1992 circa.

Year	Research Team	Method	Noise Level
1985–1986			
	Slusher *et al.* [8]	$\chi^{(3)}$, Na beam	0.75
	Shelby *et al.* [9]	$\chi^{(3)}$, fibre	0.87
	Wu *et al.* [10]	$\chi^{(2)}$, subthreshold OPO	0.37
	Maeda *et al.* [11]	$\chi^{(3)}$, Na vapour	0.96
	Raizen *et al.* [12]	$\chi^{(3)}$, Na beam	0.70
1987			
	Schumaker *et al.* [13]	$\chi^{(3)}$, 4-mode	0.80
	Grangier *et al.* [14]	$\chi^{(2)}$, subthreshold OPO	0.63
	Slusher *et al.* [15]	$\chi^{(2)}$, pulsed squeezing	0.87
1988 $\simeq$ 1992			
	Pereira *et al.* [16]	$\chi^{(2)}$, frequency doubling	0.87
	Pereira *et al.* [17]	$\chi^{(2)}$, subthreshold NOPO	0.45
	Ho *et al.* [18]	$\chi^{(3)}$, Na vapour	0.75
	Kumar *et al.* [19]	$\chi^{(2)}$, incoherent pulse	0.83
	Sizmann *et al.* [20]	$\chi^{(2)}$, frequency doubling	0.60
	Movshovich *et al.* [21]	$\chi^{(2)}$, Josephson paramp	0.99896
	Hirena *et al.* [22]	$\chi^{(2)}$, pulsed	0.78
	Rosenbluh *et al.* [23]	$\chi^{(3)}$, solitons	0.68
	Bergman *et al.* [24]	$\chi^{(3)}$, pulsed	0.32
	Hope *et al.* [25]	$\chi^{(2)}$, Ba beam	0.87
	Polzik *et al.* [5, 6]	$\chi^{(2)}$, subthreshold OPO	0.25

Table 1. *Status of observed noise levels compared to the vacuum state limit of 1.0*

4 Use of squeezed light for sensitivity beyond the standard quantum limit

Quantum mechanics induces limits on our ability to make precise measurements because every time we extract information from the system, fluctuations from the environment invade the system. However, by suitably preparing the state of the system with small fluctuations for the observable of interest (*e.g.* with a squeezed state), one can increase the precision of measurement below the standard quantum limit. Inevitably, the price to be paid for low noise in one observable is that noise will increase in an another. In the following we briefly review the reductions in phase and amplitude fluctuations of a squeezed state and discuss their use in interferometry and spectroscopy.

4.1 Phase and amplitude fluctuations

For a coherent state it is well known that

$$\delta\phi_v \sim \frac{1}{\sqrt{N}},$$

(20)

where the index v stands for vacuum. A squeezed state, however, may exhibit reduced phase fluctuations

$$\delta\phi_s \sim \sqrt{\frac{1 + \xi S_-}{N}} < \delta\phi_v \qquad (21)$$

i.e. sensitivity apparently tending to zero in the limit of ξ (the overall system efficiency) and S_- tending to +1 and -1 respectively. Indeed, a 3.0dB improvement relative to the vacuum state limit has been observed experimentally by Xiao, Wu and Kimble [26] and subsequently by Grangier et al. [14].

Analogously, for the amplitude fluctuations of a coherent state it is well known that

$$\frac{\delta A_v}{A} \sim \frac{1}{\sqrt{N}} \qquad (22)$$

while a squeezed state may exhibit reduced fluctuations

$$\frac{\delta A_s}{A} \sim \sqrt{\frac{1 + \xi S_+}{N}} < \frac{\delta A_v}{A}. \qquad (23)$$

A 2.5dB improvement relative to the vacuum state limit has been observed experimentally by Xiao, Wu and Kimble [27].

There is however a fundmental limit to the sensitivity enhancement allowed with squeezed light. By introducing again the overall efficiency $\xi = \gamma\rho\beta\alpha\eta^2$ where $(1 - \gamma)$ is the intrinsic 'loss' associated with the measurement itself (e.g. due to absorption), ρ is the escape efficiency, β is the propagation efficiency, α is the detector quantum efficiency and η is the homodyne efficiency, we find that

$$\frac{\delta A_s}{A} \sim \frac{[1 + \xi S_+]^{\frac{1}{2}}}{\sqrt{N}}. \qquad (24)$$

Within the limits

$$\rho\beta\alpha\eta^2 \to 1 \qquad 1 - \gamma \to 0 \qquad S_+ \to -1 \qquad (25)$$

it follows that

$$\frac{\delta A_s}{A} \sim \frac{1}{N}. \qquad (26)$$

Thus, the limiting enhancement is given by

$$\frac{\delta A_s}{\delta A_v} \sim \sqrt{N} \gg 1. \qquad (27)$$

With these ideas in mind, we now briefly review two simple applications of amplitude and phase squeezed light.

4.2   Interferometry beyond the vacuum state limit

It was first suggested by Caves [28] that squeezed light could have interferometric applications due to its reduced phase fluctuations. A schematic diagram of an experimental arrangement is shown in Figure 3.

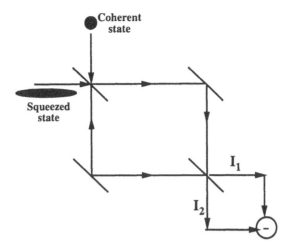

Figure 3. *Schematic diagram of the use of squeezed light in interferometry.*

It is possible to show that the signal-to-noise ratio, ψ, is dependent upon the depth of squeezing, S, present at the input port and the overall efficiency, ξ,

$$\psi \propto \frac{1}{1 + \xi S}. \tag{28}$$

An experiment employing squeezed light to improve interferometric devices beyond the vacuum state limit was first carried out by Xiao, Wu and Kimble [27].

4.3 Spectroscopy with squeezed light

Figure 4 depicts a two-level atom coupled to the 'normal' vacuum. In quantitative terms, we can consider the Lamb shift, δ, and the spontaneous decay rate, γ, as resulting from the quantum nature of the vacuum fluctuations driving the atom.

By contrast, Figure 5 depicts a two-level atom coupled to a squeezed vacuum. This system was first investigated by Gardiner [29]. A recent review article outlines the work which has been carried out on the effect of squeezed light on diverse optical systems [30].

The main problems encountered in the experimental investigation of atoms coupled with a squeezed vacuum (see Figure 5) are: a) the coupling of 'three'-dimensional atoms to a 'one'-dimensional squeezed field, b) the lack of atomic localisation due to atomic motion.

One way of solving these problems is by employing 'brute force' to localise the atom within $\bar{\lambda}$ of the focus of the squeezed radiation. Suitable wavefront engineering leading to tight focusing of the squeezed light beam incident on the atom is then required to match phase fronts. Another technique is to alter the atomic emission pattern by introducing a mirror ('optics on a half shell') in the vicinity of the atoms. This modifies the atomic 'antenna pattern' to match better the squeezed wave front. Finally,

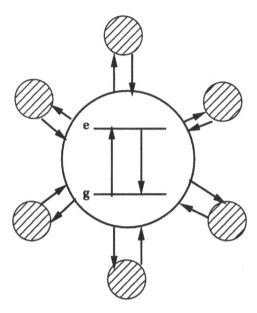

Figure 4. *Atom coupled to the 'normal' vacuum with the fluctuations giving rise to the Lamb shift, δ, and spontaneous decay, γ.*

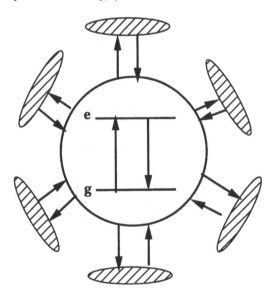

Figure 5. *Atom coupled to a squeezed vacuum.*

a third method analysed by Rice and Pedrotti [31] suggests the reduction of a three dimensional atom to one dimension by strongly coupling the atom to a cavity. This method requires the coupling of atom to cavity to be much larger than γ. It has been shown [31, 32] that squeezed light affects fundamental radiative processes into the cavity mode. The phase-dependent nature of the squeezed field results in line-narrowing of the incoherent spectrum for particular phase choices while other choices of phase result in line-broadening. Furthermore, we have demonstrated in experiments with Cesium atomic beams that probe absorption spectra may be altered in the presence of squeezed light.

As an alternative to considering the effect of squeezed light on resonance fluorescence spectra or probe absorption spectra, it is also worth investigating higher order phenomena such as two photon excitation or higher order field correlations functions like $g^{(2)}(\tau)$. Indeed, we have recently made the first demonstration of the alteration of fundamental radiation processes with squeezed light [33].

5 Correlations for spatially separated optical fields

Consider now the nondegenerate optical parametric amplifier (NOPA) shown in Figure 6.

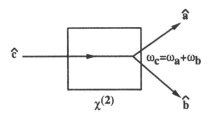

Figure 6. *Schematic of parametric down-converter.*

A treatment similar to the degenerate OPO below threshold can be extended to this case by employing the following interaction Hamiltonian for the three modes (a, b, c)

$$\hat{H}_I = i\hbar\chi^{(2)}\left[\hat{a}^\dagger\hat{b}^\dagger\hat{c} + \hat{a}\hat{b}\hat{c}^\dagger\right]. \tag{29}$$

In this system, it is interesting to study either the correlation of the intensities

$$\hat{I}_a \leftrightarrow \hat{I}_b, \quad \text{e.g.} \quad \langle(\hat{I}_a - \hat{I}_b)^2\rangle \to 0 \tag{30}$$

or the correlations of quadrature-phase amplitudes

$$\hat{X}_a, \hat{Y}_a \leftrightarrow \hat{X}_b, \hat{Y}_b. \tag{31}$$

With reference to Figure 7, we analyse the correlations of quadrature-phase amplitudes for spatially separated optical fields. By defining the quadrature variances $\Delta^2 X$ and $\Delta^2 Y$ as

$$\Delta^2 X = \langle(X_1 - X_2)^2\rangle \tag{32}$$
$$\Delta^2 Y = \langle(Y_1 - Y_2)^2\rangle \tag{33}$$

it is easy to show that for classical fields

$$\Delta^2 X \geq 2 \qquad \Delta^2 Y \geq 2 \tag{34}$$

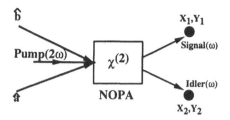

Figure 7. *Schematic of non-degenerate parametric down-converter and the signal and idler modes.*

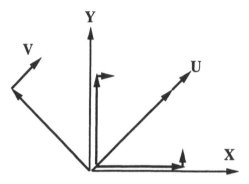

Figure 8. *Correlated fluctuations in (X,Y) projected onto (U,V) modes.*

while for quantum fields

$$\Delta^2 X \geq 0 \qquad \Delta^2 Y \geq 0. \tag{35}$$

The squeezing resulting from Type II non-degenerate parametric down-conversion may be simply explained. Correlated fluctuations in (X,Y) are projected onto (U,V) modes at $\pm 45°$ as shown in Figure 8. The correlations in (X,Y) lead to squeezing of opposite phase in (U,V), as displayed in Figure 9.

Note that the noise statistics of signal or idler beam alone are indistinguishable from a thermal field [34]. Instead, it is the correlations of the signal and idler fields taken

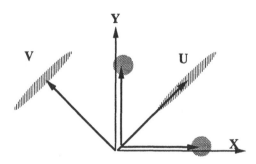

Figure 9. *The correlations in (X,Y) lead to squeezing of opposite phase in (U,V).*

together which are squeezed. Note also that the Wigner distribution W for signal and idler can be easily computed to be [34]

$$W(X_1,Y_1;X_2,Y_2) = \frac{4}{\pi^2}\exp\left\{-\left[(X_1+X_2)^2+(Y_1-Y_2)^2\right]e^{-2r}-\left[(X_1-X_2)^2+(Y_1+Y_2)^2\right]e^{2r}\right\}$$
(36)

where r is the squeezing parameter. In the limit of $r \to \infty$, $W \to \delta(X_1-X_2)\delta(Y_1+Y_2)$ leading to perfect correlations of the two separate beams.

5.1 The Einstein-Podolsky-Rosen paradox

Another application of the setup displayed in Figure 7 is to test the 'gedanken' experiment of Einstein-Podolsky-Rosen (EPR) [35]. Our system is formed by a pair of 'colliding' (interacting) modes of the electromagnetic field. Given that the two outgoing beams are correlated, we measure $(\hat{X}_2,-\hat{Y}_2)$ for the idler in order to infer at a distance $(\hat{X}_1,\hat{Y}_1)$ for the signal. The measurements are obviously affected by errors which can be quantified by the variances over the inference

$$\Delta^2 X = \langle(\hat{X}_1-\hat{X}_2)^2\rangle \qquad\qquad \Delta^2 Y = \langle(\hat{Y}_1+\hat{Y}_2)^2\rangle$$
(37)

where we assume fields of mean zero. If the concept of reality advanced by EPR is adopted, objective values of the canonically conjugate variables $(\hat{X}_1,\hat{Y}_1)$ can be obtained within the errors specified by the variances. However, quantum mechanics requires

$$\left(\Delta^2 X_1\right)\left(\Delta^2 Y_1\right) \geq 1$$
(38)

and the EPR paradox arises for [36]

$$\left(\Delta^2 X\right)\left(\Delta^2 Y\right) < 1.$$
(39)

Note that the Wigner phase-space function provides a basis for a local, realistic description of the EPR 'gedanken' experiment. Hence, ironically, EPR correlations are not manifestly quantum in the sense of a Bell inequality (although, they are nonclassical in the sense of Glauber-Sudarshan phase-space function) but rather "are precisely those between two classical particles" – J.S. Bell.

We refer the reader to Reference [1] for details about the relevance of the OPO gain in EPR measurements. The Caltech group has been working on EPR experiments involving squeezing in nondegenerate parametric down conversion during recent years [17, 37]. Technical problems such as the nonideal nature of the OPO containing two KTP crystals and polarisation mixing in the cavity initially limited progresses towards (39). However, by employing an a-cut KTP crystal at 1.079μm, we have been able to achieve the first realisation of the EPR experiment to continuous variables [37]. We have as well carried out an experiment in the realm of quantum communication as illustrated in Figure 10.

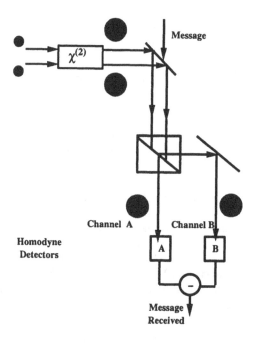

Figure 10. *A scheme for quantum communication.*

5.2   Back-action evading and non-demolition measurements

Until now we have focused on the parametric amplification in $\chi^{(2)}$ nonlinearities neglecting frequency conversion. We now consider the following interaction Hamiltonian

$$H_I \cong i\hbar[\kappa(\tilde{a}^\dagger \tilde{b}^\dagger - \tilde{a}\tilde{b}) + \sigma(\tilde{a}^\dagger \tilde{b} - \tilde{a}\tilde{b}^\dagger)] \tag{40}$$

where the pump mode c has been treated semiclassically, κ and σ describe parametric
amplification and frequency conversion, respectively. This interaction Hamiltonian describes the arrangement of Figure 11 which can be used as a first step towards Quantum
Non-Demolition (QND) measurements.

Hee we set $\kappa = \sigma$ to simplify the calculations. Following Reference [1], one obtains
the following input-output relations

$$\hat{X}_s^{\text{out}} = \hat{X}_s^{\text{in}} \qquad\qquad \hat{Y}_s^{\text{out}} = \hat{Y}_s^{\text{in}} - f(\kappa)\hat{Y}_m^{\text{in}} \tag{41}$$

$$\hat{X}_m^{\text{out}} = \hat{X}_m^{\text{in}} + f(\kappa)\hat{X}_s^{\text{in}} \qquad \hat{Y}_m^{\text{out}} = \hat{Y}_m^{\text{in}}. \tag{42}$$

where the indexes m and s label 'meter' and 'signal' respectively. $\hat{X}_s^{\text{out}}$ and $\hat{X}_m^{\text{out}}$ form a
back-action evading (BAE) pair because the signal $\hat{X}_s$ is unaffected by measurement, and
yet information from the signal is written to meter $\hat{X}_m$ with high fidelity for $f(\kappa) \gg 1$.
A guiding light on BAE measurements is the work of Yurke [38].

An experimental scheme for BAE measurement (originally suggested in [39, 40, 41])
has recently been implemented and is schematically shown in Figure 12. With paramet-

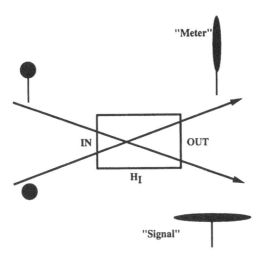

Figure 11. *Two input beams interact in a $\chi^{(2)}$ nonlinear crystal together with polarisation rotation and then are separated as 'Meter' and 'Signal' outputs.*

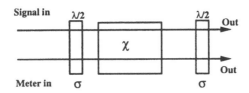

Figure 12. *A scheme for implementing a back-action evading measurement.*

ric gain, G, and polarisation rotation, σ, the back-action evading condition is obtained to be

$$G \cos 4\sigma = 1. \tag{43}$$

When this condition is satisfied we have

$$\hat{X}_s^{\text{out}} = \hat{X}_s^{\text{in}} \qquad\qquad \hat{X}_m^{\text{out}} = \hat{X}_m^{\text{in}} + f(G)\hat{X}_s^{\text{in}}, \tag{44}$$

where

$$f(G) = 2\sqrt{G^2 - 1}. \tag{45}$$

The first BAE experiment to use an arrangement similar to Figure 12 working close to the condition (43) is described in Reference [42]. Other experimental work on back-action evading measurements is reported in [43, 44, 45].

The very essence of QND detection is the ability to measure a given quantum state *repeatedly* and to do so beyond the context of (small) Gaussian fluctuations. Using two historical quotations: "We define a QND measurement of $\hat{A}$ as a *sequence* of precise measurements ..." [46] and "The key feature of such a non-demolition measurement is *repeatability* – once is not enough" [47]. Indeed an unambiguous quantum non-demolition measurement requires at least two back-action evading detectors as schematically shown in Figure 13.

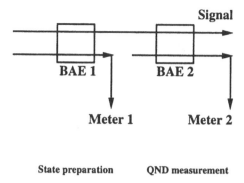

Figure 13. *A scheme for a quantum non-demolition measurement.*

Since we rarely know the exact state of any incident field, the first BAE detector prepares the the field state for the second BAE detector, which then attempts a measurement of a known state of the field. Meters 1 and 2 are compared to each other and to the signal to establish the QND measurement. Note however that the two BAE detectors could be replaced by just one if the signal input can be repeatedly circulated by using for example short pulses. Whatever the scheme, the fundamental requirement remains the ability of making repeated measurements with increased sensitivity beyond the standard quantum limit resulting from a sequence of such measurements. Success cannot be claimed from a single BAE measurement in the same way that a set of necessary conditions can never prove sufficient for a cause and effect relationship.

6   Conclusions

The scope of this contribution has been to review briefly some of the recent experimental configurations for the generation and the application of squeezed light with emphasis on the work of the Caltech Quantum Optics Group. Many exciting topics have been sacrificed for brevity and pedagogic reasons. It is important to note, however, that quantum optics is a fast growing field of research and that many of the results achieved during the last decade represent mile–stones in the way to future developments. For example, we are witnessing at present the genesis of quantum state synthesis [48]. Although it has been possible to achieve measurements beyond the vacuum-state limit, many open questions remain about the ultimate limit of quantum measurements and about the precision necessary to engineer quantum states. Interferometry with squeezed light as well as quantum non-demolition measurements are attempting to assess these limits in order to further extend the experimental manipulation of light.

Acknowledgements

We would like to thank G. Yeoman for preparing the figures. The work of HJK is supported by the National Science Foundation and the Office of Nasal Research.

References

[1] H J Kimble, *Fundamental Systems In Quantum Optics*, edited by J Dalibard, J M Raimond and J Zinn-Justin (Elsevier Science, Amsterdam, 1992)

[2] M J Collett and C W Gardiner, *Phys Rev A*, **30**, 2123 (1976)

[3] M J Collett and C W Gardiner, *Phys Rev A*, **31**, 3761 (1985)

[4] E S Polzik and H J Kimble, *Opt Lett* , **16**, 1400 (1991)

[5] E S Polzik, J Carri and H J Kimble, *Phys Rev Lett* , **68**, 3020 (1992)

[6] E S Polzik, J Carri and H J Kimble, *App Phys* , **B 55**, 279 (1992)

[7] R Loudon, *The Quantum Theory Of Light*, (John Wiley, New York, 1973)

[8] R E Slusher, L W Hollberg, B Yurke, J C Mertz, and J F Valley, *Phys Rev Lett* , **55**, 2409 (1985); R E Slusher, B Yurke, P Grangier, A LaPorta, D F Walls, and M Reid, *J Opt Soc Am B*, **4**, 1453 (1987)

[9] R M Shelby, M D Levenson, S H Perlmutter, R G DeVoe, and D F Walls, *Phys Rev Lett* **57**, 691 (1986)

[10] L A Wu, H J Kimble, J L Hall, and H Wu, *Phys Rev Lett* , **57**, 2520 (1986); L A Wu, M Xiao, and H J Kimble, *J Opt Soc Am B*, **4**, 1465 (1987)

[11] M W Maeda P Kumar and J H Shapiro, *Opt Lett* , **12**, 161 (1987)

[12] M G Raizen, L A Orozco, M Xiao, T L Boyd, and H J Kimble, *Phys Rev A*, **59**, 198 (1987)

[13] B L Schumaker, S H Perlmutter, R M Shelby, and M D Levenson, *Phys Rev Lett* , **58**, 357 (1987)

[14] P Grangier, R E Slusher, B Yurke, and A LaPorta, *Phys Rev Lett* , **59**, 2153 (1987)

[15] R E Slusher, P Grangier, A LaPorta, B Yurke, and M J Potasek, *Phys Rev Lett* , **59**, 2566 (1987)

[16] S F Pereira, M Xiao, H J Kimble, and J L Hall, *Phys Rev A*, **38**, 4931 (1988)

[17] S F Pereira, K C Peng, and H J Kimble, in *Coherence and Quantum Optics V*, edited by J H Eberly, L Mandel, and E Wolf (Plenum Press, New York, 1990) p 889

[18] S-T Ho, N C Wong, and J H Shapiro, in *Coherence and Quantum Optics V*, edited by J H Eberly, L Mandel, and E Wolf (Plenum Press, New York, 1990) p 497

[19] P Kumar, O Aytür, and J Huang, *Phys Rev Lett* , **64**, 1015 (1990)

[20] A Sizmann, R J Horowicz, E Wagner, and G Leuchs, *Opt Comm* , **80**, 138 (1990)

[21] R Movshovich *et al.* , *Phys Rev Lett* , **65**, 1419 (1990)

[22] T Hirano and M Matsuoka, *Opt Lett* , **15**, 1153 (1990)

[23] M Rosenbluh and R M Shelby, *Phys Rev Lett* , **66**, 153 (1991)

[24] K Bergman and H A Haus, *Opt Lett* , **16**, 663 (1991)

[25] D M Hope *et al.* , Tenth International Conference on Laser Spectroscopy, Font-Romeu (17-21 June 1991)

[26] M Xiao, L A Wu and H J Kimble, *Phys Rev Lett* , **59**, 278 (1987)

[27] M Xiao, L A Wu and H J Kimble, *Opt Lett* , **13**, 176 (1988)

[28] C Caves, *Phys Rev D*, **23**, 1693 (1981)

[29] C Gardiner, *Phys Rev Lett* , **56**, 1917 (1986)

[30] A S Parkins, in *Modern Nonlinear Optics, Part 2*, edited by M Evans and S Kielich (John Wiley, New York, 1993)

[31] P R Rice and L Pedrotti, *J Opt Soc Am B*, **9**, 2008 (1992)

[32] J I Cirac, *Phys Rev A*, **46**, 4354 (1992)

[33] N Ph Georgiades, E S Polzik, K Edamatsu, H J Kimble, and A S Parkins, *Phys Rev Lett* **75**, 3426 (1995)

[34] S M Barnett and P L Knight, *J Opt Soc Am B*, **2**, 467 (1985)

[35] A Einstein, B Podolsky and N Rosen, *Phys Rev* , **47**, 777 (1935)

[36] M Reid and P Drummond, *Phys Rev Lett* , **60**, 2731 (1988)

[37] Z Y Ou, S F Pereira, H J Kimble, and K C Peng, *Phys Rev Lett* , **68**, 3663 (1992); and *App Phys* , **B 55**, 265 (1992)

[38] B Yurke, *J Opt Soc Am B*, **2**, 732 (1985)

[39] S F Pereira, H J Kimble, P Alsing, and D F Walls, *J Opt Soc Am B*, **4**, 24 (1987)

[40] P Alsing, G J Milburn, and D F Walls, *Phys Rev A*, **37**, 2970 (1988)

[41] R Shelby and M Levenson, *Opt Comm* , **64**, 553 (1987)

[42] A LaPorta, R E Slusher, and B Yurke, *Phys Rev Lett* , **62**, 18 (1989)

[43] P Grangier J F Roch, and G Roger, *Phys Rev Lett* , **66**, 1418 (1991)

[44] J Ph Poizat and P Grangier, *Phys Rev Lett* , **70**, 271 (1993)

[45] S F Pereira, Z Y Ou, and H J Kimble, *Phys Rev Lett* , **72**, 214 (1994)

[46] V Braginsky, Y I Vorontsov, and K S Thorne, *Science*, **209**, 547 (1980)

[47] C M Caves, K S Thorne, R W P Drever, V D Sandberg, and M Zimmerman, *Rev Mod Phys* **52**, 341 (1980)

[48] A S Parkins, P Marte, P Zoller, and H J Kimble, *Phys Rev Lett* , **71**, 3095 (1993); and *Phys Rev A*, **51**, 1578 (1995)

Quantum superpositions in dissipative environments: decoherence and deconstruction

P L Knight and B M Garraway

Imperial College, London, UK

1   Introduction

It is difficult to construct a satisfying definition of a *nonclassical* state of light: in a sense *all* states of light contain some quantum features deriving from the discreteness of the energy quanta of that field. In practice, nonclassical features are hard to measure [1]. Many fields can be modelled accurately as *classical* fluctuating objects and their measurable properties are influenced hardly at all by their underlying quantum nature. Coherent and thermal fields are examples whose quantum characteristics are, for most practical purposes, invisible. Nevertheless, these fields *are* inherently quantum mechanical and their underlying quantum properties are recoverable at least in principle, through careful scrutiny of their statistics (for example using a micromaser). However, there *are* classes of light field states which are generally agreed to be wholly nonclassical: their *gross* properties (for example the photon number or field quadrature variances) are *below* that of a perfectly-coherent field. These sub-Poissonian and squeezed states have been the subject of a minor industry in quantum optics in the past two decades, and Meystre and Walls have recently collected together a selection of important papers [1] on nonclassical fields in quantum optics, and to which the interested reader is directed. In these lectures we will demonstrate how nonclassical field states may be constructed by superposing component states in phase-space [2]. We will then describe how these nonclassical states are peculiarly sensitive to dissipation when the coupling of the field modes to their larger environment [3] is taken into account. Well before the field energy is significantly dissipated, the coherences responsible for the nonclassical behaviour will decay, at a 'decoherence rate' [4] determined by the square of the separation of the constituent components of the nonclassical superpositions in phase-space. The traditional view of decoherence deals with the ensemble: but in quantum optics it is quite feasible to examine the behaviour of *individual* members of the ensemble, a question we address in the next part of these lectures, using newly-developed ideas of 'quantum trajectories' (see *e.g.* [5]). In doing so, we show how the dissipative environment 'deconstructs' the field into its appropriate constituent parts.

2 Quantum interference and nonclassical states

2.1 Nonclassical light, quadratures and Wigner functions

We commence our discussion by examining how a single light field mode is quantised in a box of volume V. This is described by the field operator [6]

$$\hat{E} = \left(\frac{2\omega^2}{\epsilon_0 V}\right)^{1/2} \hat{q} \sin(kz) \tag{1}$$

where ω is the field mode frequency, ϵ_0 the permittivity of free space and $k = \omega/c$ the wavevector; $\hat{q}$ is an Hermitian operator having the dimension of a length. This can be expressed in terms of annihilation ($\hat{a}$) and creation operators ($\hat{a}^\dagger$) as

$$\hat{E} = \sqrt{2}\,\mathcal{E}_0 \sin(kz) \left(\hat{a}\,e^{-i\omega t} + \hat{a}^\dagger e^{i\omega t}\right) \tag{2}$$

where

$$\mathcal{E}_0 = \left(\frac{\hbar\omega}{2\epsilon_0 V}\right)^{1/2}$$

is the 'electric field per photon' and describes the quantum unit of electric field strength and

$$\hat{a} = (2\hbar\omega)^{-1/2} \left(\omega\hat{q} + i\hat{p}\right), \qquad \left[\hat{a}, \hat{a}^\dagger\right] = 1 \tag{3}$$

where $\hat{p}$ is the conjugate field variable to $\hat{q}$. The number states of the field $|n\rangle$ may be described by the number of excitations n which govern the energy of the n-quantum state $E_n = (n + 1/2)\,\hbar\omega$ [7].

The time-dependence $\exp(\pm i\omega t)$ in Equation (2) may be regarded as a *phase* dependence: we can write $\omega t = \chi$, and imagine χ to describe the general phase-dependence of the electric field through the generalisation of Equation (2) [8],

$$\hat{E}(\chi) = \sqrt{2}\,\mathcal{E}_0 \sin(kz) \left(\hat{a}\,e^{-i\chi} + \hat{a}^\dagger e^{i\chi}\right) \tag{4}$$

where the phase χ may be that due to temporal evolution or due to the insertion of optical elements in an experimental situation to retard the field. Because the field annihilation and creation operators do not commute, we see that the total electric field operators evaluated at different phase-angles also do not commute:

$$\left[\hat{E}(0), \hat{E}(\chi)\right] = 4\,i\,\mathcal{E}_0^2 \sin\chi \tag{5}$$

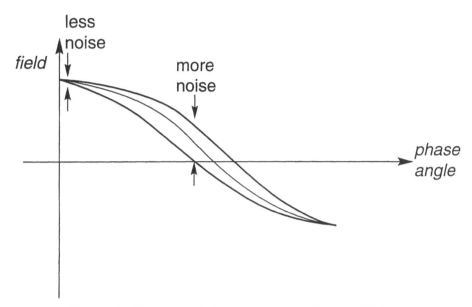

Figure 1. *Phase-dependent quantum noise of squeezed light.*

so that the associated field uncertainties $\Delta E(0)$ and $\Delta E(\chi)$ must satisfy the uncertainty relation

$$\Delta E(0)\,\Delta E(\chi) \geq 2\mathcal{E}_0^2 \,|\sin \chi|. \tag{6}$$

For most electromagnetic field states, the field uncertainty is *independent* of the field phase χ : the quantum noise is phase-independent. But if the field uncertainty at one phase (say $\chi = 0$) is manipulated to be small (compared with $\mathcal{E}_0$), then at a later field phase angle the uncertainty is correspondingly increased (Figure 1): this is the basis of *squeezed light* [8]. Note that those phase angles where the quantum noise is reduced below $\mathcal{E}_0$ occur periodically to satisfy the sinusoid of Equation (6).

It is convenient to parametrise the field in terms of the Hermitian *quadrature* operators $\hat{X}, \hat{Y}$ instead of the non-Hermitian annihilation and creation operators $\hat{a}, \hat{a}^\dagger$ so that

$$\hat{E} = 2^{3/2}\mathcal{E}_0 \sin(kz) \left(\hat{X} \cos(\omega t) + \hat{Y} \sin(\omega t)\right) \tag{7}$$

where

$$\hat{X} = \frac{1}{2}\left(\hat{a} + \hat{a}^\dagger\right), \qquad \hat{Y} = \frac{1}{2i}\left(\hat{a} - \hat{a}^\dagger\right) \tag{8}$$

$$\left[\hat{X}, \hat{Y}\right] = \frac{i}{2} \tag{9}$$

and

$$\Delta X\,\Delta Y \geq \frac{1}{4}. \tag{10}$$

For the vacuum state, or any coherent state $|\alpha\rangle$ formed by a displacement of the vacuum by an amplitude α from the origin in *phase-space*, $\Delta X = \Delta Y = 1/2$. The c-number variables X, Y characterise the phase-space of the field mode. The coherent state $|\alpha\rangle$ has mean quadratures X_0, Y_0 and uncertainties $\Delta X = \Delta Y = 1/2$ and is

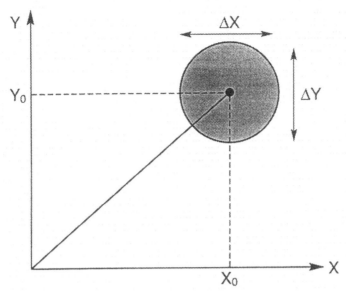

Figure 2. *Phase-space portrait of a coherent state of amplitude* (X_0, Y_0) *and fluctuations* $\Delta X, \Delta Y$.

portrayed in phase-space in Figure 2. In what follows, we will relate the shaded 'error contour' in Figure 2 to the Wigner function phase-space quasi-probability $W_\Psi(q,p)$, where

$$\hat{X} = \left(\frac{\omega}{2\hbar}\right)^{1/2} \hat{q} \tag{11}$$

$$\hat{Y} = (2\hbar\omega)^{-1/2} \hat{p}. \tag{12}$$

The phase-space description of an oscillator such as the field mode implies a joint probability for the measurement of the canonical coordinates q and p. At this point, we need to deal with the lack of commutativity of position and momentum in quantum mechanics and the concomitant non-existence of a positive joint probability distribution. The Wigner function, $W_\Psi(q,p)$, introduced in 1932, is defined as [9]

$$W_\Psi(q,p) = \int \frac{dq'}{2\pi\hbar} \, \Psi^* \left(q + q'/2\right) \, \Psi \left(q - q'/2\right) \, e^{ipq'/\hbar}. \tag{13}$$

Although this can only be regarded at best as a quasi-probability (it may be negative in some areas of phase-space), it does generate the measurable marginal distributions $|\Psi(q)|^2$ and $|\Psi(p)|^2$ on integration over p and over q, respectively. Indeed, the negativity of the Wigner functions for some states can be regarded as a signature of their non-classicality. However, the only Wigner functions which are positive everywhere are Gaussian [10]: *i.e.* restricted to coherent states and the squeezed states (generated by the quadratic squeeze operator which preserves the minimum uncertainty product $\Delta X \Delta Y = 1/2$). In Figure 3, we show the Wigner function for the nine-photon Fock state : this is a rotationally-symmetric Laguerre polynomial, with a large outer 'rim' situated at a radius $\sqrt{n} = 3$ and a sequence of fringes including substantial negative regions.

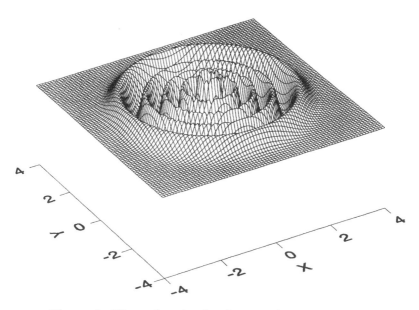

Figure 3. *Wigner function for the nine-photon Fock state .*

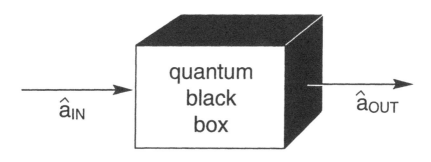

Figure 4. *Input-output relation, where an input field $\hat{a}_{IN}$ is transformed into an output field $\hat{a}_{OUT}$ by a quantum black box.*

2.2   Quantum noise as a stochastic process

We next consider the modification of a quantum-mechanical light field mode, described by an input field annihilation operator $\hat{a}_{IN}$, by a (possibly nonlinear) system which we idealise as a 'black box' as in Figure 4. Input-output theory must provide an algorithm for deriving all the statistical properties of the OUT fields from those of the IN fields. We write this as [11]

$$\hat{a}_{\mu} = f_{\mu}\left(\hat{a}_{\nu}\right) . \tag{14}$$

In a classical stochastic approach, a noisy input signal $\bar{\alpha}_\nu + \delta\alpha_\nu$ (where $\bar{\alpha}_\nu$ is the mean value and $\delta\alpha_\nu$ the 'noise') leads to a modified output $\bar{\alpha}_\mu + \delta\alpha_\mu$

$$\alpha_\mu + \delta\alpha_\mu = f_\mu(\bar{\alpha}_\nu + \delta\alpha_\nu) . \tag{15}$$

If the input fluctuations are described by a probability distribution P_{IN}, then for *classical* stochastic processes

$$P_{\text{OUT}}(\alpha_\mu) = P_{\text{IN}}\left(f_\mu^{-1}(\alpha_\nu)\right) . \tag{16}$$

A natural question to ask is whether such a transformation holds for quantum fluctuations [11]. In quantum physics, the input-output transformation should preserve the basic commutation relations: that is, it should be *canonical*:

$$\left[\hat{a}_i, \hat{a}_j^\dagger\right] = \delta_{ij} . \tag{17}$$

If we restrict ourselves to *linear* canonical transformations through the action of some function $\hat{f}$ so that

$$\hat{b}_i = f_i\left(\hat{a}_j, \hat{a}_k^\dagger\right) = u_{ij}\hat{a}_j + v_{ik}\hat{a}_k^\dagger + c_i \tag{18}$$

then the basic commutation relation is preserved provided

$$u_{nj}u_{mj}^* - v_{nj}v_{mj}^* = \delta_{nm} , \tag{19}$$

$$u_{nj}v_{mj} - v_{nj}u_{mj} = 0 . \tag{20}$$

A canonical transformation preserves volumes when acting in phase-space. We can represent this canonical transformation through the action of a unitary transformation $\hat{U}$ (the Stone–von Neumann theorem)

$$\hat{b}_i = \hat{U}^\dagger \hat{a}_i \hat{U} \tag{21}$$

which can be written in terms of the generators $\hat{J}_k$ as

$$\hat{U} = \exp\left(-i\alpha_k \hat{J}_k\right) . \tag{22}$$

Then by expansion of Equation (22) in Equation (21), we find [11]

$$\hat{b}_i = \hat{U}^\dagger \hat{a}_i \hat{U} = \hat{a}_i - i\,\alpha_k\left[\hat{a}_i, \hat{J}_k\right] + \cdots \tag{23}$$

If we insist on a linear canonical transformation, we require

$$\left[\hat{a}_i, \hat{J}_k\right] \propto \text{ either } \hat{a}_i , \text{ or } \hat{a}_j^\dagger \text{ or } \hat{I} . \tag{24}$$

In other words, a linear canonical transformation $\hat{f}$ requires a generator $\hat{J}_k$ which is made up of the identity matrix $\hat{I}$, or a *linear* combination of $\hat{a}$ and $\hat{a}^\dagger$ (the displacement operators) or *bilinear* combinations of $\hat{a}$ and $\hat{a}^\dagger$ (the squeezing operators). The bilinear combinations are

$$J^{(a)} = -i\left(\hat{a}^2 - \hat{a}^{\dagger 2}\right) = \frac{i}{2}\left(q\frac{\partial}{\partial q} - p\frac{\partial}{\partial p}\right) \tag{25}$$

$$J^{(b)} = \left(\hat{a}^2 + \hat{a}^{\dagger 2}\right) = \frac{i}{2}\left(q\frac{\partial}{\partial q} + p\frac{\partial}{\partial p}\right) \tag{26}$$

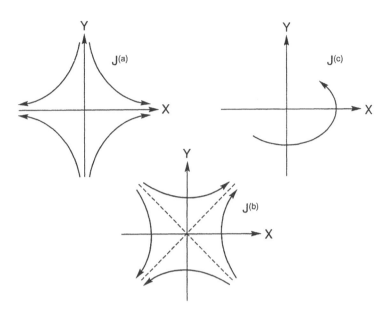

Figure 5. *Bilinear canonical transformations : stretches and rotations of patches in phase-space.*

and

$$J^{(c)} = \frac{1}{2} \left(\hat{a}^\dagger \hat{a} + \hat{a} \hat{a}^\dagger \right) \tag{27}$$

which generate the stretches and rotations shown in Figure 5. At this point, we can address the question posed earlier: which phase-space quantum quasi-probability transforms like the classical version given by Equation (16)? Only the Wigner function has this transformation property [11]

$$W_{\text{OUT}} \left(\alpha_\mu \right) = W_{\text{IN}} \left(f_\mu^{-1} \left(\alpha_\nu \right) \right) . \tag{28}$$

Other quasi-probabilities used in quantum optics, such as the diagonal coherent-state Glauber-Sudarshan P distribution or the Husimi Q-function do not transform in this way [11], and this emphasises the special rôle the Wigner function plays in quantum mechanics. The Wigner function was constructed to have simple properties under Galilean transformations from the outset; we note here that this transformation property is retained in dissipative dynamics. Given a linear canonical transformation, quantum noise can be simulated as a complex random process. The procedure is simple: first change all operators $\hat{a}_\mu, \hat{a}_\mu^\dagger$ into complex random variables α_μ, α_ν^*, and secondly use the Wigner functions for a statistical description of the fluctuations. Whether such simulations are truly semiclassical [12] is to some extent a matter of taste. We should emphasise that this procedure only works for linear transformations, and that nonlinear effects strongly perturb the phase-space and render this simple transformation worthless.

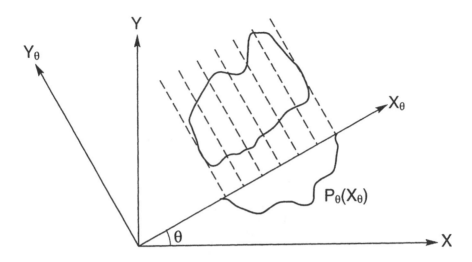

Figure 6. *Marginal distributions $P_\theta(X_\theta)$ obtained by integrating the Wigner function at fixed rotated quadrature X_θ over all conjugate quadratures Y_θ.*

2.3   Marginal distributions: reconstructed Wigner functions

The measurable marginal distributions $|\Psi(q)|^2$ and $|\Psi(p)|^2$ are not sufficient, in general, to allow a full reconstruction of the Wigner function $W_\Psi(q,p)$. However, such a reconstruction *is* possible if the marginal distributions for the *rotated quadratures*

$$X_\theta = X\cos\theta + Y\sin\theta \qquad (29)$$

$$Y_\theta = -X\sin\theta + Y\cos\theta \qquad (30)$$

are measured, as shown in Figure 6. The Wigner function can be written in terms of the quadratures X, Y as $W_\Psi(X,Y)$; or in rotated coordinates as $W_\Psi(X_\theta, Y_\theta)$. A marginal distribution is obtained by fixing X_θ and integrating over all Y_θ :

$$P_\theta(X_\theta) = \int dY_\theta\, W(X_\theta, Y_\theta) \qquad (31)$$

and is measurable through homodyne detection . If this is done for a range of different angles θ, then as shown by Risken and Vogel [13], the Wigner function can be obtained from the measured marginals by a Radon transformation familiar from tomographic imaging:

$$W(X,Y) = \frac{1}{4\pi^2}\int_{-\infty}^{\infty} dX_\theta \int_{-\infty}^{\infty} d\xi\, |\xi| \int_0^\pi d\theta\, P_\theta(X_\theta)\, e^{i\xi\,(X_\theta - X\cos\theta - Y\sin\theta)}\,. \qquad (32)$$

Raymer and his colleagues at the University of Oregon [14] have shown in a remarkable series of experiments using this 'optical tomography' technique how the Wigner functions for various field states can be reconstructed, including a squeezed state generated by a travelling-wave optical parametric oscillator. An element of caution is required in the interpretation of the tomographic reconstruction of Wigner functions: we know the

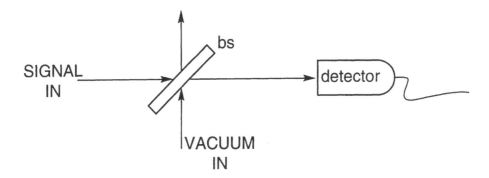

Figure 7. *The effect of imperfect detection efficiency modelled as a beam-splitter* (bs) *which diverts flux away from a perfect detector. But associated with this diversion is the introduction of vacuum noise into the detector: the detected mode Wigner function is a convolution of the input with the vacuum noise input, and this convolution broadens and smooths the measured Wigner function.*

simultaneous description of X and Y is fraught with danger [15]. In reality the measurement process, as always, perturbs the system. A tomographic detector will have a finite efficiency, and this may be modelled as a beam splitter of finite transmissivity in front of a perfect detector (Figure 7) [16]. As we show in Figure 7, the diversion of flux from the detector is associated with the addition of noise, and the measured Wigner function is a convolution of that of the input and of the vacuum. This smooths the Wigner function [17, 18]: indeed the convolution with the smooth Gaussian vacuum noise is sufficient to render the measured quasi-probability positive everywhere [18]. An inefficient photodetector is responsible for sufficient smoothing to generate the Husimi, or 'Q-function ' $Q(\beta) = (1/\pi)\langle\beta|\hat{\rho}|\beta\rangle$ [16, 19] where $\hat{\rho}$ is the field density operator and $|\beta\rangle$ a coherent state. This suggests that Wigner functions with negative regions in phase-space are not directly detectable (rendering negativity a rather indirect measure of non-classicality). Extraction by de-convolution seems intrinsically unstable, although this point is still under investigation at the time of writing.

2.4 Building nonclassical states by quantum interference

Having described how a quantum field is characterised in phase-space by the Wigner function, and how this may be measured, we turn next to the question of how such states can be created in terms of component states which may be Fock states or coherent states. We can construct nonclassical states as superpositions of Fock states $|n\rangle$ as

$$|\Psi\rangle = \sum_n c_n |n\rangle. \tag{33}$$

The Wigner functions for each Fock state component are given by Laguerre polynomials. These interfere to generate the superposition of interest. The Fock state components are radially symmetric, and tile phase-space in annulae. As an alternative basis, we can construct our state from superpositions of coherent states

$$|\Psi\rangle = \sum_{\alpha} c(\alpha) |\alpha\rangle \qquad (34)$$

which tile phase-space in circular patches. The coherent states, of course, form an over-complete set and are not orthogonal but form a perfectly respectable basis set. A finite number of states $|\alpha\rangle$ may be used for the construction, or a continuous superposition may be employed in which case the sum in Equation (34) becomes an integral. The particular basis which we employ is not entirely a matter of taste as we shall see.

The simplest example of a nonclassical superposition is the superposition of the vacuum and one photon state [20] so that

$$|\Psi\rangle = c_0 |0\rangle + c_1 |1\rangle . \qquad (35)$$

This state can exhibit squeezing (in a sense it is the simplest squeezed state), with quadrature variances

$$(\Delta X)^2 = \frac{1}{4} + \frac{1}{2} \left(|c_1|^2 - 2 |c_1|^2 \cos^2(\Delta\phi) + 2 |c_1|^4 \cos^2(\Delta\phi) \right) \qquad (36)$$

$$(\Delta Y)^2 = \frac{1}{4} + \frac{1}{2} \left(|c_1|^2 - 2 |c_1|^2 \sin^2(\Delta\phi) + 2 |c_1|^4 \sin^2(\Delta\phi) \right) \qquad (37)$$

where $\Delta\phi$ is the relative phase between c_0 and c_1. It is obvious that this state is *not* a minimum uncertainty state (M.U.S.). We find the mean value of the quadrature is non-zero, reflecting the coherence in the superposition, and for the variances we find

$$(\Delta X)^2_{\Delta\phi=0,\pi} = (\Delta Y)^2_{\Delta\phi=\pi/2,3\pi/2} = \frac{1}{4} + |c_1|^2 \left(|c_1|^2 - \frac{1}{2} \right) \qquad (38)$$

with $(\Delta X)^2$ being reduced below its coherent state (or vacuum state) value of $(1/4)$ by a maximum of $(1/16)$ when $|c_1| = 1/2$. The Wigner function for the state given by Equation (35) is shown in Figure 8, where both negativity *and* squeezing are visible. The squeezing is not of M.U.S. form.

The superposition of the vacuum and the two photon state

$$|\Psi\rangle = c_0 |0\rangle + c_2 |2\rangle \qquad (39)$$

generates rather more squeezing, although this time the mean values of the quadratures are zero, reflecting the two-photon coherence. We find for the quadrature variances,

$$(\Delta X)^2 = \frac{1}{4} \left(1 + 4 |c_2|^2 + 2\sqrt{2} |c_2| \left(1 - |c_2|^2 \right)^{1/2} \cos(\Delta\phi) \right) , \qquad (40)$$

$$(\Delta Y)^2 = \frac{1}{4} \left(1 + 4 |c_2|^2 - 2\sqrt{2} |c_2| \left(1 - |c_2|^2 \right)^{1/2} \cos(\Delta\phi) \right) , \qquad (41)$$

so that again the squeezing is not of minimum uncertainty form, but is more pronounced than the one-photon superposition: at $|c_2|^2 = (1/2) - (1/\sqrt{6})$ we find 45% maximum squeezing below the vacuum value of the quadrature variance. The Wigner function for this two-photon superposition is shown in Figure 9. This process of squeezing by superposition is improved by superposing *all* of the even number states as in the classic squeezed vacuum [8], although the complicated interference between the infinite number of Fock components actually reduces to a simple Gaussian squeezed state, which *is* of M.U.S. form.

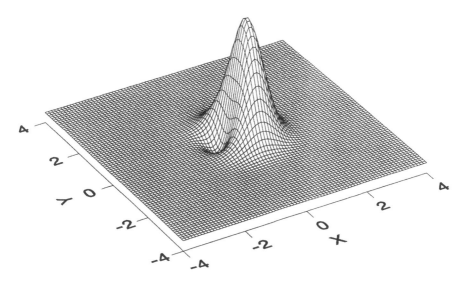

Figure 8. *Wigner function for a superposition of the vacuum and one photon state.*

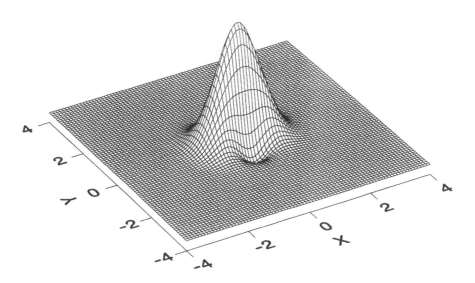

Figure 9. *Wigner function for a superposition of the vacuum and two-photon state, Equation (39).*

More interesting nonclassical states can be generated by superposing coherent states $|\alpha\rangle$, and the 'Schrödinger cat state'

$$|\Psi\rangle = \frac{1}{N_+(\alpha)} \{|\alpha\rangle + |-\alpha\rangle\} , \qquad (42)$$

where $N_+(\alpha)$ is a normalisation factor, is the best studied example [21]. It is also interesting to examine the nature of the Wigner function for the state formed by a superposition of a number of coherent states as that number N increases, whilst maintaining each component on a ring of radius R [22]. We write for this N component superposition on a ring of radius R in phase-space

$$|\Psi\rangle = \frac{1}{W} \sum_{j=0}^{N-1} \exp\left\{-\frac{2\pi i}{N} Kj\right\} \left| R \exp\left(\frac{2\pi i Kj}{N}\right) \right\rangle \qquad (43)$$

where K is something we will call the 'target' number, for reasons which will become apparent very quickly, and W is a normalisation factor. For $N = 2$, we see a Gaussian hump for each constituent state, together with fringes whose period scales inversely with the separation between the components in phase-space: the more macroscopic the superposition, the more rapid being the fringes. For $N = 4$ we see four constituent 'humps' together with complicated interference structures between them. When the number of states is increased so that the individual components nearly, but not quite, fill the circumference of the ring we see a sequence of interference structures, large amounts of negativity in the Wigner functions and large fluctuations in the photon number. But when the number of coherent states is sufficiently increased to cover entirely the circumference of the ring (of radius R), then a dramatic change occurs [22] with a collapse of the number variance as the superposition approaches ... a number state! This is seen for increasing N in Figure 10. A continuous superposition of coherent states on a ring with an appropriate choice of phases is in fact a representation of a Fock state $|n\rangle$ [21, 23]

$$|n\rangle = \frac{\exp(|\alpha|^2/2)\sqrt{n!}}{2\pi|\alpha|^n} \int_0^{2\pi} d\phi\, e^{-in\phi} \left| |\alpha| e^{i\phi} \right\rangle . \qquad (44)$$

It is amusing that coherent states, the most classical of all states, can be used to build Fock states using quantum interference. We have discussed elsewhere [24] how a sequence of conditional measurements in cavity QED may be used to realise the progression from coherent state to Fock state .

If coherent states are continuously superposed on a line

$$|\Psi\rangle = C_F \int_{-\infty}^{\infty} F(\alpha, s)|\alpha\rangle \, d\alpha , \qquad (45)$$

with $F(\alpha, s)$ chosen to be a Gaussian function of α along the line but centred on the origin, so that α is here taken as real, it is easy to demonstrate that the resultant state is a sum over even Fock states only and is in fact the squeezed vacuum [23, 25]: to do so we first write the coherent state $|\alpha\rangle$ in a Fock basis as a Poisson superposition

$$|\alpha\rangle = \sum_{m=0}^{\infty} \frac{\alpha^m}{\sqrt{m!}} e^{-\alpha^2/2}|m\rangle \qquad (46)$$

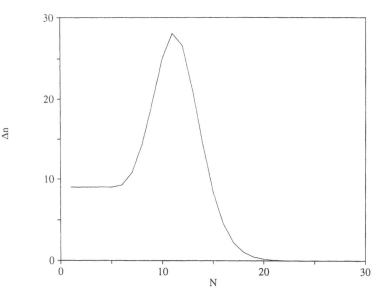

Figure 10. *Dependence of the number uncertainty Δn of the superposition of N coherent states on a ring of radius $R=3$ in phase-space, with a phase-factor $\exp\{-2\pi i K j/N\}$ for each component j to ensure a target Fock state $|K\rangle$ is generated as N increases. Here $K=4$, so the initial radius is not optimal. Nevertheless, after complicated interference at intermediate N with large fluctuations, we see a sharp reduction in Δn.*

and take as an explicit form of $F(\alpha, s)$ the Gaussian centred on the origin

$$F(\alpha, s) = \exp\left\{-\frac{(1-s)}{2s}\alpha^2\right\} \tag{47}$$

so that $C_F^{-2} = \pi (1 - s^2)^{1/2}/ 2s$, then

$$|\Psi\rangle = C_F \int_{-\infty}^{\infty} F(\alpha, s)\, e^{-\alpha^2/2} \sum_m \frac{\alpha^m}{\sqrt{m!}}|m\rangle\, d\alpha . \tag{48}$$

Then if we interchange the order of integration and summation we find

$$|\Psi\rangle = C_F \sum_{m=0}^{\infty} \frac{1}{\sqrt{m!}}|m\rangle \int_{-\infty}^{\infty} d\alpha\, \alpha^m\, e^{-\alpha^2/2s} \tag{49}$$

and perform the Gaussian integral, we find the result is the squeezed vacuum

$$\begin{aligned}
|\Psi\rangle &= \left(1 - s^2\right)^{1/4} \sum_{m=0}^{\infty} \frac{\sqrt{2m!}}{2^n\, n!}\, s^n\, |2n\rangle \\
&= \hat{S}(s)\,|0\rangle \tag{50}
\end{aligned}$$

where $\hat{S}(s)$ is the quadratic squeeze operator [8]. For this state the variances are

$$(\Delta X)^2 = \frac{1}{4}\left(\frac{1+s}{1-s}\right) \tag{51}$$

$$(\Delta Y)^2 = \frac{1}{4}\left(\frac{1-s}{1+s}\right) \tag{52}$$

so that this state *is* of M.U.S. form.

3 Dissipation versus coherent evolution: decoherence

So far, we have discussed our superposition states as if they were entirely isolated: in reality, all systems are in contact to some extent with an external world which we can model as an infinite reservoir, or heat-bath . The system dissipates both coherence and energy irreversibly upon interaction with such a reservoir, and in turn the reservoir feeds back Langevin noise into the microsystem. If the system and reservoirs are described by Hamiltonians H_R and H_S, and the interaction between them denoted $\hat{V}$, then the total system-plus-reservoir density matrix evolves in the interaction picture as

$$\dot{\rho} = -\frac{i}{\hbar} \left[\hat{V}, \rho \right] . \tag{53}$$

Initially the system and the reservoir are uncorrelated, so the initial density operator factorises into a product of subsystem reduced density operators

$$\rho(0) = \rho_S(0) \otimes \rho_R(0) . \tag{54}$$

We may formally integrate the Liouville equation of motion Equation (53) and iterate, to give

$$\rho(t) = \rho(0) - \frac{i}{\hbar} \int_0^t dt_1 \left[\hat{V}(t_1), \rho(t_1) \right] . \tag{55}$$

If we iterate one more time, and trace over the reservoir states assuming that (i) $\mathrm{Tr}_R(\hat{V}(t)\rho_R) = 0$ and (ii) $\hat{V}$ is 'weak', then we find

$$\dot{\rho}_S = \left(-\frac{1}{\hbar} \right)^2 \int_0^t dt_1 \, \mathrm{Tr}_R \left[\hat{V}(t), \left[\hat{V}(t_1), \rho_R \otimes \rho_S(t) \right] \right] . \tag{56}$$

If the system of interest is a harmonic oscillator described by annihilation ($\hat{a}$) and creation operators ($\hat{a}^\dagger$), and is coupled to a bath of oscillators described by $\hat{b}_j$ and $\hat{b}_j^\dagger$, then [19]

$$\hat{V}(t) = \hbar \, \hat{a}^\dagger \, \hat{\Gamma}(t) \, e^{i\omega_0 t} + \hbar \, \hat{a} \, \hat{\Gamma}^\dagger(t) \, e^{-i\omega_0 t}$$

where the reservoir operators are

$$\hat{\Gamma}(t) = \sum_j g_j \hat{b}_j \, e^{-i\omega_j t} , \qquad [\hat{b}_j, \hat{b}_k^\dagger] = \delta_{jk} \tag{57}$$

and ω_0 is the resonance frequency of the system oscillator. The integrals in Equation (56) can then be expressed as

$$I = \int_0^t dt_1 \left\langle \hat{\Gamma}(t)\hat{\Gamma}^\dagger(t_1) \right\rangle e^{i\omega_0 (t-t_1)} \tag{58}$$

which can be broken into real (widths) and imaginary (shift) parts as

$$I \rightarrow \frac{\gamma}{2} \left(N(\omega_0) + 1 \right) - i\Delta \tag{59}$$

(for details see [3, 19]), where $N(\omega_0)$ is given by the mean number of reservoir excitations at frequency ω_0 :

$$2\pi \, N(\omega) \, \delta(\omega - \omega') = \left\langle \hat{b}^\dagger(\omega)\hat{b}(\omega') \right\rangle . \tag{60}$$

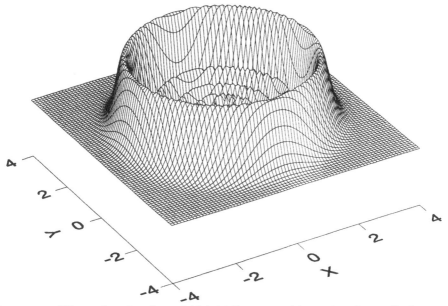

Figure 11. *Wigner function for a field initially prepared in a nine-photon Fock state , after interacting with a zero-temperature reservoir for a time $\gamma t = 0.1$.*

For a normal unsqueezed reservoir, we finally obtain the Born-Markov master equation

$$\dot{\rho}_S = \frac{\gamma}{2}(N+1)\left(2\hat{a}\rho_S\hat{a}^\dagger - \hat{a}^\dagger\hat{a}\rho_S - \rho_S\hat{a}^\dagger\hat{a}\right) + \frac{\gamma}{2}N\left(2\hat{a}^\dagger\rho_S\hat{a} - \hat{a}\hat{a}^\dagger\rho_S - \rho_S\hat{a}\hat{a}^\dagger\right) \quad (61)$$

At zero temperature, $N = 0$, so that

$$\dot{\rho}_S = \frac{\gamma}{2}\left(2\hat{a}\rho_S\hat{a}^\dagger - \hat{a}^\dagger\hat{a}\rho_S - \rho_S\hat{a}^\dagger\hat{a}\right) . \quad (62)$$

In Figure 11, we show the Wigner function for a nine-photon Fock state , but including dissipation to a zero-temperature heat-bath , computed from Equation (62) at time $\gamma t = 0.1$. We note that the fringes which were so prominent in Figure 3 have almost competely disappeared, and the prominent negativity has been much reduced (even though the initial energy is hardly depleted).

The master equation emphasises the evolution of an entire ensemble [3]. Coherences are destroyed at a rate which is proportional to the degree of excitation of the system (much faster than energy decay) as the system relaxes to a statistical mixture of relevant 'pointer bases' [4, 26, 27]. For example, a field mode initially prepared in a Fock state relaxes very rapidly to a statistical mixture of Fock states , and the fringing structure in the Wigner function disappears at a rate $\sim n\gamma$. Such an ensemble-averaged approach does not easily address the question of how an individual member of an ensemble evolves in a dissipative environment, even though this issue is accessible to experiment [28]. A number of methods have been developed recently to simulate the evolution of single realisations. One centres on quantum jumps, reflecting the information gained from the detection of decay quanta, and involves a continuous evolution which is randomly interrupted by 'instantaneous' jumps as the system state vector is conditioned by the information gained from the register of counts in detection of decay [29, 5].

4   Unravelling the master equation

Carmichael [5] has given a precise relationship between the conditioned density operator contingent on a precise sequence of detection events (a 'record') and the ensemble averaged density operator. He shows for example that if the zero temperature boson damping master equation (Equation (62)) is written in Liouvillian form

$$\frac{d\rho}{dt} = \mathcal{L}\rho \tag{63}$$

and we split the Liouvillian action $\mathcal{L}$ as a sum of two terms, an anticommutator and a 'sandwich' term,

$$\mathcal{L}\rho = -\frac{\gamma}{2}\left[\hat{a}^\dagger\hat{a}, \rho\right]_+ + \gamma\hat{a}\rho\hat{a}^\dagger = (\mathcal{L} - \mathcal{S}) + \mathcal{S} \tag{64}$$

then we may identify the 'sandwich' term $\mathcal{S}$ as a jump operator. Equation (63) can be integrated formally as

$$\begin{aligned}
\rho(t) &= \exp\left\{\left[(\mathcal{L} - \mathcal{S}) + \mathcal{S}\right]t\right\}\rho(0) \\
&= \sum_{m=0}^{\infty} \int_0^t dt_m \int_0^{t_m} dt_{m-1} \dots \int_0^{t_2} dt_1 \\
&\quad \times \left\{ e^{(\mathcal{L}-\mathcal{S})(t-t_m)}\mathcal{S}\,e^{(\mathcal{L}-\mathcal{S})(t_m-t_{m-1})}\mathcal{S}\dots\mathcal{S}\,e^{(\mathcal{L}-\mathcal{S})t_1}\rho(0)\right\}
\end{aligned} \tag{65}$$

where the quantity in curly brackets in Equation (65) is labelled $\overline{\rho_c}(t)$ by Carmichael and is the conditioned density operator describing a specific 'trajectory' or detection sequence. We can write $\overline{\rho_c}(t)$ in terms of the conditioned pure state projectors

$$\overline{\rho_c}(t) = \left|\overline{\Psi}_c(t)\right\rangle \left\langle\overline{\Psi}_c(t)\right| . \tag{66}$$

The component $\exp[(\mathcal{L} - \mathcal{S})\Delta t]$ propagates $\overline{\rho_c}(t)$ for a time Δt without a decay being recorded: for the conditioned state $\left|\overline{\Psi}_c(t)\right\rangle$

$$\left|\overline{\Psi}_c(t + \Delta t)\right\rangle = \exp\left[-iH_{\text{eff}}\Delta t/\hbar\right]\left|\overline{\Psi}_c(t)\right\rangle \tag{67}$$

where the non-Hermitian effective Hamiltonian

$$H_{\text{eff}} = H - i\hbar\frac{\gamma}{2}\hat{a}^\dagger\hat{a} \tag{68}$$

derives from the anticommutator in Equation (64). Once a decay is registered, the gain in information is responsible for the jump

$$\left|\overline{\Psi}_c(t)\right\rangle \longrightarrow \hat{a}\left|\overline{\Psi}_c(t)\right\rangle . \tag{69}$$

The procedure adopted in quantum jump simulations can then be summarised as follows:

1. Determine the current probability of an emission:

$$\Delta P = \gamma\,\Delta t\,\langle\Psi|\hat{a}^\dagger\hat{a}|\Psi\rangle . \tag{70}$$

2. Obtain a random number r between zero and one, compare with ΔP and decide on emission as follows:

3. Emit if $r < \Delta P$, so that the system jumps to the renormalised form:

$$|\Psi\rangle \longrightarrow \frac{\hat{a}|\Psi\rangle}{\sqrt{\langle\Psi|\hat{a}^\dagger\hat{a}|\Psi\rangle}}. \tag{71}$$

4. No emission if $r > \Delta P$, so that the system evolves under the influence of the non-Hermitian form

$$|\Psi\rangle \longrightarrow \frac{\left\{1 - (i/\hbar)\, H\Delta t - (\gamma/2)\, \Delta t\, \hat{a}^\dagger\hat{a}\right\}|\Psi\rangle}{(1 - \Delta P)^{1/2}}. \tag{72}$$

5. Repeat to obtain an individual trajectory, or history.

6. Average observables over many such trajectories.

To reassure ourselves that this is all true, we note the history splits into two alternatives in a time Δt :

$$|\Psi\rangle \quad \begin{array}{c} \xrightarrow{\;\Delta P\;} \quad |\Psi_{\text{emit}}\rangle \\[2mm] \xrightarrow{\;1-\Delta P\;} \quad |\Psi_{\text{no emit}}\rangle \end{array}$$

Then in terms of the density matrix, the evolution for a step Δt becomes a sum of the two possible outcomes,

$$|\Psi\rangle\langle\Psi| \longrightarrow \Delta P\,|\Psi_{\text{emit}}\rangle\langle\Psi_{\text{emit}}| + (1-\Delta P)\,|\Psi_{\text{no emit}}\rangle\langle\Psi_{\text{no emit}}| \tag{73}$$

$$\begin{aligned} = \;& \gamma\Delta t\,\hat{a}|\Psi\rangle\langle\Psi|\hat{a}^\dagger \\ & + \left\{1 - \frac{i}{\hbar}H\Delta t - \frac{\gamma}{2}\Delta t\,\hat{a}^\dagger\hat{a}\right\}|\Psi\rangle\langle\Psi|\left\{1 + \frac{i}{\hbar}H\Delta t - \frac{\gamma}{2}\Delta t\,\hat{a}^\dagger\hat{a}\right\} \end{aligned}$$

$$\begin{aligned} \sim \;& |\Psi\rangle\langle\Psi| - \frac{i}{\hbar}\Delta t\,[H, |\Psi\rangle\langle\Psi|] \\ & + \frac{\gamma}{2}\Delta t\left\{2\hat{a}|\Psi\rangle\langle\Psi|\hat{a}^\dagger - \hat{a}^\dagger\hat{a}|\Psi\rangle\langle\Psi| - |\Psi\rangle\langle\Psi|\hat{a}^\dagger\hat{a}\right\} \tag{74} \end{aligned}$$

so that

$$\frac{\Delta\rho}{\Delta t} = -\frac{i}{\hbar}[H, \rho] + \frac{\gamma}{2}\left\{2\hat{a}\rho\hat{a}^\dagger - \hat{a}^\dagger\hat{a}\rho - \rho\hat{a}^\dagger\hat{a}\right\} \tag{75}$$

as in the original master equation (63). We see that the form of the Monte-Carlo simulation scheme above is of the Euler form, and as such is a rather crude numerical approach. More accurate methods involve both Runge-Kutta and so-called 'higher-order' unravellings or 'mini-trajectories'. We have addressed this point elsewhere [30].

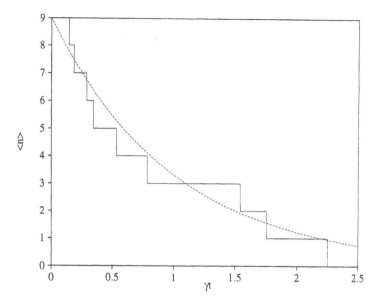

Figure 12. *Quantum jumps in the decay of a Fock state : evolution of the photon number from an initial value n = 9 due to dissipative coupling. The dashed curve shows the ensemble average:* $9\exp(-\gamma t)$.

The average over many trajectories reproduces the density matrix solution of Equation (61).

The quantum jump method applied to an initial Fock state interacting with a zero-temperature heat-bath is particularly simple to describe. The non-Hermitian no-jump evolution is described by

$$H_{\text{eff}} = -i\hbar\,\frac{\gamma}{2}\hat{a}^\dagger\hat{a} \tag{76}$$

which does, of course, not change the photon number. On the other hand, a jump from $|n\rangle$ to $|n-1\rangle$ occurs with probability $\gamma\langle\hat{a}^\dagger\hat{a}\rangle = \gamma_n$, so individual trajectories 'deconstruct' in a decoherence time $(\gamma n)^{-1}$, as shown in Figure 12. The n-photon Wigner function, the Laguerre polynomial $\mathcal{L}_n$ jumps to $\mathcal{L}_{n-1}$ at random times, with probability γ_n.

So far, we have discussed the quantum jumps in a single-component field state. We now turn our attention to superpositions. In Figure 13 we illustrate the time-evolution of a single ensemble member of a Schrödinger cat state of the form $(|\alpha\rangle + |-\alpha\rangle)/N_+(\alpha)$, where $N_+(\alpha)$ is the normalisation, and $|\alpha\rangle$ and $|-\alpha\rangle$ are two coherent states, as it undergoes two quantum jumps. During the time interval when there are no jumps, the Schrödinger cat shrinks as $\alpha(t) = \alpha(0)\,e^{-\gamma t/2}$ when the evolution (Equation (72)) is considered. This no-jump process has taken place from the initial state up to Figure 13(a). When there is a jump, then according to Equation (71), we find that the cat jumps to a different kind of cat, of the form $(|\alpha(t)\rangle - |-\alpha(t)\rangle)/N_-(\alpha(t))$. This jump is seen as we go from Figure 13(a) to Figure 13(b). This jump from an even to an odd cat is particularly visible in the Wigner functions because the interference fringes change sign. Subsequently the cat shrinks again as we evolve under the no-jump evolution again. This takes us to Figure 13(c) as the components shrink towards the

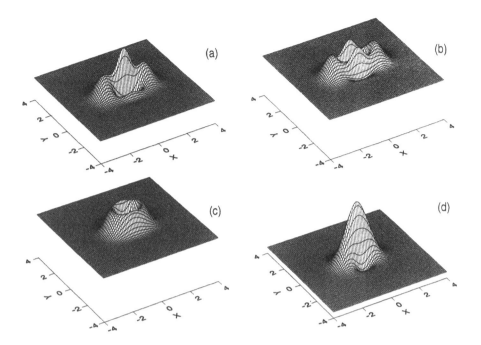

Figure 13. *A 'jumping cat' which was initially comprised of the coherent states* $|\alpha\rangle$ *and* $|-\alpha\rangle$ *with* $\alpha = 2$. *We show the Wigner functions before and after the first jump (at* $\gamma t = 0.668$*) in (a) and (b). In (c) and (d) we show the Wigner functions before and after the second jump (at* $\gamma t = 1.884$*).*

origin. Finally we portray the effect of a second jump, where once again the interference fringes change sign.

5   Quantum jumps

5.1   Intrinsic or extrinsic jumps

So far, we have discussed the evolution of *open* systems, that is of microsystems in contact with Markovian reservoirs such as the bath of vacuum field modes responsible for spontaneous emission. The quantum jump concept within an open system context has to do with the gain in information about the microsystem which is accessible from the record available in the dissipative environment. Such jump processes do *not* require an extension or modification of conventional quantum mechanics, and we will refer to these as 'extrinsic' jumps. A very different jump mechanism has been studied by a number of authors [31] and especially by Ghirardi, Rimini and Weber [32]. In these approaches the Schrödinger equation is modified in such a way that quantum coherences are automatically destroyed in a *closed* system by an *intrinsic* stochastic jump mechanism. This should be distinguished from the extrinsic mechanisms we are concerned with in the bulk of these lectures.

To see how an intrinsic jump mechanism works, we need a concrete realisation which we can apply to a specific time evolution. Milburn has proposed just such a realisation [33], in which standard quantum mechanics is modified in a simple way to generate intrinsic decoherence. He assumes that on sufficiently short time steps, the system does not evolve continuously under normal unitary evolution, but rather in a *stochastic* sequence of identical unitary transformations. This assumption leads to a modification of the Schrödinger equation which contains a term responsible for the decay of quantum coherence in the energy eigenstate basis, without the intervention of a reservoir and therefore without the usual energy dissipation associated with normal decay [34]. The decay is entirely of phase-dependence only, akin to the dephasing decay of coherences produced by impact-theory collisions or by fluctuations in the phase of a laser in laser spectroscopy, but here of intrinsic origin.

It is interesting to apply Milburn's model of intrinsic decoherence to a problem of dynamical evolution: that is, the interaction of two subsystems and the coherences which establish themselves as a consequence of their interaction. We consider the interaction between a single two-level atom and a quantised cavity mode in the Jaynes-Cummings model [35, 36] with Hamiltonian:

$$H = \hbar\omega\left(\hat{a}^\dagger\hat{a} + \frac{1}{2}\right) + \frac{\hbar\omega_0}{2}\hat{\sigma}_3 + \hbar\lambda\left(\hat{a}^\dagger\hat{\sigma}_- + \hat{a}\hat{\sigma}_+\right) \qquad (77)$$

where $\hat{\sigma}_\pm$ are the two-level spin-flip operators, $\hat{\sigma}_3$ the inversion and λ the atom-field coupling constant. For simplicity we consider exact resonance between atom and field ($\omega = \omega_0$), so that atom and field periodically exchange excitation in a sequence of Rabi oscillations. If the cavity field is prepared in a number state, $|n\rangle$, then the Rabi frequency for this exchange, given the atom is initially excited, is $\lambda\sqrt{n+1}$. If the field is prepared in a coherent state $|\alpha\rangle$, the Poissonian superposition of photon numbers in such a state leads to a Poissonian superposition of Rabi oscillations: the spread leads to a *collapse* of the Rabi oscillations in a time λ^{-1} but the discreteness of n leads to a series of partial revivals at later times. These revivals are very sensitive to dissipation [37]. The relevant coherences in the Jaynes-Cummings model are such that at the half revival time a macroscopic superposition of coherent states is established to good approximation (see *e.g.* [36]). It is the coherent recombination of the components of this superposition at the revival time which is responsible for the restoration of oscillations in the inversion. So any intrinsic decoherence will not only reduce the superpositions to a statistical mixture, it will also eliminate the revival consequent upon the survival of superpositions. We next study the Jaynes-Cummings system governed by the Milburn equation and we will show how the intrinsic decoherence modifies the time evolution of the atomic inversion.

In standard quantum mechanics the dynamics of a conservative system described by the density operator $\hat{\rho}$ is governed by the evolution operator $\hat{U} = \exp\left(-itH/\hbar\right)$, where H is the corresponding Hamiltonian. The change in the state of the quantum system in a time interval $(t, t+\tau)$ is given by the following unitary transformation

$$\rho(t+\tau) = \hat{U}(\tau)\rho(t)\hat{U}^\dagger(\tau) = \exp\left(-\frac{i}{\hbar}\tau H\right)\rho(t)\exp\left(\frac{i}{\hbar}\tau H\right), \qquad (78)$$

which is valid for arbitrarily large or small values of τ. Milburn has replaced the above paradigm with three new postulates:

1. For sufficiently short steps the system does not evolve continuously under the unitary transformation (78) but it rather changes stochastically. The probability that the state of the system is changed is $p(\tau)$, reflecting quantum jumps in the state of the system.

2. Given the state of the system is undergoing some changes, then the density operator is changed according to the relation:

$$\rho(t+\tau) = \exp\left(-\frac{i}{\hbar}\Theta(\tau)H\right)\rho(t)\exp\left(\frac{i}{\hbar}\Theta(\tau)H\right),\qquad(79)$$

where $\Theta(\tau)$ is some function of τ. In standard quantum mechanics we have $p(\tau)=1$ and $\Theta(\tau) = \tau$. In the generalised model proposed by Milburn we only require that $p(\tau) \to 1$ and $\Theta(\tau) \to \tau$ for values of τ that are sufficiently large.

3. In Milburn's model it is postulated that

$$\lim_{\tau\to 0}\Theta(\tau) = \Theta_0.\qquad(80)$$

This last postulate effectively introduces a minimum time step in the Universe. The inverse of this step is equal to the mean frequency of the unitary step, $\gamma = 1/\Theta_0$.

The rate of change of $\hat{\rho}(t)$ in Milburn's model is given by the Equation (for details see the original paper by Milburn [33]):

$$\frac{d}{dt}\rho(t) = \gamma\left\{\exp\left[-\frac{i}{\hbar\gamma}H\right]\rho(t)\exp\left[\frac{i}{\hbar\gamma}H\right] - \rho(t)\right\},\qquad(81)$$

which is equivalent to the assumption that on a very short time scale the probability that the system evolves is $p(\tau) = \gamma\tau$. Equation (81) is the proposed generalised equation which alters Schrödinger dynamics. In the limit $\gamma \to \infty$ (*i.e.* when the fundamental time step goes to zero) Equation (81) reduces to the ordinary von Neumann equation for the density operator. The stochastic element introduced by the effective jumps in Equation (80) is responsible for the appearance of an 'arrow of time' in the evolution.

Expanding Equation (81) to first order in γ^{-1}, Milburn obtained the following dynamical equation:

$$\frac{d}{dt}\rho(t) = -\frac{i}{\hbar}[H,\rho] - \frac{1}{2\hbar^2\gamma}[H,[H,\rho]].\qquad(82)$$

As shown by Milburn the first-order correction in Equation (82) leads to the diagonalisation of the density operator in the energy eigenstate basis. Moreover this term induces diffusion in variables that do not commute with the Hamiltonian. However *all* constants of motion commute with the Hamiltonian and thus remain unaffected. In the Jaynes-Cummings model this will cause the excitation number (and hence the energy) to be preserved whilst the revivals which depend on coherences are dephased. In the following section we present the exact solution [34] of Equation (82) with the Jaynes-Cummings Hamiltonian (77).

We assume that initially the field is prepared in a coherent state $|\alpha\rangle$ given by

$$|\alpha\rangle = \exp\left(\alpha\hat{a}^\dagger - \alpha^*\hat{a}\right)|0\rangle = \sum_{n=0}^{\infty}\exp\left(-|\alpha|^2/2\right)\frac{\alpha^n}{\sqrt{n!}}|n\rangle \equiv \sum_{n=0}^{\infty}Q_n|n\rangle\qquad(83)$$

and the atom was prepared in its excited state $|e\rangle$ so that

$$\hat{\rho}(0) = |\alpha\rangle\langle\alpha| \otimes |e\rangle\langle e|. \tag{84}$$

We have solved the Milburn equation (82) using super-operator techniques to give an exact solution for the density operator [34]. Using this solution we can evaluate the explicit expression for the time evolution of the atomic inversion $W(t) = \text{Tr}[\hat{\rho}(t)\hat{\sigma}_3]$ and the expression for the photon number distribution at time t : $P_n(t) = \langle n|\text{Tr}_A\,\hat{\rho}(t)|n\rangle$. For the functions $W(t)$ and $P_n(t)$ we find:

$$W(t) = \sum_{n=0}^{\infty} |Q_n|^2 \exp\left[-2(n+1)\,\lambda^2\,t/\gamma\right]\cos\left(2\sqrt{n+1}\,\lambda t\right), \tag{85}$$

and

$$
\begin{aligned}
P_n(t) = \ & |Q_n(t)|^2 \left\{ C_n^2(t)\exp\left[-2(n+1)\,\lambda^2\,t/\gamma\right] + \frac{1 - \exp\left[-2(n+1)\,\lambda^2\,t/\gamma\right]}{2} \right\} \\
& + |Q_{n-1}|^2 \left\{ S_{n-1}^2(t)\exp\left[-2n\,\lambda^2\,t/\gamma\right] + \frac{1 - \exp\left[-2n\,\lambda^2\,t/\gamma\right]}{2} \right\},
\end{aligned}
\tag{86}
$$

where the probability amplitudes Q_n are given by Equation (83) and the functions $C_n(t)$ and $S_n(t)$ are given by $C_n(t) = \cos\left(\sqrt{n+1}\lambda t\right)$ and $S_n(t) = \sin\left(\sqrt{n+1}\lambda t\right)$. We see that both Equations (85) and (86) in the limit $\gamma \to \infty$ reduce to the well known expressions for the atomic inversion and the photon number distribution in the standard Jaynes-Cummings model governed by the von Neumann equation.

It is well known that the revivals of the atomic inversion as well as oscillations in the photon number distribution arise as a consequence of quantum interference in phase-space (see e.g. [36] and references therein). In other words, these nonclassical effects have their origin in quantum coherences established during the interaction between the atom and the cavity field. From our solution (Equation (85)) it follows that the additional term in the evolution, Equation (82), which destroys quantum coherences leads to appearance of 'decay' factors $\exp\left[-2(n+1)\,\lambda^2\,t/\gamma\right]$ in Equation (85) which are responsible for the deconstruction of revivals of the atomic inversion. As the parameter γ is decreased, i.e. with a more rapid suppression of quantum coherences, we can observe a rapid deterioration of revivals of the atomic inversion. In Figure 14 we plot the time evolution of the atomic inversion for three values of the parameter λ^2/γ. From these figures it follows that the larger is the 'fundamental' time step (i.e. the smaller is the parameter γ) the more pronounced is the suppression of the first revival. These figures illustrate the decay of quantum coherences due to the very specific time evolution described in Equation (82), i.e. due to the intrinsic decoherence. Of course the system remains conservative, so there is no dissipation of energy and the inversion 'relaxes' to a zero value appropriate to an equal mixture of upper and lower levels. It is interesting to compare this dephasing with the effects of genuine dissipation to an external reservoir, either due to field damping or spontaneous emission decay [37]. Obviously in this case the energy of the atom-field system is not a constant of motion. Such external dissipation affects both coherences and energy, and both will decay. Nevertheless, the decay of quantum coherences is much faster than the energy decay: even normal dissipation wipes out coherences in the Jaynes-Cummings model (and hence revivals) well before

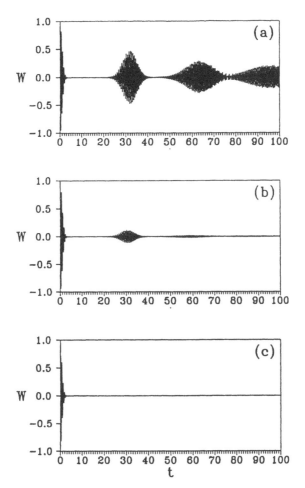

Figure 14. *Time evolution of the atomic inversion $W(t)$ of the atom initially prepared in an excited state interacting with a coherent field $|\alpha\rangle$ $(|\alpha|^2 = 25)$ for various values of the parameter (λ^2/γ) : (a) $\lambda^2/\gamma = 10^{-4}$, (b) $\lambda^2/\gamma = 10^{-3}$ and (c) $\lambda^2/\gamma = 10^{-2}$ (we have set $\lambda = 1$; the quantities t and W are dimensionless).*

any visible effect is made on the energy (*i.e.* on the value of the inversion in the collapse region).

From Equation (86) it follows that the intrinsic decoherence leads to a suppression of oscillations in the photon number distribution. In the standard Jaynes-Cummings model these oscillations appear as a consequence of quantum coherences which are dynamically established during the atom-field interaction [36]. If the dynamics of the atom-field system is governed by the Milburn equation (82) then the intrinsic decoherence suppresses quantum coherences and the oscillations in the photon number distribution. Moreover the field at one half of the revival time, $t = t_r/2 \simeq \pi\bar{n}^{1/2}/\lambda$ (where $\bar{n} = |\alpha|^2$) is not in a pure quantum-mechanical superposition state, but in a statistical mixture state.

To obtain a clearer understanding of the nature of the Milburn equation we have to stress here that the revivals of the atomic inversion depend on the establishment of quantum coherences in the energy (Fock) basis of the cavity field (*i.e.* the revivals depend on the off-diagonal terms of the density operator in the Fock basis). The Milburn equation describes perfectly well the decay of the off-diagonal terms in the Fock basis and therefore describes well the suppression of quantum effects which are related to the existence of the off-diagonal terms of the density operator in the Fock basis. Percival [38] has recently re-opened the question of quantum state diffusion and localisation for similar closed (rather than open) systems.

5.2 Observation of quantum jumps

Having dealt with the idea of *intrinsic* quantum jumps produced by modified, non-orthodox, quantum mechanics we return to *conventional* quantum mechanics but of *open* systems: systems interacting with a Markovian environment, in which jumps are the result of irreversible emissions into the external bath which can be detected to give a record of the open system dynamics. In a real sense, the jumps in an open system are jumps in our knowledge of the state of the system. Individual, repeatable jumps were in the past obscured by the statistical averaging necessary when observing an ensemble of very many identical systems. But developments in experimental techniques, especially in quantum optics have enabled us to observe the dynamics of single quantum systems: fields can be trapped in high Q cavities and manipulated by interaction with individual atoms, and single electrons and ions can be trapped and shielded for very long periods of time. In 1975, Dehmelt [39] suggested the use of trapped three-level ions to make visible quantum jumps in a configuration he termed an atomic amplifier. The configuration is shown in Figure 15. The 1−0 transition is strongly allowed and corresponds to an emission of order 10^7 photons per second when is strongly-driven. The state 2 is chosen to be metastable: lifetimes of seconds have been used. The rare (because of the transition being weak) excitation events to state 2 will terminate the strong fluorescence and lead to a period of 'darkness' which persists until the ion decays back to the ground state 0. The probability of remaining in the 'shelving' state 2 introduces an element of randomness in the duration of the dark periods. The *lack* of fluorescence constitutes a measurement of the 0−2 transition (a *null* result of a measurement is still a measurement and leads to a gain in information about the system [40]). The further study of this shelving process was stimulated by a paper by Cook and Kimble [41] who described how the three-level ion shelving scheme leads to a random telegraph fluorescence signal, where the observed fluorescence turns on and off abruptly as the ion is shelved and decays at random. This triggered a substantial experimental effort and resulted in a number of observations of the telegraphic quantum jumps in three-level ion fluorescence [42]. It is worth noting that jumps in the excitation levels of a single electron trapped in a Penning trap were already observed in 1977 [43] although this does not involve fluorescence detection, of course. Shelving is an essential building block of trapped ion frequency standards [44].

It is worth studying the dynamics of a three-level ion in this shelving situation to see how the bright and dark periods are determined by the excitation dynamics. Provided the 0−1 and 0−2 Rabi frequencies are small compared with the decay rates, we find

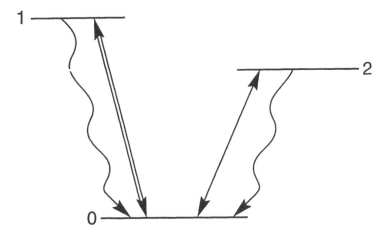

Figure 15. *Dehmelt's atomic amplifier, in which a three-level trapped ion is excited by two incident light fields; level 1 is readily excited and is a normal state which strongly fluoresces back to the ground state 0, whereas state 2 is metastable. Excitation to state 2 will interrupt the 1−0 fluorescence for a period until the ion decays back to 0 and enables the strong 0−2 resonance fluorescence to recommence. The lack of strong fluorescence during the temporary shelving induced by a single photon absorption to state 2 is the 'amplification'.*

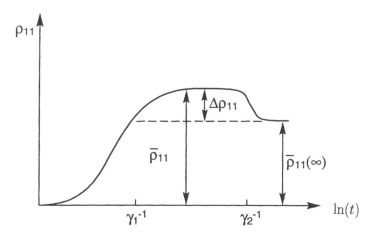

Figure 16. *Time-evolution of the population in the strongly-fluorescing level 1 of the three-level ion shown in Figure 15. The lifetimes γ_1^{-1} and γ_2^{-1} are marked on the time axis. What is crucial here is the 'hump' $\Delta\rho_{11}$: this is a signature of the telegraphic nature of the fluorescence.*

for the population in the strongly-fluorescing level 1 as a function of time something like the behaviour shown in Figure 16. We choose for this figure the values $\gamma_1 \gg \gamma_2$, reflecting the metastability of level 2. For times short compared with the metastable lifetime γ_2^{-1}, then of course the atomic dynamics is hardly aware of level 2 and evolves

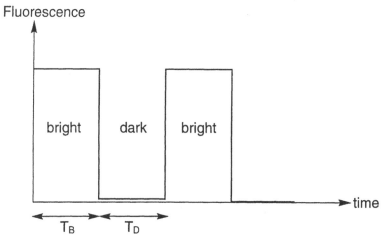

Figure 17. *A few periods of bright and dark sequences in the fluorescence intensity I from a three-level system. The bright periods last on average T_B, and the dark periods T_D.*

as a 0–1 *two-level* system. After a time γ_2^{-1}, the metastable state has an effect and the (ensemble-averaged) population in level 1 reduces to the appropriate *three-level* equilibrium values. The 'hump' $\Delta\rho_{11}$ shown in Figure 16 is actually a signature of the telegraphic fluorescence. To show this, consider a few sequences of bright and dark periods in the telegraph signal as shown in Figure 17. The total rate of emission R is proportional to the bright rate times the fraction of the evolution made up of bright periods:

$$R = \left(\frac{T_B}{T_B + T_D}\right)\bar{\rho}_{11} \tag{87}$$

but this has to be equal to the true average,

$$R = \rho_{11}(\infty), \tag{88}$$

so that

$$\frac{T_D}{T_B} = \frac{\bar{\rho}_{11} - \rho_{11}(\infty)}{\rho_{11}(\infty)} = \frac{\Delta\rho_{11}}{\rho_{11}(\infty)}, \tag{89}$$

and the ratio of the period of bright to dark intervals is governed, as we claimed, by the 'hump' $\Delta\rho_{11}$.

So far, we have concentrated on situations where the Rabi frequencies are small (or for incoherent excitation). What happens for *coherent* resonant excitation with larger Rabi frequencies? The answer to this naive question is *nothing*: there are essentially *no* quantum jumps for coherently-driven resonantly excited three-level systems! But this is because of the idea of resonance is tricky: the strong 0–1 Rabi frequency dresses the atom and AC-Stark splits the transition, forcing the system substantially out of resonance. If this is recognised and the probe laser driving the 0–2 transition is detuned from the bare resonance until it matches the dressed atom resonance, then the jumps and telegraphic fluorescence return. As far as we know, the dependence of the telegraph fluorescence on detuning for coherently excited transitions has yet to be investigated experimentally.

Let us return to the idea of a null measurement. We imagine that we observe the fluorescence from a driven three-level ion over a time scale which is long compared with the strongly fluorescing state lifetime γ_1^{-1} but very short compared with the shelf state lifetime γ_2^{-1}, so that $\gamma_1^{-1} \ll \Delta t \ll \gamma_2^{-1}$. Pegg and Knight [45] have shown that the average period of brightness and darkness in the telegraphic fluorescence can be obtained from considerations of null detection. During such an interval Δt, the population in the shelf state, $P_2(t)$, hardly has time to evolve, but population can be rapidly cycled from the ground state $|0\rangle$ to the strongly-fluorescing state $|1\rangle$ and back. Detection of a photon at the beginning of a Δt interval implies a survival in the $0-1$ sector for the whole interval and a bright period, whereas a *null* detection is sufficient for us to be confident that the atom is shelved for the whole Δt interval and a dark period ensues.

If we take our origin of time to be after an interval Δt in which we see a photon, then $P_2(0) = 1$. We can introduce the 'life expectancy' T_B as the time the atom spends in the $0-1$ sector *continuously*. If the atom is still in this sector at a time t_1 (known from an observation of another fluorescence photon just prior to t_1), then the life expectancy will be also T_B. So we can partition the outcomes into the case where at t_1 it has survived in the $0-1$ sector with probability $P_{10}(t_1)$ and the case where the ion did not survive the whole interval t_1 continuously in the $0-1$ sector [45]

$$T_B = P_{10}(t_1)\,(t_1 + T_B) + (1 - P_{10}(t_1))ft_1 \,, \tag{90}$$

where f is a fraction (< 1). Then for small t_1,

$$T_B = \frac{t_1 P_{10}(t_1)}{1 - P_{10}(t_1)} = \frac{t_1}{1 - P_{10}(t_1)} - t_1 \tag{91}$$

and if t_1 is small so we may neglect the possibility of a *return* from state $|2\rangle$ back in to the $0-1$ sector, $(1 - P_{10}(t_1)) \approx P_2(t_1)$, so that

$$T_B^{-1} = \left.\frac{dP_2}{dt}\right|_{t=0} \quad \text{given} \quad P_2(0) = 0\,. \tag{92}$$

This is finite, so we know that the fluorescence *will* terminate. To obtain a value for T_B, we merely need to know the evolution equation (not its solution) for the population in state $|2\rangle$: this would be the Bloch equation for coherent excitation, or the Einstein rate equation for incoherent excitation.

The calculation of the mean period of darkness proceeds along similar lines: if *no* photons are detected in an interval Δt just before $t = 0$, we find

$$T_D^{-1} = -\left.\frac{dP_2}{dt}\right|_{t=0} \quad \text{given} \quad P_2(0) = 1\,. \tag{93}$$

Estimates of T_B and T_D are readily obtainable if the $0-1$ transition is saturated. If $P_2(0)$ is found to be zero, then $P_1 = 1/2 = P_0$, then

$$\left.\frac{dP_2}{dt}\right|_{t=0} \approx \left(\begin{array}{c} \text{stimulated rate} \\ \text{out of 0 to 2} \end{array}\right) \times P_0(0)$$

and $T_B^{-1} = R/2$, where R is this stimulated rate. The average dark period is found to be $T_D = 1/R$ by similar arguments [45]. This shows that the random telegraph signal for saturated transitions has bright periods lasting two times longer on average than the dark periods.

Now reconsider the Monte-Carlo simulation of quantum jumps. For purely spontaneous emission from a two-level atom, the relevant density operator master equation is

$$\dot{\rho}_S = -\frac{i}{\hbar}[H, \rho_S] - \frac{\gamma}{2}\left(\hat{\sigma}_+\hat{\sigma}_-\rho_S + \rho_S\hat{\sigma}_+\hat{\sigma}_-\right) + \gamma\hat{\sigma}_-\rho_S\hat{\sigma}_+. \tag{94}$$

To propagate the conditioned state vector from time t to $(t + \Delta t)$ we need to calculate the current probability of emission, given by

$$\Delta p = \gamma\Delta t\langle\Psi(t)|\hat{\sigma}_+\hat{\sigma}_-|\Psi(t)\rangle. \tag{95}$$

Then the jump probability Δp is compared with a computer-generated random number r. If $r < \Delta p$, then as before, we re-set the state according to

$$|\Psi\rangle \longrightarrow N^{-1}\hat{\sigma}_-|\Psi\rangle \tag{96}$$

where N^{-1} is a normalisation factor. If $\Delta p < r$, we take the no-jump possibility and propagate $|\Psi\rangle$ under the influence of the non-Hermitian effective Hamiltonian $H_{\text{eff}} = H - i\hbar\gamma\hat{\sigma}_+\hat{\sigma}_-/2$, where H is the unperturbed, atomic Hamiltonian, so that

$$|\Psi\rangle \longrightarrow N^{-1}(1 - iH_{\text{eff}}\Delta t/\hbar)|\Psi\rangle. \tag{97}$$

A single trajectory evolves in a smooth evolution under H_{eff}, interrupted by the jumps. Bouwmeester and colleagues [46] have discussed the physical interpretation of the no-jump evolution and connected this to the decay behaviour in Jaynes 'neoclassical theory' [47]. When many trajectories are averaged, the traditional exponential decay law is recovered.

Resonance fluorescence can be described by a modest extension of the above: all one needs to do is to supplement H by a term describing the coupling of an external laser driving field to the atomic dipole, so that

$$H = \Delta\hat{\sigma}_+\hat{\sigma}_- + (g\hat{\sigma}_+ + g^*\hat{\sigma}_-) \tag{98}$$

where the detuning between the atomic frequency ω and the laser frequency Ω is given by $\Delta = \omega - \Omega$, and the coupling constant g is given by one half of the Rabi frequency $(-\wp\mathcal{E}_\Omega/\hbar)$ where $\wp$ is the atomic dipole moment and $\mathcal{E}_\Omega$ is the amplitude of the laser driving field. Given that the laser field can re-excite the atom, a single trajectory now evolves with many quantum jumps, and an ensemble average over many trajectories reproduces the behaviour predicted from the Bloch equations. This procedure can be generalised to multilevel atoms, and in particular to the three-level V system traditionally used to generate the quantum telegraphic 'shelving' fluorescence so characteristic of quantum jumps. In Figure 18 we indicate the level scheme under consideration. The $0-1$ transition is strongly driven, whereas state 2 is assumed to be a metastable state weakly probed by a much less intense laser field; we allow a small detuning of the probe laser from the $0-2$ transition. The strong driving of the $0-1$ transition Stark shifts the position of state 0, and hence the $0-2$ transition is shifted from its bare transition to a

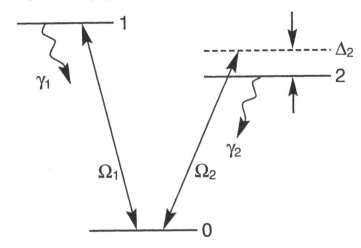

Figure 18. *Three-level V scheme to illustrate quantum telegraph signal in fluorescence from a three-level system. The 0–1 transition is strongly allowed and resonantly excited; state 1 decays at a rate γ_1. State 2 is metastable, with a long lifetime γ_2^{-1}, so that $\gamma_2 \ll \gamma_1$, and is excited by a field which may be detuned by an amount Δ_2 from the bare 0–2 resonance.*

dressed position; the detuning can be adjusted to compensate for this. If this detuning is ignored and bare resonance is maintained, essentially *no* telegraphic signal is generated as we discussed earlier, whereas if the detuning appropriate to the Stark-shifted resonance is maintained, the telegraphic excitation can proceed [48]. In Figure 19 we show the result of a Monte-Carlo wavefunction simulation of these quantum jumps. We see periods where the population resides in state 1 (and can fluoresce) interspersed with periods where the atom is 'shelved' in state 2. The rival effects of coherent excitation and of jumps are quite interesting but will be discussed elsewhere [49].

6   Quantum state diffusion

The second simulation method we consider is quantum state diffusion in which the evolution of $|\Psi(t)\rangle$, representing one member of the ensemble, is determined by a non-linear stochastic differential equation [50, 51]. We write the equation (for the specific case of one-photon damping of a field state) in the Itô form involving drift and diffusion processes so that

$$
\begin{aligned}
|\Psi(t+\delta t)\rangle &= (1 - i\delta t H)|\Psi(t)\rangle + \gamma\left[\langle\hat{a}^\dagger\rangle\hat{a} - \frac{1}{2}\hat{a}^\dagger\hat{a} - \frac{1}{2}\langle\hat{a}^\dagger\rangle\langle\hat{a}\rangle\right]|\Psi(t)\rangle\,\delta t \\
&\quad + \sqrt{\gamma}\,[\hat{a} - \langle\hat{a}\rangle]\,|\Psi(t)\rangle\,\mathrm{d}\xi_t
\end{aligned}
\tag{99}
$$

where $\mathrm{d}\xi_t$ is a random complex Wiener variable which perturbs the system. It varies randomly between each time step and each sample run so that when averaged:

$$
\overline{\mathrm{d}\xi_t} = 0\,, \qquad \overline{\mathrm{d}\xi_t\mathrm{d}\xi_{t'}} = 0\,, \qquad \overline{\mathrm{d}\xi_t^*\mathrm{d}\xi_{t'}} = \delta_{t,t'}\,\delta t\,.
\tag{100}
$$

In consequence the time evolution is not smooth on a very small scale because of the discretised fluctuations of the state vector in its Hilbert space. This distinguishes the

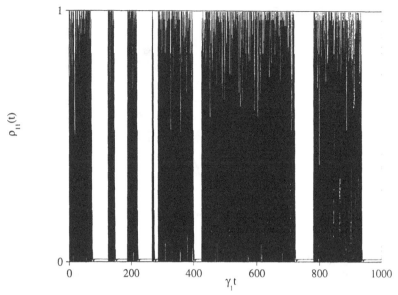

Figure 19. *Results of a simulation of the population ρ_{11} in the strongly-fluorescing state 1 in the three-level shelving scheme. For this simulation $\Omega_1/\gamma_1=2.5$, $\Omega_2/\gamma_1=0.1$, $\Delta/\gamma_1=2.5$, $\gamma_2/\gamma_1 = 0.005$.*

quantum state diffusion calculations from the state vector Monte-Carlo approach in a characteristic way; in the latter case the state vector (and hence the observables) jump only at certain places rather than at every time point. This manifestation of the constant jiggling of the state vector is clearly seen in the single trajectory examples of Figure 20 (solid and dotted curves). However, as in the state vector Monte-Carlo the results for an ensemble average smooth out as we increase the number of samples.

It is interesting to note that if it happens that $|\Psi(t)\rangle$ is an eigenstate of the Lindblad operator $\hat{a}$, then the random fluctuations do not contribute to the system evolution in (99). This results in a vanishing of the random perturbations as can be seen in Figure 20 as equilibrium is established. That $|\Psi(t)\rangle$ is an eigenstate of $\hat{a}$ is not a sufficient condition for the steady state; an inspection of Equation (99) shows that we also require $|\Psi(t)\rangle$ to be an eigenstate of $H - i\hat{a}^{\dagger}\hat{a}/2$ as is found in the state vector Monte-Carlo case.

The stochastic equation (99) amounts to a Langevin equation for the state vector (rather than the usual case of observables). As in the case of state vector Monte-Carlo simulations we can recover the master equation (62) when we average over the ensemble. Equation (99), constructed by Gisin *et al.* , has recently been interpreted as the state vector evolution in a balanced heterodyne detection scheme [52]. A related equation is found for the case of balanced homodyne detection by Carmichael [5]. In both cases a quantum jump theory is developed for the measurement process in the presence of a strong classical local oscillator. This quantum jump theory, unlike the single mode case presented above, contains two jump operators, *i.e.* one for each of the detectors. Using the fact that the local oscillator is strong and creates many jumps in the detector, the

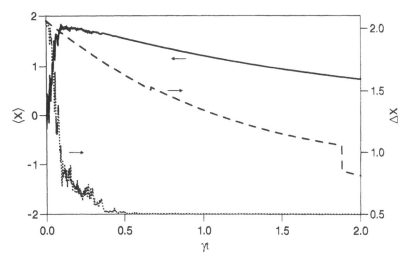

Figure 20. *Localisation of a Schrödinger 'cat' state which is initially comprised of the coherent states $|\alpha\rangle$ and $|-\alpha\rangle$ with $\alpha = 2$. We show $\langle \hat{X} \rangle$ (solid curve—left-hand scale) and ΔX (dotted curve—right-hand scale). The localisation is seen in the development of the non-zero value of $\langle \hat{X} \rangle$ and in the rapid reduction of the uncertainty. The dashed curve shows ΔX for the quantum jump trajectory of Figure 13. The value of $\langle \hat{X} \rangle$ is zero in this case because there is no localisation.*

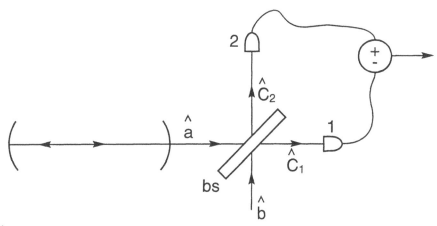

Figure 21. *Heterodyne measurement of cavity field $\hat{a}$ by mixing at a beam-splitter a local oscillator field $\hat{b}$ (assumed to be in an intense coherent state $|\beta\rangle$, $|\beta| \gg 1$) followed by detection of the two output fields described by the operators $\hat{c}_1, \hat{c}_2$.*

number of jumps in a short time interval can be determined by a stochastic function $d\xi$ as found in Equation (99) above. Imagine, for definiteness, the heterodyne situation shown in Figure 21. The two heterodyne output modes $\hat{c}_1, \hat{c}_2$ are superpositions of input modes $\hat{a}$ (describing the quantum system of interest) and $\hat{b}$ (describing the local oscillator, which we imagine to be in a coherent state $|\beta\rangle$ of amplitude $|\beta| \gg 1$). Then for a 50% transmitting beam-splitter [8],

$$\hat{c}_1 = \frac{1}{\sqrt{2}} \left(\hat{a} + \hat{b} \right) , \qquad \hat{c}_2 = \frac{1}{\sqrt{2}} \left(\hat{a} - \hat{b} \right) . \tag{101}$$

The two detectors count photons in the output modes, so we introduce jump operators $\hat{J}_1$ and $\hat{J}_2$ for these detectors

$$\hat{J}_1 = \gamma^{1/2} \hat{c}_1 , \qquad \hat{J}_2 = \gamma^{1/2} \hat{c}_2 . \tag{102}$$

Within a short time Δt we will count many jumps, because of the large amplitude of the local oscillator field β; the total number of counted jumps in Δt will fluctuate in number to lead to a stochastic evolution. Note that

$$\hat{c}_1^\dagger \hat{c}_1 = \frac{1}{2} \left(\hat{a}^\dagger \hat{a} + \hat{b}^\dagger \hat{b} + \hat{a}^\dagger \hat{b} + \hat{b}^\dagger \hat{a} \right) \tag{103}$$

and the local oscillator is in the coherent state $\left| \beta e^{i\phi} \right\rangle$. Then

$$\left\langle \hat{c}_1^\dagger \hat{c}_1 \right\rangle = \frac{1}{2} \left\langle \hat{a}^\dagger \hat{a} + \beta^2 + \hat{a}^\dagger \beta e^{i\phi} + \hat{a} \beta e^{-i\phi} \right\rangle \tag{104}$$

and similarly

$$\left\langle \hat{c}_2^\dagger \hat{c}_2 \right\rangle = \frac{1}{2} \left\langle \hat{a}^\dagger \hat{a} + \beta^2 - \hat{a}^\dagger \beta e^{i\phi} - \hat{a} \beta e^{-i\phi} \right\rangle . \tag{105}$$

In balanced heterodyne measurement, the difference in counts of the two detectors is measured so that the terms $\langle \hat{a}^\dagger \hat{a} \rangle$ and β^2 cancel, and we see that the difference is given by the mean value of the quadrature operator $\hat{x}_\phi = (1/2)(\hat{a}^\dagger e^{i\phi} + \hat{a} e^{-i\phi})$:

$$\left\langle \hat{c}_1^\dagger \hat{c}_1 \right\rangle - \left\langle \hat{c}_2^\dagger \hat{c}_2 \right\rangle = \beta \left\langle \hat{a}^\dagger e^{i\phi} + \hat{a} e^{-i\phi} \right\rangle \equiv 2\beta \left\langle \hat{x}_\phi \right\rangle . \tag{106}$$

To simulate the heterodyne measurement, we need to define the effective non-Hermitian evolution H_{eff} which governs the no-jump events. We set

$$H_{\text{eff}} = -\frac{i\hbar}{2} \left(\hat{J}_1^\dagger \hat{J}_1 + \hat{J}_2^\dagger \hat{J}_2 \right) \approx -i\hbar \frac{\gamma}{2} \left(\hat{a}^\dagger \hat{a} + \beta^2 \right) \tag{107}$$

where the second form follows if we assume the local oscillator to be in an intense coherent state $\left| \beta e^{i\phi} \right\rangle$ and neglect its quantum nature. Then the no-jump evolution is given by

$$\exp \left(-i H_{\text{eff}} \Delta t / \hbar \right) \approx 1 - \frac{\gamma}{2} \left(\hat{a}^\dagger \hat{a} + \beta^2 \right) \Delta t . \tag{108}$$

The jump operators in the same approximation are

$$\hat{J}_1 = \left(\frac{\gamma}{2} \right)^{1/2} \left(\hat{a} + \beta e^{i\phi} \right) , \qquad \hat{J}_2 = \left(\frac{\gamma}{2} \right)^{1/2} \left(\hat{a} - \beta e^{i\phi} \right) \tag{109}$$

so that the probabilities of jumps per unit time are given by

$$\begin{aligned}
\left\langle \hat{J}_1^\dagger \hat{J}_1 \right\rangle &\sim \frac{\gamma}{2} \beta^2 \left(1 + \frac{2}{\beta} \langle \hat{x}_\phi \rangle \right) \approx \gamma \beta^2 / 2 , \\
\left\langle \hat{J}_2^\dagger \hat{J}_2 \right\rangle &\sim \frac{\gamma}{2} \beta^2 \left(1 - \frac{2}{\beta} \langle \hat{x}_\phi \rangle \right) \approx \gamma \beta^2 / 2 .
\end{aligned} \tag{110}$$

We choose the interval time Δt to be large enough to contain many detected counts, but small compared with the system evolution time. Then the total number of detected jumps is $\langle \hat{J}_i^\dagger \hat{J}_i \rangle \Delta t$ (where $i = 1, 2$) on average, though it fluctuates.

If we divide up the interval Δt into n segments of length δt, then the probability of m counts in one of the detectors is

$$P(m) = \frac{n!}{(n-m)!\, m!}(P\delta t)^m (1 - P\delta t)^{n-m}, \tag{111}$$

where $P = \langle \hat{J}^\dagger \hat{J} \rangle$ for the given detector. But since $P\Delta t \gg 1$, we can approximate this by the Gaussian form

$$P(m) = \left((2\pi)^{1/2}\sigma \right)^{-1} \exp\left[-(m - P\Delta t)^2 / 2\sigma^2 \right] \tag{112}$$

where

$$\sigma^2 = P\Delta t(1 - P\delta t) \approx P\Delta t. \tag{113}$$

In other words,

$$P(m) = \frac{1}{\left(2\pi \left\langle \hat{J}^\dagger \hat{J} \right\rangle \Delta t \right)^{1/2}} \exp\left[-\frac{\left(m - \left\langle \hat{J}^\dagger \hat{J} \right\rangle \Delta t \right)^2}{2\left\langle \hat{J}^\dagger \hat{J} \right\rangle \Delta t} \right] \tag{114}$$

where

$$m(t) = \left\langle \hat{J}^\dagger \hat{J} \right\rangle \Delta t + \sqrt{\left\langle \hat{J}^\dagger \hat{J} \right\rangle} \Delta W \tag{115}$$

and the Wiener noise term is such that

$$\left\langle \Delta W^2 \right\rangle = \Delta t. \tag{116}$$

To use the jump method, we need to evaluate powers of the jump operators: if there are m_i detected counts in detector i over time Δt, then we will need

$$\left(\hat{J}_1 \right)^{m_1} = \left[\beta \left(\frac{\gamma}{2} \right)^{1/2} \left(1 + \frac{\hat{a}}{\beta} e^{-i\phi} \right) \right]^{m_1} \propto \left(1 + \frac{m_1}{\beta} \hat{a} e^{-i\phi} \right), \qquad \beta \gg 1. \tag{117}$$

Similarly

$$\left(\hat{J}_2 \right)^{m_2} \propto \left(1 - \frac{m_2}{\beta} \hat{a} e^{-i\phi} \right), \tag{118}$$

where

$$m_i = \left\langle \hat{J}_i^\dagger \hat{J}_i \right\rangle \Delta t + \sqrt{\langle \hat{J}_i^\dagger \hat{J}_i \rangle}\, \Delta W_i \qquad (i = 1, 2). \tag{119}$$

The un-normalised evolution over Δt is given by

$$|\Psi\rangle \longrightarrow \exp\left(-i H_{\text{eff}} \Delta t / \hbar \right) \left(\hat{J}_1 \right)^{m_1} \left(\hat{J}_2 \right)^{m_2} |\Psi\rangle, \tag{120}$$

where the operator ordering is not important with the approximations used. Using the results above, we find

$$|\Psi\rangle \longrightarrow \left[1 - i H_{\text{eff}} \Delta t / \hbar + \beta^{-1} \hat{a} e^{-i\phi} (m_1 - m_2) \right] |\Psi\rangle \tag{121}$$

so that

$$|\Psi\rangle \longrightarrow \left[1 - iH_{\text{eff}}\Delta t/\hbar + e^{-i\phi}\frac{\hat{a}}{\beta}\left(2\gamma\beta\langle\hat{x}_\phi\rangle\Delta t + \beta\left(\frac{\gamma}{2}\right)^{1/2}(\Delta W_1 - \Delta W_2)\right)\right]|\Psi\rangle.$$

(122)

We will add together the two Wiener increments as $\Delta W = (\Delta W_1 - \Delta W_2)/\sqrt{2}$, and from now on write all increments in terms of differentials, e.g. dW. Now the local oscillator and the quantum system have different frequencies, so the time dependent phase $\phi(t)$ can be incorporated into the definition of the Wiener increment, as $e^{-i\phi(t)}\,dW \to d\xi_t$. We now drop a counter-rotating term from equation (122) and we find, after all this, the stochastic evolution equation

$$|d\tilde{\Psi}\rangle \sim \left[-\frac{\gamma}{2}\left(\hat{a}^\dagger\hat{a} + \beta^2\right)dt + \gamma\hat{a}\langle\hat{a}^\dagger\rangle\,dt + \sqrt{\gamma}\hat{a}\,d\xi_t\right]|\Psi\rangle$$

(123)

where $|\tilde{\Psi}\rangle$ is the state $|\Psi\rangle$ before normalisation. This is one form of the nonlinear stochastic evolution equation characteristic of the Quantum State Diffusion model [50].

To obtain the equation of motion for the normalised state vector one has to be careful about differentiation because of the normalisation of the noise term (Equation (100)). The Itô calculus may be used under which, for example,

$$d(\langle\tilde{\Psi}(t)|\tilde{\Psi}(t)\rangle) = \langle\tilde{\Psi}(t)|d\tilde{\Psi}\rangle + \langle d\tilde{\Psi}|\tilde{\Psi}(t)\rangle + \langle d\tilde{\Psi}|d\tilde{\Psi}\rangle.$$

(124)

Furthermore, to obtain a connection with Equation (99) it is necessary to include a phase factor as well as renormalising the state vector so that [53]

$$|\Psi(t)\rangle = \frac{e^{i\zeta(t)}|\tilde{\Psi}(t)\rangle}{\sqrt{\langle\tilde{\Psi}(t)|\tilde{\Psi}(t)\rangle}}.$$

(125)

The phase factor is stochastic because

$$id\zeta = \frac{\langle\tilde{\Psi}(t)|\hat{a}^\dagger|\tilde{\Psi}(t)\rangle d\xi^* - \langle\tilde{\Psi}(t)|\hat{a}|\tilde{\Psi}(t)\rangle d\xi}{2\langle\tilde{\Psi}(t)|\tilde{\Psi}(t)\rangle}$$

(126)

resulting in the equation for the normalised $|\Psi(t)\rangle$:

$$\begin{aligned}
|d\Psi\rangle &= \gamma\left[-\frac{1}{2}\hat{a}^\dagger\hat{a} + \hat{a}\langle\Psi(t)|\hat{a}^\dagger|\Psi(t)\rangle + \frac{1}{2}\langle\Psi(t)|\hat{a}^\dagger\hat{a}|\Psi(t)\rangle\right. \\
&\quad \left.- \langle\Psi(t)|\hat{a}^\dagger|\Psi(t)\rangle\langle\Psi(t)|\hat{a}|\Psi(t)\rangle\right]|\Psi(t)\rangle dt \\
&\quad + \frac{1}{2}\gamma\left[-\langle\Psi(t)|\hat{a}^\dagger\hat{a}|\Psi(t)\rangle + \langle\Psi(t)|\hat{a}^\dagger|\Psi(t)\rangle\langle\Psi(t)|\hat{a}|\Psi(t)\rangle\right]|\Psi(t)\rangle d\xi^* d\xi \\
&\quad + \sqrt{\gamma}\left[\hat{a} - \langle\Psi(t)|\hat{a}|\Psi(t)\rangle\right]|\Psi(t)\rangle d\xi.
\end{aligned}$$

(127)

This equation yields Equation (99) given $d\xi^* d\xi \longrightarrow dt$. The only difference between these equations is that the state vector in Equation (99) is only normalised for the ensemble average whilst in Equation (127) it is always normalised. In practice this seems to result in no noticeable differences because whilst our computer program integrates Equation (99) it renormalises the state vector at every time step.

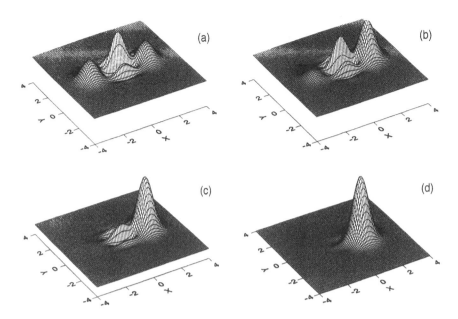

Figure 22. *Wigner functions from the quantum state diffusion of a single sample. We see $\gamma t = 0, 0.05, 0.2$ and 0.8 in* (a) $-$ (d)*. Localisation is seen as the increasing dominance of one component of the 'cat'.*

We should emphasise that the non-linear equations we consider here in no way violate the principles of quantum mechanics. Each kind of simulation is non-linear because it is conditional on a specific unravelling of the master equation (62) and when we ensemble average over a large number of state vectors we always recover the master equation (62) with the appropriate reservoir operators [5, 29, 50, 54].

In Figure 22 we demonstrate the process by which the state localises in quantum state diffusion from an initial superposition by plotting the Wigner function. Interference fringes can be seen in the initial state, Figure 22(a), which consists of a Schrödinger 'cat' coherent superposition of two coherent states $|\alpha\rangle$ and $|-\alpha\rangle$ (with $\alpha = 2$ in this case), with the two components representing the live 'cat' and the dead 'cat'. If we now examine the time sequence in Figures 22(a)-(d) we observe that one of the components of the 'cat' diminishes as the other component grows. This is accompanied by a rapid loss of the fringes. The Wigner function clearly shows localisation to one of the coherent state components of the original 'cat', the choice of component being determined randomly according to the stochastic input. In Figure 20 we showed the values of the quadrature $\langle \hat{X} \rangle = \langle \Psi(t)|(\hat{a} + \hat{a}^\dagger)|\Psi(t)\rangle/2$ and uncertainty $\Delta X = (\langle \hat{X}^2 \rangle - \langle \hat{X} \rangle^2)^{1/2}$ as a function of time and including the times at which the Wigner functions of Figure 22 are appropriate. The localisation manifests itself in $\langle \hat{X} \rangle$ through the rapid development of a non-zero value of the displacement which takes place as one of the component coherent states is chosen. Subsequently, the value of $\langle \hat{X} \rangle$ decays as the coherent state itself approaches the ground state. The localisation is seen in ΔX as a rapid reduction (at essentially the decoherence rate $\gamma |\alpha|^2$) to a value of one half, the uncertainty value for a coherent state. The value of one half is maintained as the coherent state dissipates

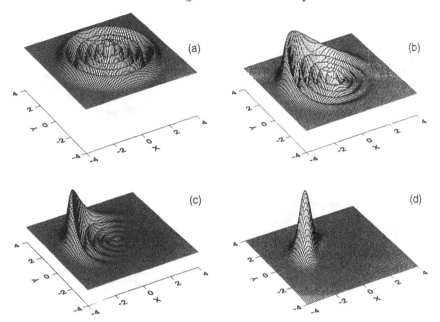

Figure 23. *Localisation of a Fock state $|n\rangle$ with $n = 9$. These results are for a single sample undergoing quantum state diffusion. The Wigner functions are shown for γt =0, 0.1, 0.5 and 1.0 in (a)–(d).*

to the origin. The picture would be quite different if we were to consider the density matrix approach. In that case the fringes in the Wigner function would also rapidly disappear, but we would be left with both of the components reflecting a statistical mixture of outcomes [55].

We have so far presented the heterodyne experiment with a simple choice of two components (outcomes) of a Schrödinger 'cat'. Any state of the oscillator can be represented as a superposition of coherent states, and the quantum state diffusion will describe localisation to a single coherent state component. We next consider a complex superposition of coherent states: a Fock state . As mentioned earlier a continuous superposition of coherent states with any radius r can form a Fock state provided the correct phase factors $e^{-in\theta}$ are present. However, when localisation takes place a coherent state is selected from that circle for which each component coherent state has nearly the same energy as the original Fock state . The process is illustrated in Figure 23. The initial state in (a) contains nine quanta and its Wigner function, formed from a Laguerre polynomial, is made up of an outer rim with inner concentric fringes. As time progresses, (b)-(c), we observe that localisation starts with the preferential orientation of the rim and fringes in one direction. This chosen direction is different for each realisation. As the outer rim localises, the inner fringes wash out with the vanishing of quantum interference between coherent state components. In (d) we see that the fringes have almost disappeared and the remaining localised state approaches a coherent state. In this case, the localisation rate is much less than the decoherence rate: in order to localise, sufficient decay time has to elapse to generate a a range of photon numbers to support the full Poisson width of a coherent state component. During the subsequent

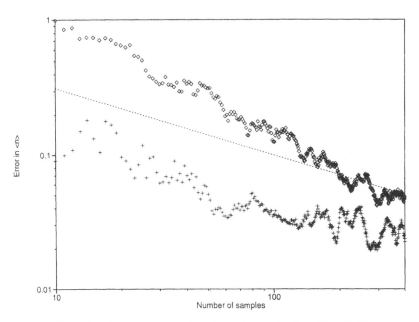

Figure 24. *Error in simulation as a function of sample size. The R.M.S. error is time averaged and shown for quantum jump simulations ($\Diamond$) and quantum state diffusion ($+$). The straight line indicates the trend of the central limit theorem, though the scaling factor is not known. In all the simulations the initial state was a 9-photon Fock state subjected to purely dissipative (1-photon) processes.*

time evolution the Wigner function becomes quite rounded and moves in towards the phase-space origin. We have also investigated the localisation from initially squeezed states, and again we see localisation into components of Equation (45) (and under some conditions, the formation of *Schrödinger 'cats'*).

7   Comparison of ensemble averaged results

We have seen in the previous two sections how the state vector Monte-Carlo method and quantum state diffusion method of simulation lead to very different behaviour for individual realisations and reflect the gain in information from specific conditionings, whilst for an average over large samples they realise the results of the master equation (62). In the case where we are only interested in ensemble averages, we can treat each simulation merely as a computational algorithm en route to an ensemble behaviour and ignore the subtleties of how each simulation reflects a specific conditioning. Then in all practical cases the number of samples must be finite and given that we have a choice of the two methods for performing the simulation, the question arises as to which to use: when ensemble averaged, both methods seem to give results with about the same accuracy. In order to check this we have calculated the mean fluctuations between simulated, and sample averaged, observables and the computed observables from direct integration of the master equation (62) in a similar fashion to the two photon processes examined in [54]. Here we consider one-photon decay and the fluctuations are averaged over the time evolution and are shown in Figure 24 for the observable $\langle \hat{n} \rangle$ where the symbols $\diamond$ indicate the fluctuations with the state vector Monte-Carlo approach and the symbols $+$ show fluctuations in quantum state diffusion. The fluctuations decrease in size with sample number roughly as expected from the central limit theorem. However, these results also depend on the integration step size δt and can only be taken as representative. For example, if the step size is reduced, then greater accuracy may be achieved with larger numbers of samples.

8   Conclusions

In these lectures we have presented a pedagogical approach to the construction of non-classical light through quantum interference amongst component parts, and their 'deconstruction' through the irreversible interaction with a dissipative environment. This allows us to investigate the evolution of single realisations of unstable quantum systems and to study the quantum jumps and the localisation characteristic of particular measurement schemes, reflecting the conditioned nature of information gain from observation of individual realisations.

Acknowledgements

We thank V Bužek, A Ekert, L Gilles, M S Kim, H Moya-Cessa, I C Percival, D T Pegg, W Power and J Steinbach for discussions. This work was supported in part by the European Community and by the UK Engineering and Physical Sciences Research Council.

References

[1] See for example the papers reproduced in *Nonclassical Effects in Quantum Optics*, edited by P Meystre and D F Walls (American Institute of Physics, New York, 1991)

[2] K Vogel and W P Schleich, in *Fundamental Systems in Quantum Optics*, edited by J Dalibard, J M Raimond, and J Zinn-Justin (Elsevier, Amsterdam, 1991), page 713

[3] W H Louisell, *Quantum Statistical Properties of Radiation*, (Wiley, New York, 1974)

[4] W H Zurek, *Phys Today*, **44**, 36 (1991); W H Zurek, S Habib, and J P Paz, *Phys Rev Lett*, **70**, 1187 (1993) and references therein

[5] H J Carmichael, *An Open Systems Approach to Quantum Optics*, Lecture Notes in Physics (Springer, Berlin, 1993)

[6] M Sargent III, M O Scully, and W E Lamb Jr, *Laser Physics*, (Addison-Wesley, Reading, 1975)

[7] R Loudon, *Quantum Theory of Light*, (Oxford University Press, Oxford, 1983)

[8] R Loudon and P L Knight, *J Mod Opt*, **34**, 209 (1987)

[9] E P Wigner, *Phys Rev*, **40**, 749 (1932); see also M Hillery, R F O'Connell, M O Scully, and E P Wigner, *Phys Rep*, **106**, 121 (1984)

[10] R L Hudson, *Rep on Math Phys*, **6**, 249 (1974)

[11] A K Ekert and P L Knight, *Phys Rev A*, **42**, 487 (1990); *ibidem*, **43**, 3934 (1991) See also the original papers, M J Collett and C W Gardiner, *Phys Rev A*, **30**, 1386 (1984); C W Gardiner and M J Collett, *Phys Rev A*, **31**, 3761 (1985); and also C W Gardiner, *Quantum Noise*, (Springer, Berlin, 1991)

[12] S Reynaud, A Heidmann, E Giacobino and C Fabre, *Progress in Optics*, **30**, edited by E Wolf, (Elsevier, Amsterdam, 1992) and references therein

[13] K Vogel and H Risken, *Phys Rev A*, **40**, 2847 (1989); see also H Kühn, D -G Welsch and W Vogel, *J Mod Opt* , **41**, 1607 (1994) and references therein

[14] D T Smithey, M Beck, M G Raymer, and A Faridani, *Phys Rev Lett* , **70**, 1244 (1993); M G Raymer, M Beck and D F McAlister, *Phys Rev Lett* , **72**, 1137 (1994)

[15] U Leonhardt, PhD thesis, Humboldt University, Berlin (1993)

[16] The *Q*-function is related to the Wigner function by a Gaussian convolution When an imperfect detector is modelled as a beam-splitter followed by a perfect detector, the added noise (the complementary part of the dissipation) is responsible for a smoothing of the quasi-probability, which approaches the *Q*-function positivity as the detector imperfections increase

[17] A Bandilla and H Paul, *Ann Phys* (Leipzig) **23**, 323 (1969); H Paul, *Fortschr Phys* , **22**, 424 (1974); U Leonhardt, H Paul, and G M D'Ariano, *Phys Rev A*, **52**, 4899 (1995); U Leonhardt, *Phys Rev A*, **49**, 1231 (1994)

[18] V Bužek, C Keitel, and P L Knight, *Phys Rev A*, **51**, 2575 (1995)

[19] D F Walls and G J Milburn, *Quantum Optics*, (Springer, Berlin, 1994)

[20] K Wódkiewicz, P L Knight, S J Buckle, and S M Barnett, *Phys Rev A*, **35**, 2567 (1987)

[21] V Bužek and P L Knight, *Progress in Optics*, edited by E Wolf, (Elsevier, Amsterdam, 1994) and references therein

[22] J Janszky, P Domokas and P Adam, *Phys Rev A*, **48**, 2213 (1993)

[23] V Bužek and P L Knight, *Opt Comm* , **81**, 331 (1991)

[24] B M Garraway, B Sherman, H Moya-Cessa, P L Knight, and G Kurizki, *Phys Rev A*, **49**, 535 (1994); B Sherman and G Kurizki, *Phys Rev A*, **45**, 7674 (1992); K Vogel, V Akulin, and W P Schleich, *Phys Rev Lett*, **71**, 1816 (1993)

[25] J Janszky and An V Vinogradov, *Phys Rev Lett* , **64**, 2771 (1990)

[26] J P Paz, S Habib, and W H Zurek, *Phys Rev D*, **47**, 488 (1993); W H Zurek, *Phys Today*, **44**, 36 (1991); *ibidem*, **46**, 84 (1993); W H Zurek, *Prog Theor Phys* , **89**, 281 (1993)

[27] E Joos and H D Zeh, *Z Phys B*, **59**, 223 (1985); D F Walls and G J Milburn, *Phys Rev A*, **31**, 2403 (1985); S J D Phoenix, *Phys Rev A*, **41**, 5132 (1990); G J Milburn, *Phys Rev A*, **36**, 5271 (1987)

[28] See, for example: N Gisin, P L Knight, I C Percival, R C Thompson and D C Wilson, *J Mod Opt* , **40**, 1663 (1993) and references therein; M Brune, S Haroche, J M Raimond, L Davidovich and N Zagury, *Phys Rev A*, **45**, 5193 (1992); S Haroche and J M Raimond, *Advances in Atomic, Molecular and Optical Physics Supplement 2*, (1994, Academic Press), page 123

[29] J Dalibard, Y Castin, and K Mølmer, *Phys Rev Lett* , **68**, 580 (1992); K Mølmer, Y Castin, and J Dalibard, *J Opt Soc Am B*, **10**, 524 (1993)

[30] J Steinbach, B M Garraway, and P L Knight, *Phys Rev A*, **51**, 3302 (1994)

[31] C M Caves and G J Milburn, *Phys Rev D*, **36**, 3543 (1987); L Diosi, *Phys Rev A*, **40**, 1165 (1989); G C Ghirardi, P Pearle, and A Rimini, *ibidem*, **42**, 78 (1990); I C Percival, *Proc Roy Soc* , **451**, 503 (1995); J S Bell, *Physics World*, **3**, 33 (1990)

[32] G C Ghirardi, A Rimini, and T Weber, *Phys Rev D* **34**, 470 (1986)

[33] G J Milburn, *Phys Rev A*, **44**, 5401 (1991)

[34] H Moya-Cessa, V Bužek, M S Kim and P L Knight, *Phys Rev A*, **48**, 3900 (1993)

[35] E T Jaynes and F W Cummings, *Proc IEEE*, **51**, 89 (1963)

[36] B W Shore and P L Knight, *J Mod Opt* , **40**, 1195 (1993)

[37] S M Barnett and P L Knight, *Phys Rev A*, **33**, 2444 (1986); T Quang, P L Knight, and V Bužek, *Phys Rev A*, **44**, 6092 (1991)

[38] I C Percival, see [31]

[39] H G Dehmelt, *Bull Am Phys Soc* , **20**, 60 (1975)

[40] R H Dicke, *Am J Phys* , **49**, 925 (1981)

[41] R J Cook and H J Kimble, *Phys Rev Lett* , **54**, 1023 (1985)

[42] W Nagourney, J Sandberg, and H G Dehmelt, *Phys Rev Lett* , **56**, 2797 (1986); Th Sauter, W Neuhauser, R Blatt, and P E Toschek, *Phys Rev Lett* , **57**, 1696 (1986); J C Bergquist, R G Hulet, W M Itano, and D J Wineland, *Phys Rev Lett* , **57**, 1699 (1986)

[43] R S Van Dyck, Jr, P B Schwinberg, and H G Dehmelt, *Phys Rev Lett* **38**, 310 (1977)

[44] R Blatt, P Gill, and R C Thompson, *J Mod Opt* , **39**, 193 (1992) and references therein

[45] D T Pegg and P L Knight, *Phys Rev A*, **37**, 4303 (1988)

[46] D Bouwmeester, R J C Spreeuw, G Nienhuis, and J P Woerdman, *Phys Rev A*, **49**, 4170 (1994)

[47] E T Jaynes, in *Coherence in Quantum Optics IV*, edited by L Mandel and E Wolf, (Plenum, New York, 1978)

[48] M S Kim and P L Knight, *Phys Rev A*, **36**, 5265 (1987); *ibidem*, **40**, 215 (1989)

[49] M S Kim, B M Garraway, and P L Knight, in preparation

[50] N Gisin and I C Percival, *J Phys A*, **25**, 5677 (1992); *ibidem*, **26**, 2233 (1993); *ibidem*, **26**, 2245 (1993)

[51] N Gisin and I C Percival, *Phys Lett A*, **167**, 315 (1992)

[52] H M Wiseman and G J Milburn, *Phys Rev A*, **47**, 1652 (1993); see also [5] and Y Castin, J Dalibard, and K Mølmer, in *Atomic Physics 13*, edited by H Walther, T W Hänsch, and B Neizert (American Institute of Physics, New York, 1993)

[53] B M Garraway and P L Knight, *Phys Rev A*, **50** 2548 (1994)

[54] B M Garraway and P L Knight, *Phys Rev A*, **49**, 2785 (1994)

[55] V Bužek, A Vidiella-Barranco, and P L Knight, *Phys Rev A*, **45**, 6570 (1992)

Optical physics of quantum wells

David A B Miller

AT&T Bell Laboratories
Holmdel, USA

1   Introduction

Quantum wells are thin layered semiconductor structures in which we can observe and control many quantum mechanical effects. They derive most of their special properties from the quantum confinement of charge carriers (electrons and 'holes') in thin layers (e.g. 40 atomic layers thick) of one semiconductor 'well' material sandwiched between other semiconductor 'barrier' layers. They can be made to a high degree of precision by modern epitaxial crystal growth techniques. Many of the physical effects in quantum well structures can be seen at room temperature and can be exploited in real devices. From a scientific point of view, they are also an interesting 'laboratory' in which we can explore various quantum mechanical effects, many of which cannot easily be investigated in the usual laboratory setting. For example, we can work with 'excitons' as a close quantum mechanical analogue for atoms, confining them in distances smaller than their natural size, and applying effectively gigantic electric fields to them, both classes of experiments being difficult to perform on atoms themselves. We can also carefully tailor 'coupled' quantum wells to show quantum mechanical beating phenomena that we can measure and control to a degree that is difficult with molecules.

In this article, we will introduce quantum wells, and will concentrate on some of the physical effects that are seen in optical experiments. Quantum wells also have many interesting properties for electrical transport, though we will not discuss those here. We will briefly allude to some of the optoelectronic devices, though again we will not treat them in any detail.

2   Introduction to quantum wells

First we will introduce quantum wells by discussing their basic physics, their structure, fabrication technologies, and their elementary linear optical properties.

2.1   Semiconductor band structure and heterostructures

All of the physics and devices that will be discussed here are based on properties of direct gap semiconductors near the center of the Brillouin zone. For all of the semiconductors of interest here, we are concerned with a single, S-like conduction band, and two P-like valence bands. The valence bands are known as the heavy and light hole bands. Importantly for quantum wells, the electrons in the conduction band, and the (positively charged) 'holes' in the valence band behave as particles with effective masses different from the free electron mass. The simplest 'k.p.' band theory says that the electron effective mass, m_e, and the light hole effective mass, m_{lh} are approximately equal and proportional to the band gap energy. For GaAs, which has a band gap energy ~ 1.5 eV, the actual values are $m_e \sim 0.069\, m_0$ and $m_{lh} \sim 0.09\, m_0$, where m_0 is the free electron mass. The heavy hole effective mass, m_{hh}, is typically more comparable to the free electron mass ($m_{hh} \sim 0.35\, m_0$ for the most common situation in quantum wells), and does not vary systematically with the band gap energy.

Quantum wells are one example of heterostructures—structures made by joining different materials, usually in layers, and with the materials joined directly at the atomic level. When two semiconductors are joined, it is not clear in advance how the different bands in the two materials will line up in energy with one another, and their is no accurate predictive theory in practice. Hence, an important experimental quantity is the 'band offset ratio' ; this is the ratio of the difference in conduction band energies to the difference in valence band energies. For GaAs/AlGaAs heterostructures, for example, approximately 67% of the difference in the band gap energies is in the conduction band offset, and 33% is in the valence band offset, giving a ratio 67:33. In this particular material system, both electrons and holes see higher energies in the AlGaAs than in the GaAs, giving a so-called 'Type I' system. Heterostructures in which electrons and holes have their lowest energies in different materials are called 'Type II', but such structures will not be considered further here.

Heterostructures in general have many uses. They can be used for advanced electronic devices (*e.g.* modulation-doped field-effect transistors, heterojunction bipolar transistors, resonant tunneling devices), optical components (*e.g.* waveguides, mirrors, microresonators), and optoelectronic devices and structures (*e.g.* laser diodes, photodetectors, quantum well and superlattice optical and optoelectronic devices). Although heterostructures may be useful in electronics, they are crucial in many optoelectronic devices (*e.g.* lasers). Perhaps their most important technological aspect may be that they can be used for all of these electronic, optical, and optoelectronic purposes, and hence may allow the integration of all of these.

2.2   Quantum well structures and growth

A quantum well is a particular kind of heterostructure in which one thin 'well' layer is surrounded by two 'barrier' layers. Both electrons and holes see lower energy in the 'well' layer, hence the name (by analogy with a 'potential well'). This layer, in which both electrons and holes are confined, is so thin (typically about 100 Å , or about 40 atomic layers) that we cannot neglect the fact that the electron and hole are both waves. In fact, the allowed states in this structure correspond to standing waves in the direction perpendicular to the layers. Because only particular waves are standing waves, the system is quantised, hence the name 'quantum well'.

There are at least two techniques by which quantum well structures can be grown, molecular beam epitaxy (MBE) (Cho [1]), and metal-organic chemical vapour deposition (MOCVD) (Furuya and Miramoto [2]). Both can achieve a layer thickness control close to about one atomic layer. MBE is essentially a very high vacuum technique in which beams of the constituent atoms or molecules (*e.g.* Ga, Al, or As) emerge from ovens, land on the surface of a heated substrate, and there grow layers of material. Which material is grown can be controlled by opening and closing shutters in front of the ovens. For example, with a shutter closed in front of the Al oven, but open shutters in from of the Ga and As ovens, GaAs layers will be grown. Opening the Al shutter will then grow the alloy AlGaAs, with the relative proportion of Ga and Al controlled by the temperatures of the ovens. With additional ovens and shutters for the dopant materials, structures of any sequence of GaAs, AlAs, and AlGaAs can be grown with essentially arbitrary dopings. MOCVD is a gas phase technique at low pressure (*e.g.* 25 torr). In this case the constituents are passed as gasses (*e.g.* trimethylgallium and arsine) over a heated substrate, with the resulting composition being controlled by the relative amounts of the appropriate gasses. Hybrid techniques, using the gas sources of MOCVD in a high vacuum molecular beam system, also exist, and are known variously as gas-source MBE or chemical beam epitaxy (CBE) (Tsang [3]). Which technique is best depends on the material system and the desired structure or device. Typical structures grown by these techniques might have total thickness of microns, and could have as many as hundreds of layers in them.

There are many different materials that can be grown by these techniques, and many of these have been used to make quantum well structures. One significant restriction is that it is important to make sure that the lattice constants (essentially, the spacing between the atoms) of the materials to be grown in the heterostructure are very similar. If this is not the case, it will be difficult to retain a well-defined crystal structure throughout the layers—the growth will not be 'epitaxial'. The growth is simplest when the lattice constants are identical. Fortunately, AlAs and GaAs have almost identical lattice constants, which means that arbitrary structures can be grown with high quality in this materials system. Another commonly used system is InGaAs with InP; in this case, the proportions of In and Ga are adjusted to give a lattice constant for the ternary (three-component) InGaAs alloy that is equal to InP. Use of four component (quaternary) alloys (*e.g.* InGaAsP) allows sufficient degrees of freedom to adjust both the lattice constant and the bandgap energy. Up to a certain critical thickness, which depends on the degree of lattice constant mismatch, it is possible to grow structures with materials that naturally have different lattice constants. In this case, the materials grow in a highly strained state but can adopt the local lattice constant and retain good epitaxial crystal structure. Such strained materials are of increasing technological importance, although we will not discuss them further here.

A partial list of materials used for quantum well structures includes: III-V's— GaAs/GaAlAs on GaAs (Type I), GaSb/GaAlSb on GaSb (Type I), InGaAs/InAlAs on InP (Type I), InAs/GaSb (Type II), InGaAs/GaAs (Type I, strained); II-VI's— HgCdTe/CdTe, ZnSe/ZnMnSe (semimagnetic), CdZnTe/ZnTe (Type 1, strained); IV-VI's—PbTe/PbSnTe; IV - Si/SiGe (strained).

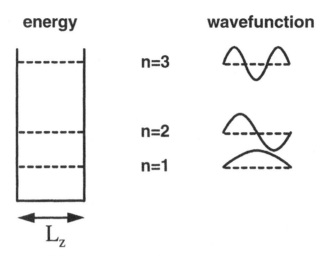

Figure 1. *'Infinite' quantum well and associated wavefunctions.*

2.3   Particle-in-a-box quantum well physics

We can understand the basic properties of a quantum well through the simple 'particle-in-a-box' model. Here we consider Schrödinger's equation in one dimension for the particle of interest (*e.g.* electron or hole)

$$-\frac{\hbar^2}{2m}\frac{d^2\phi_n}{dz^2} + V(z)\,\phi_n = E_n\phi_n\,, \tag{1}$$

where $V(z)$ is the structural potential (*i.e.* the 'quantum well' potential) seen by the particle along the direction of interest (z), m is the particle's (effective) mass, and E_n and ϕ_n are the eigenenergy and eigenfunction associated with the n-th solution to the equation.

The simplest case is shown in Figure 1. In this 'infinite well' case, we presume for simplicity that the barriers on either side of the quantum well are infinitely high. Then the wavefunction must be zero at the walls of the quantum well.

The solution is then particularly simple:

$$E_n = -\frac{\hbar^2}{2m}\left[\frac{n\pi}{L_z}\right]^2\,, \qquad \phi_n = A\sin\left(\frac{n\pi z}{L_z}\right) \qquad n = 1, 2, ... \tag{2}$$

The energy levels (or 'confinement energies') are quadratically spaced, and the wavefunctions are sine waves. In this formula, the energy is referred to the energy of the bottom of the well. Note that the first allowed energy (corresponding to $n=1$) is above the bottom of the well.

We see that the energy level spacing becomes large for narrow wells (small L_z) and small effective mass m. The actual energy of the first allowed electron energy level in a typical 100Å GaAs quantum well is about 40meV, which is close to the value that would be calculated by this simple formula. This scale of energy is easily seen, even at room temperature.

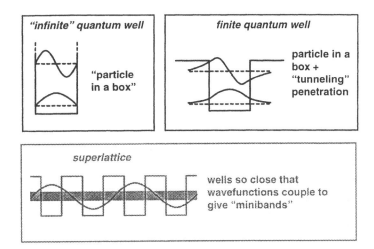

Figure 2. *Comparison of 'infinite' quantum well, 'finite' quantum well, and superlattice behaviour. For the superlattice, a wavefunction for one of the possible superlattice miniband states is shown (actually the state at the top of the miniband).*

The solution of the problem of an actual quantum well with finite height of barriers is a straightforward mathematical exercise. It does, however, require that we choose boundary conditions to match the solutions in the well and the barriers. One boundary condition is obvious, which is that the wavefunction must be continuous. Since the Schrödinger equation is a second order equation, we need a second boundary condition, and it is not actually obvious what it should be. We might think that we would choose continuity of the wavefunction derivative across the boundary; we cannot do so because the masses are different on the two sides of the boundary in general, and it can be shown that such a simple condition does not conserve particle flux across the boundary when the masses are different. One that does conserve particle flux is to choose $(1/m)d\phi/dz$ continuous; this is the most commonly used one, and gives answers that agree relatively well with experiment, but there is no fundamental justification for it. This lack of fundamental justification should not worry us too much, because we are dealing anyway with an approximation (the 'envelope function' approximation). If we were to use a proper first principles calculation, we would have no problem with boundary conditions on the actual wavefunction (Burt [4]). The solution of the finite well problem does not exist in closed form (Weisbuch [5]), requiring numerical solution of a simple equation to get the eigenenergies. The wavefunctions of the bound states are again sine waves inside the quantum well, and are exponentially decaying in the barriers. The energies are always somewhat lower than those we would calculate using the infinite well. It can be shown from the solution (Weisbuch [5]) that there is always at least one bound state of a finite quantum well. Figure 2 illustrates the differences between the idealised 'infinite' quantum well, the actual 'finite' well.

Also in Figure 2, we have illustrated a superlattice. It will be useful here to define the difference between quantum wells and superlattices. The simplest, crystallographic, definition of a superlattice is a 'lattice of lattices'. With that definition, any regular sequence of well and barrier layers would be a superlattice. A more useful definition here is the 'electronic' definition; in this definition, we call such a regular structure a superlattice only if there is significant wavefunction penetration between the adjacent wells. Otherwise, the physics of the multiple layer structure is essentially the same as a set of independent wells, and it is more useful to call the structure a multiple quantum well (MQW). If there is significant wavefunction penetration between the wells, we will see phenomena such as 'minibands', and the structure is then usefully described as a superlattice. The 'minibands' arise when quantum wells are put very close together in a regular way, just as 'bands' arise in crystalline materials as atoms are put together. Just as with quantum wells, simple models for superlattices can be constructed using envelope functions and effective masses, and such models are also good first approximations. As a rule of thumb, for well and barrier layers thicker than about 50Å in the GaAs/AlGaAs system, with a typical Al concentration of about 30% in the barriers, the structure will probably be best described as a multiple quantum well.

3   Linear optical properties of quantum wells

To understand the linear interband optical absorption in quantum wells, we will first neglect the so-called 'excitonic' effects. This is a useful first model conceptually, and explains some of the key features. For a full understanding, however, it is important to understand the excitonic effects. In contrast to bulk semiconductors, excitonic effects are very clear in quantum wells at room temperature, and have a significant influence on device performance.

3.1   Optical absorption neglecting excitons

The simplest model for absorption between the valence and conduction bands in a bulk semiconductor is to say that we can raise an electron from the valence band to a state of essentially the same momentum in the conduction band (a 'vertical' transition) by absorbing a photon. The state in the conduction band has to have essentially the same momentum because the photon has essentially no momentum on the scale usually of interest in semiconductors. In this simple model, we also presume that all such transitions have identical strength, although they will have different energies corresponding to the different energies for such vertical transitions. The optical absorption spectrum therefore has a form that follows directly from the density of states in energy, and in bulk (3D) semiconductors the result is an absorption edge that rises as the square root of energy, as shown in Figure 3.

In quantum wells, for the direction perpendicular to the layers, instead of momentum conservation we have a selection rule. The rule states that (to lowest order) only transitions between states of the same quantum number in the valence and conduction bands are allowed. This rule follows from the fact that the optical absorption strength is proportional to the overlap integral of the conduction and valence (envelope) wave functions. For sinusoidal standing waves, as shown for the infinite quantum well in Figure 2, there is only a finite overlap between identical standing waves. (This rule is somewhat weaker for finite wells, although it is still a very good starting point.) We can if we wish still view this as conservation of momentum, since we can regard the standing

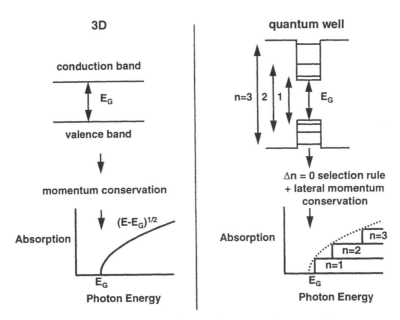

Figure 3. *Optical absorption in bulk (i.e. 3D) semiconductors and in quantum wells, in the simplest model where excitonic effects are neglected.*

waves as states of specific momenta. (The converse is also true, in that even in bulk semiconductors we can equally well view the momentum conservation rule as following from allowed overlap integrals, in that case between plane wave wavefunctions.)

In a quantum well, the electrons and holes are still free to move in the directions parallel to the layers; hence, we do not really have discrete energy states for electrons and holes in quantum wells; we have instead 'subbands' that start at the energies calculated for the confined states. The electron in a given confined state can in addition have any amount of kinetic energy for its in-plane motion in the quantum well, and so can have any energy greater than or equal to the simple confined-state energy for that subband. The density of states for motion in the plane of the quantum well layers turns out to be constant with energy, so the density of states for a given subband really is a 'step' that starts at the appropriate confinement energy. Optical transitions must still conserve momentum in this direction, and just as for bulk semiconductors, the optical absorption must still therefore follow the density of states. Hence, in this simple model, the optical absorption in a quantum well is a series of steps, with one step for each quantum number, n. It is easily shown, from the known densities of states, that the corners of the steps 'touch' the square root bulk absorption curve (when that curve is scaled to the thickness of this infinite quantum well). Thus, as we imagine increasing the quantum well thickness, we will make a smooth transition to the bulk behaviour, with the steps becoming increasingly close until they merge into the continuous absorption edge of the bulk material.

3.2   Consequences of heavy and light holes

As we mentioned above, there are two kinds of holes that are relevant here, corresponding to the heavy and light hole bands. Since these holes have different masses, there
are two sets of hole subbands, with different energy spacings. The light hole subbands,
because they have lighter mass, are spaced further apart. The consequence for optical
absorption is that there are actually two sets of 'steps'. The heavy-hole-to-conduction
set starts at a slightly lower energy and is more closely spaced than the light-hole-
to-conduction set. The heavy hole set is usually dominant when we look at optical
absorption for light propagating perpendicular to the quantum well layers.

If, however, we look in a waveguide, with light propagating along the quantum
well layers, there are two distinct optical polarisation directions: one with the optical
electric vector parallel to the quantum well layers (so-called 'transverse electric' or
TE polarisation); and the other with the optical electric vector perpendicular to the
quantum well layers (so-called 'transverse magnetic' or TM polarisation). The TE case
is essentially identical to the situation for light propagating perpendicular to the layers,
where the optical electric vector is always in the plane of the quantum wells; the optical
electric vector is always perpendicular to the direction of propagation for a plane wave.
The TM case is substantially different from the TE case, however.

Because of microscopic selection rules associated with the unit cell wavefunctions,
for the TM polarisation, the heavy-hole-to-conduction transitions are forbidden, and all
of the absorption strength goes over to the light-hole-to-conduction transitions. Hence,
at least with this simple model, there is now only one set of steps in the absorption.
The reason for loss of the heavy hole transitions is not a special property of quantum
wells; this kind of selection rule is a consequence of defining a definite symmetry axis
in the material, in this case the growth direction of the quantum well layers. Exactly
the same selection rule phenomenon will result if we apply a uniaxial stress to a bulk
semiconductor. One practical consequence of this selection rule effect is that quantum
well waveguide lasers essentially always run in TE polarisation; there are more heavy
holes than light holes in thermal equilibrium, and hence the gain associated with heavy
holes is larger, and hence the gain is larger for the TE polarisation. In general, this
simple classification of holes into 'heavy' and 'light' is only valid near the center of the
Brillouin zone. The detailed structure of the valence bands is particularly complicated
since the various subbands would actually appear to cross one another. In fact, such
crossing are avoided, and the resulting subband structure is quite involved. For many
devices, such effects are not very important (at least for a basic understanding of the
devices), and we will not discuss such valence band effects further here.

3.3   Optical absorption including excitons

Figure 4 shows an actual absorption spectrum of a quantum well sample, demonstrating
that the quantum well absorption is indeed a series of steps; simple calculations based
on the particle-in-a-box models will correctly give the approximate positions of the
steps. But it is also clear that there are sets of peaks in the spectra not predicted or
explained by the simple 'non-excitonic' model described above. These peaks are quite
a strong effect, and will be particularly important near the band-gap energy, here at

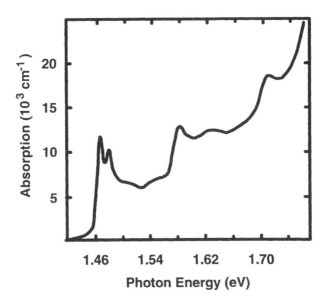

Figure 4. *Absorption spectrum of a typical GaAs/AlGaAs quantum well structure at room temperature.*

about 1.46eV photon energy. Most devices also operate in this region.

To understand these peaks, we need to introduce the concept of excitons. The key point missing in the previous discussion is that we have neglected the fact the electrons and holes are charged particles (negative and positive respectively) that attract each other. Hence, when we have an electron and a hole in a semiconductor, their wavefunctions are not plane waves; plane waves correspond to the case of uniform independent motion of the electron and the hole. Instead, we should expect that the electron and hole wavefunction should correspond to the case where the electron and hole are close to one another because of their Coulomb attraction. The formal error we made in the analysis above is therefore that we did not use the correct eigenfunctions for the electrons and holes, and hence we formally got an incorrect answer when we used those eigenfunctions to calculate the optical absorption.

Unfortunately, to include excitonic effects properly we need to use a different picture, since the whole band structure picture is a single particle picture; it essentially describes the energies seen by either a single electron or a single hole, but cannot handle both at once. Fortunately, there is a simple picture that allows us to understand the resulting excitonic effects.

The correct approach is not to consider raising an electron from the valence band to the conduction band, but rather to consider the creation of an electron-hole pair. In this picture, we find the eigenfunctions of the electron-hole pair in the crystal, and base our calculation of optical absorption on those pair states. First, then, we must understand what are the states of an electron-hole pair in a crystal. Fortunately, at least for the case where the attraction is not too strong, this problem is already solved; it is essentially the same problem as the states of the hydrogen atom, corrected for

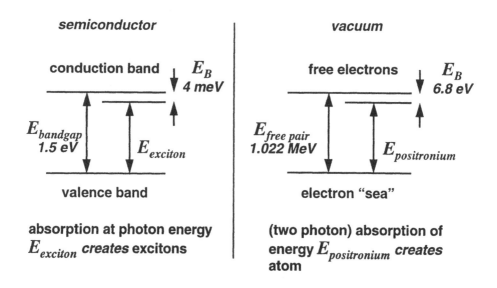

Figure 5. *Analogy between excitonic absorption and the optical creation of positronium atoms.*

the different effective masses and dielectric constants in the semiconductor. Using this model, we find, for example, the binding energy of lowest, 1S, exciton is

$$E_B = \frac{\mu e^4}{8h^2 \epsilon_R^2 \epsilon_0^2} \tag{3}$$

where h is Planck's constant, ϵ_R is the relative permittivity, ϵ_0 is the permittivity of free space, and μ is the reduced mass

$$\mu = \frac{m_e m_h}{m_e + m_h}. \tag{4}$$

For bulk GaAs (where excitons can be clearly seen at low temperature), $E_B \sim 4$ meV.

Hence, we find the first remarkable property of excitons compared to the non-excitonic model. It is possible to create an exciton with an energy E_B less than that required to create a 'free' electron-hole pair. A free electron-hole pair is analogous to an ionised hydrogen atom, and the energy required to create such a free pair in the semiconductor is actually the simple band-gap energy of the single particle picture. Hence we expect some optical absorption at photon energies just below the band-gap energy. This point is illustrated in Figure 5, on the left, showing a possible transition at an energy $E_{exciton} = E_{bandgap} - E_B$.

It is not only the absorption that creates the 1S exciton that is important, although under most conditions it is the only one that we see clearly as a distinct peak. In fact the entire absorption spectrum of these kinds of semiconductors is properly described in terms of the complete set of hydrogenic states. For example, the absorption above the band-gap energy results from the creation of excitons in the unbound hydrogenic states. In the classical sense, such states correspond to hyperbolic orbits. There is also additional absorption just below the bandgap energy from the other, bound excitonic states.

Note now that we are explaining the optical absorption in terms of the creation of a particle, the exciton (or, exactly equivalently, an electron-hole pair). It is important to understand that the absorption we see is not that associated with raising an existing hydrogenic particle to an excited state, as would be the case with conventional atomic absorption; we are instead creating the particle. An analogy that may help understand this distinction is the absorption in the vacuum that creates electron-positron pairs. This is illustrated in Figure 5, on the right.

By absorbing photons in the vacuum, we can in principle create positronium atoms, which are simply hydrogenic systems composed of an electron and a positron (instead of a proton). In this case, we are simultaneously creating a positron in the Fermi sea (the analogue of the valence band), as well as an electron. This happens to be a two-photon transition, and is therefore a rather weak effect in practice, but it illustrates the difference between the creation of an atom and the absorption between levels of the atom. A very important difference between these is that in the creation case, we have 'excitonic' absorption even when we have no excitons in the material. In the positronium case, it is very clear that there need be no positronium atoms there since we are starting with a perfect vacuum!

In the simple model where we neglected excitonic effects, the strength of a particular transition was determined by the square of the overlap integral between the 'initial' (valence subband) state and the 'final' (conduction subband) state. In principle, we have a similar result for the present excitonic model, although in this case the initial state is the 'empty' crystal, and the final state is the crystal with an exciton added. We will not formally derive the consequences of this change in model, but the net result is that, in the exciton creation case, the strength of the absorption to create an electron-hole pair in a given state is proportional to the probability that the resulting electron and hole will be in the place (strictly, the same unit cell). Now we can understand why the excitonic absorption gives such a strong peak for creation of 1S excitons. In the 1S exciton, the electron and hole are bound closely to one another (within a diameter of about 300Å in bulk GaAs), and the 1S wavefunction actually peaks at zero relative displacement. Hence the probability of finding the electron and the hole in the same place is actually very large, and so the resulting absorption is strong.

Thus far, we have discussed excitons in general, without explaining why they are particularly important in quantum wells. Excitonic effects are clear in many direct gap semiconductors at low temperature. At room temperature, however, although excitonic effects are still very important in determining the shape and strength of the interband absorption, the actual peaks corresponding to the bound states are difficult to resolve. The main reason for this lack of resolution is that the bound states are rapidly ionised by collisions with optical phonons at room temperature. In fact, they are typically ionised in times short compared with a classical orbit time. By a Heisenberg uncertainty principle argument, the associated optical linewidth is broadened, to such a point that the line is no longer clearly resolvable.

The quantum well offers two important differences compared to the bulk material, both of which stem from the confinement in the quantum well. Figure 6 shows, in a

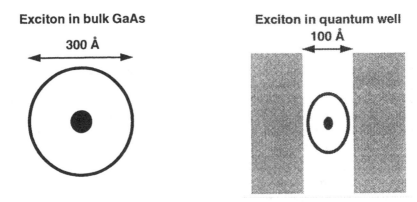

Figure 6. *Comparison of bulk and quantum well exciton sizes and shapes.*

semiclassical picture, an exciton in bulk GaAs and an exciton in a quantum well.

Hydrogenic theory for the GaAs exciton gives a diameter of about 300 Å, as mentioned above. When we create an exciton in a quantum well that is only 100Å thick, the exciton must become smaller, at least in the direction perpendicular to the quantum wells. Remarkably, however, it also becomes smaller in the other two directions (in the plane of the quantum well). This surprising conclusion can be checked, for example, by variational calculations. We can rationalise it by saying that nature prefers to keep the exciton more nearly spherical to minimise energy overall. If, for example, we allowed the exciton to become a flat 'pancake' shape, it would acquire high kinetic energies from the large second derivatives at the edges of the pancake. It is also the case that the exact solution of the two-dimensional hydrogen atom, corresponding to a very thin quantum well with very high walls, has a diameter one quarter that of the three-dimensional hydrogen atom. There are two consequences.

1. The electron and hole are even closer together than in the three-dimensional case, so the absorption strength to create such an exciton is even larger.

2. The exciton has a larger binding energy because the electron and hole are closer together, and hence it orbits 'faster'.

As a result of the faster orbiting, the exciton is able to complete a classical orbit before being destroyed by the optical phonon, and hence it remains a well-defined resonance. Equivalently, although the linewidth of the quantum well exciton is comparable to that of the bulk exciton, the binding energy is larger, and the peak is still well resolved from the onset of the 'interband' absorption at the band-gap energy. These two reasons explain both why the quantum well excitons are relatively stronger and also better resolved than the bulk excitons. The practical consequence is that in quantum wells we may be able to make some use of the remarkable excitonic peaks since we can see them at room temperature.

Note that in practice we frequently refer to the peaks in the spectrum as the exciton peaks or the exciton absorption peaks. It is important to remember always that these peaks represent the absorption that *creates* the (1S) exciton, and also that all of the rest of the interband absorption is also excitonic, although not to create the 1S state. Loosely, we often talk of 'the exciton' as meaning only the 1S exciton.

Incidentally, the two strong peaks that we typically see near the band-gap energy are the 1S exciton absorption peaks associated with the first heavy-hole to conduction transition (the stronger peak at lower photon energy) and with the first light-hole to conduction transition (the weaker peak at slightly higher photon energy).

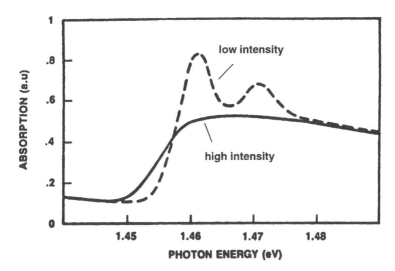

Figure 7. *Absorption spectrum without (dashed line) and with (solid line) optically created free carriers.*

4 Nonlinear optics in quantum wells

There are many possible nonlinear optical effects in quantum wells. Here we will discuss only one class of effects, namely those related to optical absorption saturation near to the band-gap energy. This particular set of effects has been strongly considered for various different kinds of devices, and is a serious candidate for applications in laser modelocking.

In the simplest case, we shine a laser beam on the material so that the resulting optical absorption generates a significant population of electrons and holes, either creating 'excitons' (*i.e.* electron-hole pairs initially in bound hydrogenic states) or 'free carriers' (*i.e.* electron-hole pairs initially in unbound hydrogenic states). Absorbing directly into the exciton (absorption) peaks will generate excitons, whereas absorbing at higher photon energies will generate free carriers. For steady state effects at room temperature with continuous laser beams, it makes little difference which we generate, since the excitons will ionise rapidly (*e.g.* in ~ 300 fs) and in thermal equilibrium we will essentially have only free carriers anyway. Figure 7 shows the effect on the absorption spectrum of generating a significant density (*e.g.* $\sim 10^{17}$ cm^{-3} free carrier density) in the quantum wells.

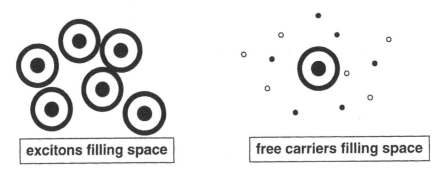

Figure 8. *Illustration of saturation by filling space with excitons (left) and filling space with free carriers. At a density of one exciton per exciton volume, or one 'cold' free carrier per exciton volume, it becomes essentially impossible to create any more excitons. This can be visualised geometrically in terms of the probability of trying to 'throw' one more exciton onto the plane without it landing on any of the other excitons or free carriers.*

As can be seen from Figure 7, with the presence of such a free carrier density, the exciton peaks have become saturated. The detailed physics of such phenomena is very complicated, and can only be analysed using many-body theory. We can, however, understand qualitatively various of the mechanisms (Chemla *et al.* [6], Schmitt-Rink *et al.* [7]). It is important in understanding these mechanisms to remember that the excitonic absorption peaks are associated with the *creation* of excitons. The simple absorption saturation mechanism that we would see in, for example, an atomic vapour, in which the absorption saturates when half of the atoms have been excited to their upper state, does not exist here; there is no density of atoms to start with since there are no excitons present and hence we cannot deduce any density of excited atoms at which saturation occurs by such an argument.

There is, however, one mechanism that is particularly easy to understand. Our physical intuition tells us that we cannot create two similar excitons in the same place; this is exactly like trying to create two atoms in the same place. Hence, as we begin to fill up space with excitons, we will start to run out of space to create more. Consequently, the probability of being able to create more excitons must reduce, and so the optical absorption associated with creating them must decrease. Therefore the exciton absorption line will saturate. This mechanism is illustrated on the left of Figure 8. For such a mechanism, we will get saturation with about one created exciton per exciton area, which works out to an exciton density of about 10^{17} cm^{-3}. This mechanism is known as 'phase-space filling'.

The real reason for this saturation mechanism is Pauli exclusion; we cannot have two electrons in the same state in the same place. It is also true that free carriers, if they are 'cold' enough to be in the states near the band center from which the exciton is comprised, can also prevent creation of more excitons, again by the Pauli exclusion principle. This is illustrated on the right of Figure 8.

Although we have discussed the saturation mechanisms so far in a qualitative and simplistic way, it turns out that these mechanisms are rigorously correct, with the only detail being the effective area of the exciton to use in the argument [7]. There are also other mechanisms that are somewhat less obvious, and some of these are of comparable size [7]. A second class of mechanisms that will change the absorption is screening effects. If we create a density of free carriers, they will tend to change the dielectric constant, and hence change the size of the exciton, typically increasing it. If the size of the exciton increases, then the probability of finding the electron and hole in the same place decreases; hence the optical absorption strength decreases, giving an effect that also behaves like saturation. This direct, classical Coulomb screening is thought to be weak in quantum wells because the walls of the wells prevent the movement of charge necessary for effective screening. There is, however, an even more subtle effect that is actually of comparable size to the other 'saturation' mechanisms, which is exchange screening. Essentially, when we include the effect of Pauli exclusion in our calculations, we find that the results we would calculate using our simple classical Coulomb effects are not correct at high density, because the Pauli exclusion forces the electrons to be further apart than we had thought (and similarly for the holes). This is described as if it were a screening effect, although it is actually a Pauli exclusion effect that causes us to correct our previous Coulomb calculation. This exchange screening also tends to increase the size of the exciton, reducing absorption.

Although in the steady state at room temperature, we see primarily the effect of free carriers on excitons, at lower temperatures or at short time scales we can see the effect of excitons on excitons. These latter effects are actually typically somewhat stronger (*e.g.* a factor of 1.5–2). Hence, if we initially create excitons and monitor the saturation of the exciton peaks, we can see a fast transient absorption saturation associated with the exciton-exciton effects, followed by a somewhat weaker saturation as the excitons ionise. This actually allows us to measure the exciton ionisation time, which is about 300fs at room temperature.

These saturation effects associated with the exciton peak in quantum wells are relatively sensitive. They have been explored for a variety of nonlinear optical switching devices. Even with the additional benefits of the strong quantum well excitons, these effects are not large enough to make low enough energy devices for current practical systems interest for information processing. The effects are important, however, for two reasons. One reason is that they set limits on the operating power of other devices, such as the modulators and electroabsorptive switches discussed below. Secondly, the effects are large enough for serious use as mode locking saturable absorber elements for lasers, and this is becoming practical now (Smith *et al.* [8], Keller *et al.* [9], Chen *et al.* [10]).

5   Quantum well electro-absorption physics

When electric fields are applied to quantum wells, their optical absorption spectrum near to the band-gap energy can be changed substantially (Miller *et al.* [11]), an effect we can call electroabsorption. Such effects have been extensively investigated for optical modulators and switches. There are two very distinct directions in which we can apply electric fields to quantum wells, either with the electric fields parallel to the layers or with the electric field perpendicular to the layers. The case of fields perpendicular to the layers is the one most peculiar to quantum wells, and it is called the Quantum-Confined Stark Effect (QCSE). Here, we will first discuss the case of fields parallel to the layers.

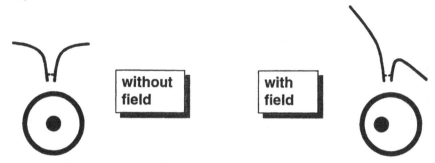

Figure 9. *Effect of field on quantum well excitons for fields parallel to the layers.*

5.1   Electric fields parallel to the layers

For the case of electric fields parallel to the layers, we get effects that are qualitatively similar to those seen in bulk semiconductors. The benefit of the quantum well here is that we can exploit the excitonic electroabsorptive effects at room temperature. The main effect we see is that the exciton absorption peak broadens with field.

Figure 9 illustrates qualitatively the Coulomb potential of the electron in the presence of the hole, and sketches how the exciton is deformed as the field is applied. In principle, when we apply an electric field to a hydrogenic system, we can shift the energy levels and the resulting transition energies, an effect known in atoms as the Stark effect. For a symmetrical state, such as the ground state of a hydrogenic system, the Stark effect can be viewed as the change in electrostatic energy caused by polarising the atom with the field, $\mathbf{E}$. The change in energy is therefore $-(1/2)\mathbf{P}\cdot\mathbf{E}$, where $\mathbf{P}$ is the induced polarisation. This corresponds to a reduction in the energy of the hydrogenic system. Hence, we might expect that as we apply an electric field to a semiconductor, we should see the 1S exciton absorption peak move to lower energies, because the energy of the resulting exciton we create is lower by this Stark shift.

It is true that there is such a Stark shift, but it is not the dominant effect that we see. The reason is that the Stark shift of a hydrogenic system is limited to about 10% of the binding energy. Since the binding energy of the exciton is only about 10meV, the shift is therefore limited to about 1meV, and hence it is not a very large effect. When we try to shift the energy by more than this, the hydrogenic system becomes field ionised (*i.e.* the electron and hole are 'ripped' apart by the field) so rapidly that the particle cannot complete even a substantial fraction of a classical orbit before being destroyed, and the whole concept of a bound state loses any useful meaning. In fact, what we see primarily as we apply the field is the broadening of the exciton absorption resonance caused by the shortening of the exciton's lifetime—again, a 'Heisenberg uncertainty principle' broadening. (It is just possible to see the shift at low fields in a carefully controlled experiment.)

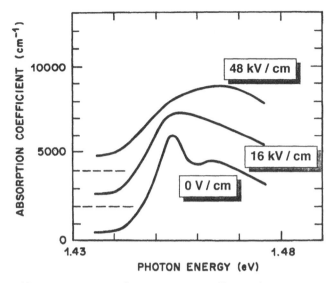

Figure 10. *Absorption spectra for a quantum well sample at room temperature with various electric fields applied in the plane of the quantum well layers. The spectra are shifted vertically for clarity. The exciton peaks broaden with field.*

It is worth noting that the fields we are capable of applying to the exciton are gigantic in a relative sense. Applying 1V / μm (= 10^4 V/cm) corresponds to a field of one binding energy ($\sim$ 10 meV) over one exciton diameter ($\sim$ 100Å). Such a field corresponds to a massive perturbation, certainly much larger than can readily be achieved with a hydrogen atom itself. It is also not surprising that such a field should cause the exciton to be field-ionised in less than a classical orbit time.

Figure 10 shows a typical set of spectra for parallel field electroabsorption in quantum wells. The broadening and disappearance of the peaks is clearly seen. One consequence is the appearance of a weak absorption tail at lower photon energies. The appearance of this tail is often referred to as the Franz-Keldysh effect. This is somewhat misleading, however, since the Franz-Keldysh effect is really a non-excitonic effect, whereas excitonic effects dominate near to the band-gap energy; the Franz-Keldysh effect, for example, cannot model the exciton broadening at all.

5.2 Electric fields perpendicular to the layers

The behaviour of the electroabsorption for electric fields perpendicular to the quantum well layers is quite distinct from that in bulk semiconductors. Figure 11 shows a typical set of spectra.

Here we can see that, instead of being broadened by the electric field, the exciton absorption peaks are strongly shifted by the field. The shifts can be tens of meV, and the applied fields here can be much larger than those shown above for the parallel field case, while still preserving the exciton peaks.

The reason for the difference in the electroabsorption in the perpendicular field case

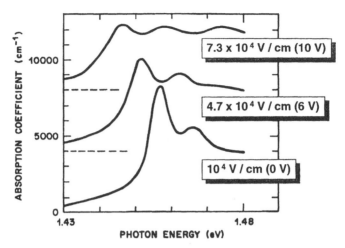

Figure 11. *Absorption spectra for electric fields applied perpendicular to the quantum well layers. The spectra are shifted vertically for clarity. The voltages correspond to those applied to a diode structure containing quantum wells.*

is a straightforward consequence of the quantum well. As we apply an electric field perpendicular to the layers, we pull the electron in an exciton in one direction (towards the positive electrode) and the hole in the other direction, just as we would expect. The difference is that the walls of the wells prevent the exciton from field ionising. Instead, the electron becomes squashed against one wall of the quantum well and the hole against the other, as illustrated in Figure 12 for the lowest electron and hole states ($n = 1$). Because the electron and hole are still relatively close to one another, they are still strongly attracted by their Coulomb attraction, and they still orbit round one another in the plane of the quantum wells, albeit in a somewhat displaced orbit. Hence the exciton can still exist as a particle for times longer than a classical orbit time, and the exciton absorption peak is not greatly broadened. Because the particle still exists even with very strong fields, we can obtain very large Stark shifts. In fact, the Stark shifts can be many times the binding energy (this is not unphysical since we are decreasing the energy, not increasing it). To see this effect, it is of course important that the quantum well is significantly smaller than the bulk exciton diameter; otherwise, the exciton can effectively be field-ionised simply by separating them by a sufficient distance within the well. For obvious reasons, this shift of the exciton absorption peaks is called the Quantum-Confined Stark Effect (QCSE). In principle, we could see similar effects with a hydrogen atom itself, but to do so we would need to confine the hydrogen atom within a distance less than 1Å, and apply fields of $\sim 10^{11}$ V/cm, both of which are currently impractical.

We can also see from Figure 11 that, as the exciton absorption peaks are shifted to lower energies, they do become weaker. The reason for this is that the electron and hole are being separated from one another by being pulled in opposite directions, and hence there is less probability of finding them in the same place. Consequently, the absorption strength decreases. It is worth mentioning here that, as absorption strength is lost in this way by the so-called 'allowed' transitions (*e.g.* the first hole state to

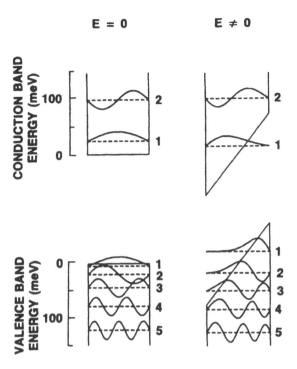

Figure 12. *Electron and hole wavefunctions for the first few states in an 'infinite' quantum well. Without field, the wavefunctions are sinusoidal; with field they are Airy functions.*

the first electron state), it is picked up by formerly 'forbidden' transitions (*e.g.* the second hole state to the first electron state) (Miller *et al.* [12]). With the applied field, the electron and hole wavefunctions in all of the levels are distorted; instead of being sinusoidal, they are Airy functions, as illustrated in Figure 12. In this case there tend to be finite overlap integrals between all possible states. The strengths of these various 'forbidden' transitions are bounded by sum rules—essentially, the electric field cannot change the total amount of absorption in the system.

We will not give details of the quantum mechanical calculation of the shifts of the peaks. It is, however, relatively straightforward. Although the above explanations for the reasons for the continued existence of the excitons are important for the effect, and the excitonic effects are also important for the strength of the absorption, the dominant part of the shift comes from the underlying shift of the single particle states. We can see this shift in Figure 12, where we see the $n = 1$ electron level moving down and the $n = 1$ hole level moving up; the net result is that the energy separation between the $n = 1$ hole and electron levels is reduced with field. The shift in the exciton binding energy itself is relatively unimportant by comparison (typically a few meV). The calculation of the shift of individual electron or hole levels with field can be done by various means. For a tilted well, the wave equation becomes Airy's differential equation, and hence the exact solutions (*e.g.* as in Figure 12) are Airy functions.

David A B Miller

It is also, incidentally, quite correct to view the quantum-confined Stark effect shifts as resulting from the polarisation of the exciton by the electric field, hence giving a $(1/2)\mathbf{P} \cdot \mathbf{E}$ shift in the exciton energy. To lowest order, the polarisation is proportional to the field (*i.e.* the induced separation of the electron and hole is initially linear with field), and so the QCSE is a quadratic effect to lowest order.

Although we have discussed effects here only for the case of simple 'rectangular' quantum wells, many other forms of quantum wells are also possible, such as coupled quantum wells, graded quantum wells, stepped quantum wells. These other structures also show related effects induced by applied electric fields, although we will not discuss these here. Also, superlattices show a class of effects induced by fields, known as Wannier-Stark localisation. This effect and effects in coupled quantum wells are closely related.

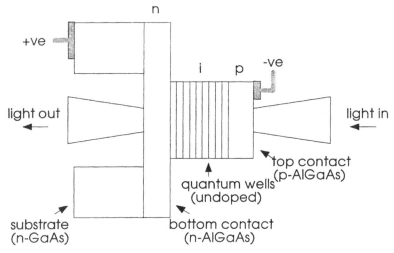

Figure 13. *Quantum well modulator diode structure.*

5.3   Quantum well electro-optic devices

The QCSE is particularly attractive for optical modulators. A simple device structure is shown in Figure 13. A quantum well region, typically containing 50 to 100 quantum wells, is grown as the undoped 'intrinsic' ('i') region in a p-i-n diode. This quantum well region will therefore have a thickness of about 1–2 microns. In operation, the diode is reverse biased. 1V across 1 micron corresponds to a field of 10^4 V/cm, so substantial QCSE shifts and changes in the absorption spectrum can be made with applied voltages of the order of 5–10V. The reverse biased diode is convenient because there is no conduction current that needs to flow in the diode in order to apply these relatively substantial fields.

The 'p' and 'n' regions of the diode are made out of a material that is substantially transparent at the wavelength of interest. In the case of GaAs quantum wells, they would most likely be made out of AlGaAs, the same material as used for the quantum well barriers. For the particular case of GaAs/AlGaAs quantum wells, the usual substrate material is GaAs. Unfortunately, GaAs is opaque at the wavelengths used with such modulators, hence the substrate has to be removed for a transmissive modulator, such as the one in Figure 13.

There are two very important features of the quantum well modulator. The first is that the electroabsorptive mechanisms (the QCSE) in the quantum well are strong enough to make a modulator that can work for light propagating *perpendicular* to the surface of the semiconductor. Despite the fact that this modulator will typically only be a few microns in thickness altogether, it can make changes in transmission of a factor of 2–3 in a single pass of a light beam. This modulation is large enough to make usable systems. This feature has the very important consequence that one can therefore make two-dimensional arrays of devices. This opens up many possibilities for novel highly-parallel optoelectronic systems.

A second key feature is that the operating energy of this modulator is small. The amount of energy required to change the optical properties in this modulator is essentially the stored electrostatic energy at the operating field. In other words, one has to charge up the capacitance of the device. The capacitance of a 1 micron square area of a semiconductor device 1 micron thick is approximately 10^{-16} F. Hence the total energy required to charge the device is $(1/2)$ $CV^2 \sim 1$–$2fJ/\mu m^2$. This energy density is comparable to the energy density in switching electronic devices, and is much smaller (*e.g.* by a factor of ~ 100) than the energy per unit optical area required to turn on a laser diode or saturate an optical absorption. Hence this device is very attractive as a potentially highly efficient optical output device for electronic circuits.

In addition, the speed limit on the use of the QCSE is probably in the range of ~ 100fs or less. Practical modulator devices will not be limited by the electroabsorption mechanism itself, seeing instead only the usual resistive/capacitive limits in applying voltage to the diodes. Because the diodes can be small, the capacitance can also be small, and hence these devices are attractive for high speed modulators.

The QCSE can also be used effectively for waveguide modulators (where the light propagates along the surface). In this case, a useful modulator can be made with only a few quantum wells because the propagation distance can be long (*e.g.* 100 microns). As a result, waveguide quantum well modulators can operate with low voltages (less than 1V). The change in absorption from the QCSE also results in changes in refractive index (through the Kramers-Krönig relations), and waveguide refractive modulators can also be made.

The Self Electro-optic Effect Device (SEED) principle is to combine a quantum well modulator (or a set of modulators) with a photodetector (or a set of photodetectors) to make an optically controlled device with an optical output (or outputs) (Miller *et al.* [13],[14]). A major reason for thinking about such a class of devices is an opportunity for efficient integration. Quantum well modulators driven directly by external electrical connections are limited in speed and operating energy by the external electrical parasitics, such as capacitance and the need to drive with impedance matched lines at high speeds. The modulators themselves can be driven with very low total energies if the drive is electrically local and not brought in by some external connection. Furthermore, since they are semiconductor devices, the prospects for integration with various optoelectronic and electronic devices are good.

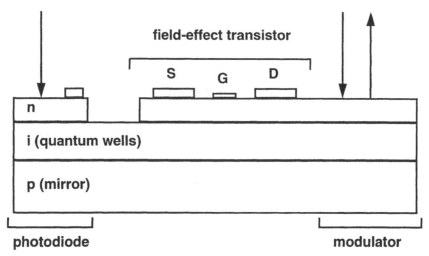

Figure 14. *Concept of the field-effect transistor* SEED *(*FET-SEED*). S - source; G - gate; D - drain.*

Although it is often the case that the conversion from optics to electronics and back to optics is inefficient and costly, this need not be the case if the devices are well integrated. In practice, the quantum well devices can be integrated effectively, and allow two-dimensional arrays of smart optoelectronic units or 'smart pixels'. Such devices offer new possibilities in information processing and switching architectures.

Various optical switching and linear analog devices are possible using combinations of quantum well diodes, without any transistors. These rely on positive or negative feedback; the generated photocurrent leads to a change in voltage across the quantum well diodes, which in turn changes the optical absorption and hence the photocurrent. The feedback can be positive or negative depending on whether the absorption decreases or increases with voltage respectively. Positive feedback can lead to bistable switching. The best-known such device is the symmetric-SEED, which consists of two quantum well diodes reverse-biased in series; this 'S-SEED' shows various digital (Lentine *et al.* [15]) and analog (Miller [16], de Souza *et al.* [17]) modes.

Quantum well diodes, operating as both photodetectors and modulators, can also be integrated with transistors. There are at least two reasons to do such integration. First, using transistors for electronic gain allows the optical energy requirements to be reduced. A second reason is that electronics is very good at performing complex logic functions, at least locally. Combining electronics with the abilities of optics for interconnection may allow the best of both worlds. Indeed there are many advantages of optics for interconnection that become apparent once a good integration technology is available (Miller [18]).

Figure 14 shows the concept of one integration, the field-effect transistor SEED (FET-SEED) (Miller *et al.* [19], D'Asaro *et al.* [20]). Here a quantum well diode is grown on top of a mirror as usual, in this case with the n-layer at the top. Then field effect transistors can be fabricated in the top layer. Hence this concept allows photodetectors for optical inputs, transistors for gain and logic, and quantum well modulators for optical outputs.

The present state of the art in this technology is that, in the laboratory, small circuits have been operated with 22fJ input optical energy, and at speeds up to 650Mb/s. In actual multistage systems, larger 'smart pixels' are operating with 100fJ optical input energy at 155Mb/s. A smart pixel array with 96 optical beams and 400 transistors was used in this system (Lentine *et al.* [21]). Such operating energies and speeds are of serious interest. For example, with 100fJ input energy and a factor of 10 loss overall in an optical system, a 1W laser has sufficient power to drive 1Tb/s of information through a system. This is a very large data rate, and one that is difficult to achieve with purely electronic systems for a number of reasons (mostly related fundamentally to the skin effect and Maxwell's equations).

Another approach to such integration is to combine quantum well devices with silicon electronics. There has been success in monolithic integration of quantum well diodes onto silicon (Goossen *et al.* [22]), and it seems likely that various other hybrid schemes might also be possible (Goossen *et al.* [23]).

6 Terahertz oscillations in coupled quantum wells

One area of current interest in optical physics of quantum wells is in measuring coherent quantum processes. A particular example of this is recent work on Terahertz oscillations excited by short optical pulses. In this work, we can create a quantum mechanical system in a particular linear superposition of states, including setting the quantum mechanical phases in the superposition. We can then watch the system evolve in time, and, importantly, we can measure the evolution of the system with a time resolution short compared to the time scale of the evolution.

One appropriate system for this work is a 'coupled' quantum well pair. It is well known, and easy to show, that a pair of quantum wells separated by a thin barrier will have two coupled states. The lower energy state of the pair has a symmetric combination of the two individual well states, and the higher energy state of the pair has an antisymmetric combination. The splitting, ΔE, between these two states becomes larger as we reduce the thickness of the barrier, and can readily be made of the order of 5–10meV in actual coupled wells. If we were to prepare this system so that there was one 'electron' in one of these wells initially, it would oscillate back and forward between the two wells with a frequency $\nu = \Delta E/h$. For ΔE's of 5–10meV, the corresponding frequencies are about 1.2–2.4THz. The system oscillates because it is not in an eigenstate, but is rather in the linear superposition of two eigenstates.

Because the electron is moving back and forth between the two wells, there is actually a current in this system at these frequencies (or equivalently an oscillating polarisation). As a result, this system will actually emit electromagnetic radiation in the Terahertz region. If we can detect the electric field of the Terahertz radiation (rather than merely its intensity), we will be able to measure the time evolution of this oscillating polarisation, including both amplitude and phase. Hence we can monitor the evolution of the phase of the quantum mechanical system, a rather unusual ability.

The actual sample structure for this experiment and a resulting Terahertz signal are shown in Figure 15. We excite the system with a short optical pulse whose frequency bandwidth covers both of the coupled states so that we excite a linear superposition

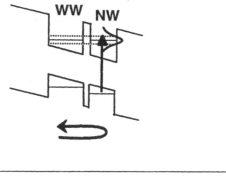

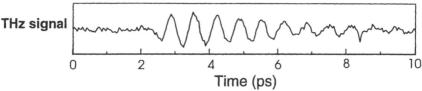

Figure 15. *Schematic diagram of a biased coupled quantum well (upper part of the figure) with both coupled states excited simultaneously by an appropriate short (and hence broad-band) optical pulse, resulting in oscillation as shown by the resulting detected Terahertz oscillation (lower part of the figure). This particular structure has a narrow well (NW) and a wide well (WW).*

of them. Note that the short optical pulse is essentially fully coherent—the frequency spectrum is not simply like a small slice out of the spectrum from a light bulb, but has a definite phase relation between the different frequency components; as a result, the two states are excited with a definite phase relation (actually the phase relation that puts the electrons initially in the right hand well). In this particular structure we have used an experimental trick. The two wells are not identical, though we have applied an electric field to 'line up' the lowest states in the two wells so we still have coupled wells. The reason for the asymmetry is so that the optical excitation only creates electrons in the right hand well. (In a symmetric structure, equal populations would be created in both wells, and there would be no net oscillation.) The Terahertz signal starts after the creation of the electrons in the right hand well, and shows the expected oscillation. The oscillation decays in time, in this case because of dephasing of the excitation from collisions with phonons, other particles, or imperfections.

The experimental apparatus to perform these experiments is shown in Figure 16. This apparatus is also capable of generating two optical pulses from the original laser pulse with a controllable delay set by the piezo translation stage. The excitation pulse (or pulses) are incident on a multiple (coupled) quantum well sample in a cryostat. As a result of this excitation, the sample emits some Terahertz electromagnetic radiation. This radiation is collimated by a parabolic mirror, sent to another parabolic mirror and focussed onto an antenna. The antenna is a classical dipole antenna. The instantaneous voltage from the antenna is sampled by a photoconductive sampling gate. This gate 'opens' for a short time after it is hit with a short optical gating pulse. The resulting charge that flows through the gate to a conventional electrical signal detection system is therefore proportional to the instantaneous Terahertz electric field. The entire Terahertz

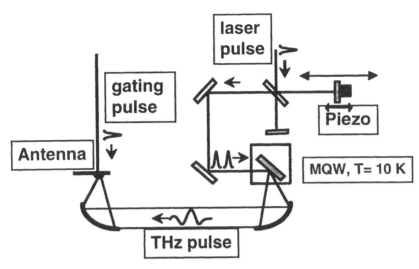

Figure 16. *Experimental setup for two pulse excitation of a quantum well sample with controlled phase between pulses, and subsequent measurement of the resulting Terahertz electric field pulse by gated photoconductive sampling of a receiving antenna.*

signal is traced out by repeating the experiment for progressively longer delays between the exciting laser pulse and the gating pulse. Such delays are controlled by simple linear translation of a mirror.

This apparatus also illustrates that such Terahertz experiments lie in an overlap region between various different regimes. The Terahertz signal is generated quantum mechanically, focussed optically and detected like a radio signal. The photon energy or frequencies involved are such that we can move between optics and radio techniques, and between quantum mechanical and classical descriptions. It is also true that in this regime we are at the boundary between classical electrical transport and coherent transport; the coupled well oscillation can be viewed as a coherent transport phenomenon, and it is also possible to observe normal classical transport on these time-scales. Finally, it is possible to view the Terahertz generation as a second order nonlinear optical processtechnically a difference frequency generation, and so we are also at the boundary between coherent transport and nonlinear optics.

When we extend these experiments to excitation by two closely spaced optical pulses, we get into a regime that exposes much more of the quantum mechanical nature of this process, and which at first sight is counter-intuitive. Figure 17 shows the resulting signals for excitation by two pulses when the time separation of the two pulses corresponds to two complete oscillation periods.

In such an experiment, we might imagine that the first pulses would set some charge oscillating, and the second pulse, being of essentially the same amplitude, would set up about the same amount again; given the carefully chosen time separation between the pulses, the second oscillation and the first would add, and we would get twice as much charge oscillating, and hence twice as much field generated. This is clearly not the case, as can be seen from the results. A key error in such an analysis is that we

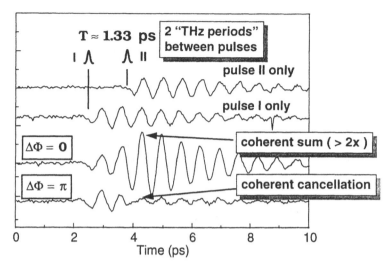

Figure 17. *Terahertz signals for excitation by two pulses separated by approximately 2 oscillation periods. The top two traces show the resulting signals for excitation by each of the two pulses separately. The bottom two traces show two cases for excitation with both pulses. One shows coherent addition of the two induced oscillations, the other shows coherent cancellation. Note that the difference between these two cases is only in the difference in the relative optical phase of the pulses.*

must add the wavefunctions first, then deduce the resulting observable. This has two kinds of consequences. First of all, if we indeed do manage to add the wavefunction created by the second pulse to the approximately equal oscillating wavefunction left over from the first pulse, we may indeed double the wavefunction amplitude that is oscillating; but the resulting Terahertz current or polarisation is proportional to the amount of charge oscillating, and the amount of charge oscillating is proportional to the *square* of the amplitude of the wavefunction. Hence the resulting Terahertz electric field could actually *quadruple*, not merely double. This kind of effect is seen in the third trace (from the top) of Figure 17. The second consequence results because we must add the complete wavefunctions, not merely the envelopes. We must remember that the actual wavefunction of an electron in a quantum well is not really just the slowly-varying envelope we have been discussing. The slowly varying envelope is just the modulation envelope multiplying the unit cell wavefunctions. If we change the optical phase of a pulse by half a cycle, we will change the phase of oscillation of the unit cell wavefunctions by half a cycle. This can mean that the second pulse, if it has the wrong optical phase, can actually completely cancel the wavefunction generated by the first pulse by coherent cancellation at the unit cell level. This phenomenon is illustrated in the fourth (bottom) trace of Figure 17; note that in going from the third to the fourth traces in Figure 17, all that has changed is the relative *optical* phase between the two pulses—a timing change of only 1.4fs!

The experiment discussed here is only an illustration of the many quantum mechanical effects that can be explored optically in quantum wells. Further details on this and other current work can be found in References [24], [25], [26], and [27].

7   Conclusions

The optical properties of quantum wells near to the optical band-gap energy have proved to be a fascinating laboratory for studying many novel physical mechanisms. In addition, several novel and practical devices have resulted that offer significantly new opportunities for optoelectronic systems. We can expect continued evolution of physics, devices, and novel information processing systems in the years to come.

Further Reading

- For an introduction to quantum well optical physics and devices, see [28].

- For a longer treatment of the physics, see [29] and [30].

- For an extensive discussion of quantum well optical physics see [31]

- For a discussion of band structure and states in quantum wells, see [32].

- For treatments of quantum well optoelectronic devices see [33], [14] and [34].

References

[1] A Y Cho, *J of Crystal Growth*, **111**, 1 (1991)

[2] K Furuya, and Y Miramoto, *Int J High Speed Electronics*, **1**, 347 (1990)

[3] T W Tsang, 1990, *J Crystal Growth*, **105**, 1 (1990)

[4] M G Burt, *J Phys: Condens Matter*, **4**, 6651 (1992)

[5] C Weisbuch, in *Semiconductors and Semimetals*, **24**, edited by R Dingle (Academic Press, New York) page 1

[6] D S Chemla, D A B Miller, P W Smith, A C Gossard, and W Wiegmann, *IEEE J Quantum Electron*, **QE-20**, 265 (1984)

[7] S Schmitt-Rink, D S Chemla, and D A B Miller, *Phys Rev B*, **32**, 6601 (1985)

[8] P W Smith, Y Silberberg, and D A B Miller D A B, *J Opt Soc Am B*, **2**, 1228 (1985)

[9] U Keller, G W 't Hooft, W H Knox, and J E Cunningham, *Opt Lett*, **16**, 1022 (1991)

[10] Y K Chen, M C Wu, T Tanbun-Ek, R A Logan, and M A Chin, *Appl Phys Lett*, **58**, 1253 (1991)

[11] D A B Miller, D S Chemla , T C Damen, A C Gossard, W Wiegmann, T H Wood, and C A Burrus, *Phys Rev B*, **32**, 1043 (1985)

[12] D A B Miller, J S Weiner, and D S Chemla, *IEEE J Quantum Electron*, **QE-22** 1816 (1986)

[13] D A B Miller, D S Chemla, T C Damen, T H Wood, C A Burrus, A C Gossard, and W Wiegmann, *IEEE J Quantum Electron*, **QE-21**, 1462 (1985)

[14] D A B Miller, *Optical and Quantum Electron*, **22**, S61 (1990)

[15] A L Lentine, H S Hinton, D A B Miller, J E Henry, J E Cunningham, and L M F Chirovsky, *IEEE J Quantum Electron*, **QE-25**, 1928 (1989)

[16] D A B Miller, *IEEE J Quantum Electron*, **QE-29**, 655 (1993)

[17] E A de Souza, L Carraresi, G D Boyd, and D A B Miller, *Opt Lett*, **18**, 974 (1993)

[18] D A B Miller, *Opt Lett*, **14**, 146 (1989)

[19] D A B Miller, M D Feuer, T Y Chang, S C Shunk, J E Henry, D J Burrows, and D S Chemla, *IEEE Phot Tech Lett*, **1**, 61 (1989)

[20] L A D'Asaro, L M F Chirovsky, E J Laskowski, S S Pei, T K Woodward, A L Lentine, R E Leibenguth, J W Focht, J Freund M, G G Guth, and L E Smith, *IEEE J Quantum Electron*, **QE-29**, 670 (1993)

[21] A L Lentine, F B McCormick, T J Cloonan, J M Sasian, R L Morrison, M G Beckman, S L Walker, M J Wojcik, S J Hinterlong, R J Crisci, R A Novotny, and H S Hinton, in *Conference on Lasers and Electro-optics, May 2-7, 1993, Baltimore, Maryland* Postdeadline Paper CPD24 (Optical Society of America, 1993)

[22] K W Goossen, G D Boyd, J E Cunningham, W Y Jan, D A B Miller, D S Chemla, and R M Lum, 1989, *IEEE Photonics Tech Lett*, **1**, 304 (1989)

[23] K W Goossen, J E Cunningham, and W Y Jan, *IEEE Photonics Tech Lett*, **5**, 776 (1993)

[24] H G Roskos, M C Nuss, J Shah, K Leo, D A B Miller, A M Fox, S Schmitt-Rink, and K Köhler, *Phys Rev Lett*, **68**, 2216 (1992)

[25] I Brener, P C M Planken, M C Nuss, L Pfeiffer, D E Leaird, and A M Weiner, *Appl Phys Lett*, **63**, 2213 (1993)

[26] M C Nuss, P C M Planken, I Brener, H G Roskos, M S C Luo, and S L Chuang, *Appl Phys B*, **58**, 249 (1994)

[27] I Brener, P C M Planken, M C Nuss, M S C Luo, S L Chuang, L Pfeiffer, D E Leaird, and A M Weiner, *J Opt Soc Am B* (special issue on 'Terahertz electromagnetic pulse generation, physics and applications')

[28] D A B Miller, *Optics and Photonics News*, **1**, 7 (1990)

[29] D A B Miller, D S Chemla, and S Schmitt-Rink, in *Optical Nonlinearities and Instabilities in Semiconductors* edited by H Haug (Academic Press, Boston, 1988)

[30] D S Chemla, S Schmitt-Rink, and D A B Miller, in *Optical Nonlinearities and Instabilities in Semiconductors* edited by H Haug (Academic Press, Boston, 1988)

[31] S Schmitt-Rink, D S Chemla, and D A B Miller, *Advances in Physics*, **38**, 89 (1989)

[32] G Bastard, *Wave mechanics applied to semiconductor heterostructures* (Les Editions de Physique, Les Ulis, France, 1988)

[33] D A B Miller, *Int J of High Speed Electronics*, **1**, 19 (1990)

[34] A L Lentine, and D A B Miller, *IEEE J Quantum Electron*, **29**, 655 (1993)

Simple quantum dynamics

Stig Stenholm

Research Institute for Theoretical Physics
University of Helsinki

1   Introduction

The technical development of short pulse laser devices has lead to a growth of the interest in time resolved processes governed by the nonrelativistic Schrödinger equation. Most conventional tests of quantum theory are based on energy calculations (eigenvalue problems) or steady state probability fluxes (scattering theory). It is thus of great fundamental interest that we are now in a position to excite and follow quantum time evolution in a range where the intricacies of Schrödinger dynamics plays a dominating role.

Zewail and his collaborators [1] have probed many molecular dynamics phenomena including elementary chemical reactions. Simple molecular systems like Na_2 have been investigated in pump-probe experiments by Gerber *et al.* [2] and in time and frequency resolved spontaneous emission by Walmsley *et al.* [3]. Laser cooling combined with optical trapping methods have created the field of cold collisions. This is a growing activity where theoretical and experimental tools are combined; Julienne *et al.* [4] provide a review. Semiconductor work is progressing rapidly as seen *e.g.* from the review by Göbel [5].

The new computer generations are able to handle large numbers of data fast and accurately. This is necessary in order to integrate the time dependent Schrödinger equation. Special integration methods like Crank-Nicholson or the split operator methods are required. With these methods, simple one-dimensional quantum problems can be solved for the dynamic time evolution, see [6].

The time and space dependent Schrödinger equation in one dimension is of the form

$$i\hbar\frac{\partial}{\partial t}\Psi(x,t) = -\frac{\hbar^2}{2M}\frac{\partial^2}{\partial x^2}\Psi(x,t) + V(x)\Psi(x,t).\tag{1}$$

If we introduce scaled variables of the form

$$\zeta = \frac{x}{R_c},\quad \tau = \frac{t}{T_c},\quad v(\zeta) = \frac{V(R_c\zeta)}{v_0},\tag{2}$$

where R_c, T_c and v_0 are the scaling parameters, we find the dimensionless Schrödinger equation

$$i\frac{\partial}{\partial\tau}\Psi(\zeta,\tau) = -\left(\frac{T_c}{R_c^2}\right)\frac{\hbar^2}{2M}\frac{\partial^2}{\partial\zeta^2}\Psi(\zeta,\tau) + v(\zeta)\left(\frac{v_0 T_c}{\hbar}\right)\Psi(\zeta,\tau).\tag{3}$$

If we now set all coefficients in Equation (3) equal to unity, we can fix the space and time scales into suitable ranges of the dimensionless parameters. Both molecular systems and semiconductor heterostructures have energies slightly below 1eV; if we set $v_0 = 0.5$eV, this determines T_c to be 1.3fs. Setting the coefficient in the kinetic energy of Equation (3) to unity determines the relation between the mass and the spatial scale. For a semiconductor electron effective mass of about 10% of the free mass, we obtain $R_c \approx 1$nm. For the mass of a light molecule, *i.e.* a few times ten proton masses, we obtain $R_c \approx 10^{-2}$Å. These energy, space and time scales are suitable for the physical situations we want to consider. Dimensionless energies are below unity, and the space and time ranges extend to about 10^3 dimensionless units. All these are convenient ranges for numerical work, and the calculational programs can be very similar both for molecular and semiconductor structures.

When quantum time evolution is followed, one has, of course, to include many relevant details of the physical systems. However, in many cases the main features can be understood from simple models and their combinations. The knowledge of solvable time dependent quantum problems is hence of interest. The peculiarities of quantum theory in the time domain and its main consequences can be understood from such models. In fitting realistic situations, these models form the framework around which one builds and develops the full theory. Once the basic physics is known, it becomes a computational problem to include the realistic features into the modelling. Then the availability of ever increasing computer resources allows the treatment of situations quite close to the laboratory experiment.

It is my aim to present a series of simple dynamical quantum systems in these lectures. Many of the results are old and well established, but their applications and actuality have greatly increased in recent years because of the tremendous progress in time resolved laser research. It is not my aim to provide detailed modelling of actual experiments, but I hope to offer a menu of simple models upon which you can build and expand in order to understand and describe the new experiments expected to emerge in this topical and rapidly growing field.

2 Time dependent Born-Oppenheimer dynamics

2.1 Setting up the problem

The usual Born-Oppenheimer approximation is presented as a time independent approach giving energy eigenvalues [7],[8]. It can, however, be extended to dynamical situations in a straightforward way. In this section we will formulate the theory for calculating the result of time resolved experiments in molecular systems.

We describe the molecule by a set of nuclear coordinates $R = \{R_1, R_2, ...\}$ and a set of electronic coordinates $r = \{r_1, r_2, ...\}$. The position of the molecule in space is determined by its center of mass R_0 (3 parameters) and its orientation in space Ω. This involves 3 angles except in the case of linear molecules. For N nuclei, there remains $3N-6$ free nuclear degrees of freedom ($3N-5$ for linear molecules). The number of electronic degrees of freedom depends on the number of valence electrons of each nucleus. In practice, the center-of-mass position R_0 can be calculated with neglect of the electrons. In the following we will not indicate the vector nature of the position variables; it is assumed to be understood.

The kinetic energy of a nucleus of mass M will be

$$T_N = -\frac{\hbar^2}{2M} \frac{\partial^2}{\partial R^2}. \tag{4}$$

This will be the generic notation for the nuclear kinetic energy; the various nuclear masses will not be labelled explicitly. The electronic kinetic energy will similarly be written

$$T_e = -\frac{\hbar^2}{2m} \frac{\partial^2}{\partial r^2}. \tag{5}$$

The Coulomb potential between the electrons will be denoted by $V_{ee}(r)$, that between nuclei by $V_{NN}(R)$, and the electron-nuclei interaction by $V_{eN}(r, R)$. The potentials depend on the particle coordinates as indicated. The molecular system is situated in the external radiation field $\mathbf{E}(R, T) = \hat{\boldsymbol{\varepsilon}} \, E(R, t)$, where $\hat{\boldsymbol{\varepsilon}}$ is a polarisation vector. The coupling is linear in the external field and we write

$$V_{int} = -\mathbf{D} \cdot \hat{\boldsymbol{\varepsilon}} \, E(R, t). \tag{6}$$

We next define the electronic Hamiltonian

$$H_e(r, R) = T_e + V_{ee}(r) + V_{NN}(R) + V_{eN}(r, R), \tag{7}$$

and its corresponding eigenstates and eigenvalues

$$H_e(r, R)|\varphi_n\rangle = \mathcal{U}_n(R)|\varphi_n\rangle. \tag{8}$$

Here the nuclear positions occur purely parametrically. The energy eigenvalues $\mathcal{U}_n(R)$ are the adiabatic potentials where the molecular dynamical motion takes place (see Figure 1).

The eigenstates $|\varphi_n\rangle$ are solutions of a Hermitian eigenvalue problem, and they thus form a complete orthonormal set

$$\langle \varphi_m|\varphi_n\rangle = \delta_{nm}. \tag{9}$$

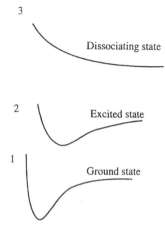

Figure 1. *Typical adiabatic potentials of molecular dynamics.*

Here we use these states as *a basis* to expand in. For the time dependent state of the total system we write

$$|\Psi(t)\rangle = \sum_n \psi_n(R,t)|\varphi_n\rangle, \tag{10}$$

This representation is still *exact*. The coefficients $\psi_n(R,t)$ can be regarded as the wave functions for the nuclear motion. When the expansion (10) is inserted in the time dependent Schrödinger equation

$$H\,|\Psi(t)\rangle = i\hbar \sum_n \dot{\psi}_n(R,t)\,|\varphi_n\rangle, \tag{11}$$

we obtain

$$\sum_n [T_N + \mathcal{U}_n(R)]\,\psi_n(R,t)|\varphi_n\rangle - \sum_n \mathbf{D}\cdot\hat{\varepsilon}\ E(R,t)\psi_n(R,t)\,|\varphi_n\rangle = i\hbar \sum_n \dot{\psi}_n(R,t)|\varphi_n\rangle. \tag{12}$$

If we now neglect the action of T_N on the electronic state $|\varphi_n\rangle$, and project Equation (12) on the state $|\varphi_m\rangle$, we find the coupled set of equations

$$[T_N + \mathcal{U}_m(R)]\psi_m(R,t) - \sum_n \langle\varphi_m|\mathbf{D}\cdot\hat{\varepsilon}\,|\varphi_n\rangle E(R,t)\psi_n(R,t) = i\hbar \frac{\partial}{\partial t}\psi_m(R,t). \tag{13}$$

The influence of the neglected terms will be discussed in Section 2.2. In the spirit of the dipole approximation, we may evaluate the electric field strength $E(R,t)$ at the molecular center-of-mass R_0. As an illustration we look at the following special case.

2.1.1 Two-atomic molecule

In this case we have only one nuclear coordinate and

$$T_N = -\frac{\hbar^2}{2M}\left[\frac{1}{R^2}\frac{\partial}{\partial R}\left(R^2\frac{\partial}{\partial R}\right) + \frac{\hat{L}^2}{R^2}\right], \tag{14}$$

where the rotational kinetic energy is in the operator

$$\hat{L}^2 = \frac{1}{\sin\theta}\frac{\partial}{\partial\theta}\left(\sin\theta\frac{\partial}{\partial\theta}\right) + \frac{1}{\sin^2\theta}\frac{\partial^2}{\partial\varphi^2}. \tag{15}$$

Its eigenstates are the angular momentum eigenfunctions, and introducing the ansatz

$$\chi_n(R,\theta,\varphi) = \psi_n(R,t)\frac{Y_{L,M}(\theta,\varphi)}{R} \tag{16}$$

into the Schrödinger equation, we obtain

$$\left[-\frac{\hbar^2}{2M}\left(\frac{\partial^2}{\partial R^2} - \frac{L(L+1)}{R^2}\right) + \mathcal{U}_n(R)\right]\psi_n(R,t) = i\hbar\frac{\partial}{\partial t}\psi_n(R,t). \tag{17}$$

In this case the problem is reduced to a genuinely one-dimensional one. For no rotation, $L = 0$, and the potential is the electronic adiabatic eigenvalue.

2.1.2 The two-level system

We look at a combination of two electronic energy levels denoted by labels 1 and 2. We take a monochromatic field of the form

$$E(t) = \mathcal{E}_0 \cos(\omega t). \tag{18}$$

The state vector is written in the form of an ansatz

$$|\psi\rangle = \chi_1(R,t)|\varphi_1\rangle + e^{-i\omega t}\chi_2(R,t)|\varphi_2\rangle. \tag{19}$$

With the Equation (13) written in the form

$$i\hbar\frac{\partial}{\partial t}\begin{bmatrix}\chi_1 \\ e^{-i\omega t}\chi_2\end{bmatrix} = \begin{bmatrix}T_N + \mathcal{U}_1 & \mu_{12}\mathcal{E}_0\cos(\omega t) \\ \mu_{12}\mathcal{E}_0\cos(\omega t) & T_N + \mathcal{U}_2\end{bmatrix}\begin{bmatrix}\chi_1 \\ e^{-i\omega t}\chi_2\end{bmatrix} \tag{20}$$

we can apply the rotating wave approximation

$$e^{\pm i\omega t}\cos(\omega t) = \frac{1}{2} \tag{21}$$

to obtain

$$i\hbar\frac{\partial}{\partial t}\begin{bmatrix}\chi_1 \\ \chi_2\end{bmatrix} = \begin{bmatrix}T_N + \mathcal{U}_1 & V \\ V & T_N + \mathcal{U}_2 - \hbar\omega\end{bmatrix}\begin{bmatrix}\chi_1 \\ \chi_2\end{bmatrix}. \tag{22}$$

Here the coupling is written

$$V = \frac{1}{2}\mu_{12}\mathcal{E}_0. \tag{23}$$

This may still contain a variation with time slow on the scale of ω^{-1}. The energy level 2 is seen shifted by one photon energy $\hbar\omega$ (see Figure 2). The Schrödinger equation (22) is the one integrated to describe the behaviour of wave packets on molecular energy levels [6].

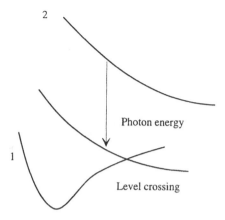

Figure 2. *One photon energy shift of the level 2.*

The two-level potential matrix

$$\mathbf{V} = \begin{bmatrix} \mathcal{U}_1(R) & V(R) \\ V(R) & \mathcal{U}_2(R) \end{bmatrix}$$ (24)

can be diagonalised by the unitary transformation

$$UVU^\dagger = \begin{bmatrix} D_1 & 0 \\ 0 & D_2 \end{bmatrix}$$ (25)

with the transformed state vector

$$|\varphi\rangle = U|\psi\rangle$$ (26)

satisfying the equation

$$i\hbar\frac{\partial}{\partial t}|\varphi\rangle = -\frac{\hbar^2}{2M}\left(U\frac{\partial^2}{\partial R^2}U^\dagger\right)|\varphi\rangle + \begin{bmatrix} D_1 & 0 \\ 0 & D_2 \end{bmatrix}|\varphi\rangle.$$ (27)

The derivative does not commute with the unitary transformation but

$$U\frac{\partial}{\partial R}U^\dagger = \frac{\partial}{\partial R} + U\frac{\partial U^\dagger}{\partial R}.$$ (28)

This makes the theory look like a gauge theory, where the coupling resides in the kinetic energy

$$i\hbar\frac{\partial}{\partial t}|\varphi\rangle = \frac{1}{2M}\left(p + i\hbar U\frac{\partial U^\dagger}{\partial R})\right)^2|\varphi\rangle + \begin{bmatrix} D_1 & 0 \\ 0 & D_2 \end{bmatrix}|\varphi\rangle.$$ (29)

For slowly varying potentials, the coupling term can be neglected and the motion can be described on the decoupled adiabatic energy surfaces.

2.1.3 Mechanical effects of light

In laser cooling and atomic optics experiments [9], [10] we need to describe the quantum motion of a few-level system under the influence of photon momentum. Then the center-of-mass variable R_0 becomes a genuine dynamical variable, and we introduce the kinetic energy corresponding to this by setting

$$T_{cm} = -\frac{\hbar^2}{2M_0} \frac{\partial^2}{\partial R_0^2}. \tag{30}$$

The intramolecular potentials are supposed constant, and we set

$$\mathcal{U}_1 = E_1, \quad \mathcal{U}_2 = E_2. \tag{31}$$

The dependence of the impinging field on the center-of-mass coordinate has to be included in the coupling V; for a standing wave field this is of the form $\cos(qR_0)$. We obtain the Schrödinger problem

$$i\hbar \frac{\partial}{\partial t} \begin{bmatrix} \chi_1(R_0) \\ \chi_2(R_0) \end{bmatrix} = \begin{bmatrix} T_{cm} + E_1 & V(R_0) \\ V(R_0) & T_{cm} + E_2 - \hbar\omega \end{bmatrix} \begin{bmatrix} \chi_1(R_0) \\ \chi_2(R_0) \end{bmatrix}. \tag{32}$$

Using this equation, one can approach many problems in the theory of light pressure experiments.

2.2 Discussion of validity

The terms neglected in getting from Equation (12) to (13) derive from neglecting the last two terms in the derivative

$$\frac{\partial^2}{\partial R^2} [\psi_n(R,t)|\varphi_n\rangle] = |\varphi_n\rangle \frac{\partial^2}{\partial R^2} \psi_n(R,t) + 2\frac{\partial}{\partial R}|\varphi_n\rangle \frac{\partial}{\partial R}\psi_n(R,t) + \psi_n(R,t)\frac{\partial^2}{\partial R^2}|\varphi_n\rangle \tag{33}$$

In order to estimate the magnitude of the derivative $\partial/\partial R$ we apply it to Equation (8) and obtain

$$\left(\frac{\partial H_e}{\partial R}\right)|\varphi_n\rangle + H_e\frac{\partial}{\partial R}|\varphi_n\rangle = |\varphi_n\rangle\frac{\partial \mathcal{U}_n}{\partial R} + \mathcal{U}_n\frac{\partial}{\partial R}|\varphi_n\rangle. \tag{34}$$

Taking the diagonal element of this relation and using Equation (8) we obtain

$$\frac{\partial \mathcal{U}_n(R)}{\partial R} = \langle\varphi_n|\left(\frac{\partial H_e}{\partial R}\right)|\varphi_n\rangle. \tag{35}$$

This relation is called the Hellmann-Feynman theorem, and it allows one to calculate the molecular force constants directly in terms of derivatives with respect to the electronic Hamiltonian.

Taking the off-diagonal terms of Equation (33) and using the orthogonality of the electronic states we obtain

$$\langle\varphi_m|\frac{\partial}{\partial R}|\varphi_n\rangle = \left[\mathcal{U}_n(R) - \mathcal{U}_m(R)\right]^{-1} \langle\varphi_m|\left(\frac{\partial H_e}{\partial R}\right)|\varphi_n\rangle. \tag{36}$$

Thus only at intramolecular level crossings, *i.e.* points where $\mathcal{U}_n(R) = \mathcal{U}_m(R)$ for $m \neq n$, is the contribution from the derivatives large. At these points of degeneracy, we must perform a diagonalisation eliminating the crossing, but as long as we consider electronic energy levels coupled through optical photons, we can safely neglect terms of the type (36).

In the projection of (12) on (13) we encounter also diagonal matrix elements. These can be obtained from the condition that the electronic state normalisation should stay invariant under a small displacement of the nuclear coordinates

$$\langle \varphi_n(R + \delta R) | \varphi_n(R + \delta R) \rangle = \langle \varphi_n(R) | \varphi_n(R) \rangle + 2\mathrm{Re}\left(\langle \varphi_n(R) | \frac{\partial}{\partial R} | \varphi_n(R) \rangle \right) \delta R. \quad (37)$$

For nondegenerate energy states, the eigenfunctions can be chosen real, and the matrix element multiplying δR must vanish exactly. Thus we can safely neglect the second term on the right-hand side of Equation (32). The off-diagonal elements of the second term constitute a coupling between the electronic energy levels. Because we are interested in levels coupled through optical photons, the static intramolecular coupling has only a minute effect, it is too far off resonance. If we wish, we can add its diagonal elements to the molecular Hamiltonian determining the time evolution in Equation (13). This gives

$$\Delta\mathcal{U}_n = -\frac{\hbar^2}{2M} \langle \varphi_n(R) | \frac{\partial^2}{\partial R^2} | \varphi_n(R) \rangle. \quad (38)$$

The derivative of the nuclear configuration R is expected to be of the same order of magnitude as that with respect to the electronic configuration r, the effect should be the same whichever charge we move. Consequently we can estimate the correction (38) to be of the magnitude

$$\langle \Delta\mathcal{U}_n \rangle \approx \frac{m}{M} \langle T_e \rangle. \quad (39)$$

Thus the correction is about 3 to 4 orders of magnitude smaller than the energy scale of the electrons. The typical molecular vibrational energy is of the order

$$\hbar\Omega \sim \sqrt{\frac{m}{M}} \langle T_e \rangle, \quad (40)$$

and the correction (39) is negligible.

We have discussed all the approximations necessary to neglect the static corrections to the Born-Oppenheimer approximation. These are all found to be negligible, when we look at optically coupled levels. We have, however, not considered the effect of the external coupling field. For very strong and very fast fields, the approximations may no longer hold. No detailed investigations about the limitations imposed by these factors have been carried out.

2.3   The Franck-Condon treatment

We can develop the theory one step further. We may introduce the vibrational eigenfunctions on the adiabatic energy surfaces by the eigenvalue equations

$$(T_N + \mathcal{U}_m(R))\, u_\nu^{(m)}(R) = \Omega_\nu^{(m)} u_\nu^{(m)}(R). \quad (41)$$

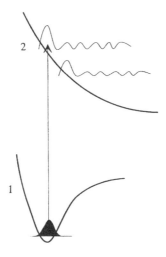

Figure 3. *The Franck-Condon factor.*

The vibrational eigenfunctions $\{u_\nu^{(m)}(R)\}$ form a complete orthonormal set for each m; for different values of m they have no simple relationships.

Next we expand the nuclear wave functions in the eigenfunctions according to

$$\psi_m(R,t) = \sum_\nu C_\nu^{(m)}(t)u_\nu^{(m)}(R).\tag{42}$$

We want to insert this expansion into the Schrödinger equation (13). Following Condon, we assume that the matrix element $\langle \varphi_m|\mathbf{D}\cdot\hat{\varepsilon}\,|\varphi_n\rangle$ does not depend on the nuclear coordinates R, this is a drastic approximation whose validity requires further attention. With this assumption, the Schrödinger equation becomes

$$i\hbar\frac{\partial}{\partial t}C_\xi^{(m)}(t) = \Omega_\nu^{(m)}C_\xi^{(m)}(t) + \sum_{\nu,n}\mu_{nm}\mathcal{E}_0\left(u_\xi^{(m)}|u_\nu^{(n)}\right)C_\nu^{(n)}(t),\tag{43}$$

where the overlap element $(u_\xi^{(m)}|u_\nu^{(n)})$ is the Franck-Condon factor determining the coupling between the levels (see Figure 3).

The Franck-Condon coupling usually requires good knowledge of the detailed structure of the energy surfaces. There are a few models where we can evaluate them explicitly [11]; we discuss them in the following subsection.

2.3.1 Harmonic oscillator models

We start with the case where the excited state is only a displaced replica of the ground state potential (see Figure 4(a)). We have

$$\begin{aligned}H_1 &= \frac{P^2}{2M} + \frac{1}{2}M\Omega_1^2 R^2 \\ H_2 &= \Delta E + \frac{P^2}{2M} + \frac{1}{2}M\Omega_1^2(R-R_0)^2\,.\end{aligned}\tag{44}$$

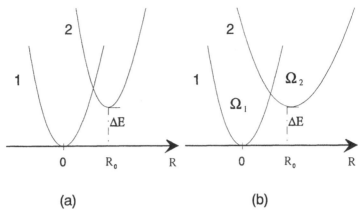

Figure 4. *Harmonic oscillator models.*

Using the harmonic oscillator creation operator $a^\dagger$ and introducing the unitary transformation

$$D(\xi) = \exp\left[-\xi(a - a^\dagger)\right] = \exp\left(-i\frac{R_0 P}{\hbar}\right) \tag{45}$$

we find

$$D(\xi)RD^\dagger(\xi) = R - R_0, \tag{46}$$

when

$$\xi = R_0 \sqrt{\frac{M\Omega_1}{2\hbar}} = \frac{R_0}{\sqrt{2}A_1}, \tag{47}$$

where A_i is the extension of the ground state wave function on level i.

Starting from the eigenvalue equation

$$H_1|u_\nu^{(1)}\rangle = \hbar\Omega_1\nu|u_\nu^{(1)}\rangle, \tag{48}$$

we obtain the relation

$$D(\xi)H_1 D^\dagger(\xi)|u_\nu^{(2)}\rangle = \hbar\Omega_1\nu|u_\nu^{(2)}\rangle = (H_2 - \Delta E)|u_\nu^{(2)}\rangle. \tag{49}$$

We have thus obtained the eigenfunctions for the vibrations on level 2 in the form

$$|u_\nu^{(2)}\rangle = D(\xi)|u_\nu^{(1)}\rangle, \tag{50}$$

and the Franck-Condon factors are now

$$|\langle u_\nu^{(2)}|u_\mu^{(1)}\rangle|^2 = |\langle u_\nu^{(1)}|D(\xi)|u_\mu^{(1)}\rangle|^2. \tag{51}$$

These states are expressible in terms of coherent states; if *e.g.* one state is the ground state we obtain the Poisson distribution

$$|\langle u_\nu^{(2)}|u_0^{(1)}\rangle|^2 = \frac{(R_0/A_1)^{2\nu}}{2^\nu \nu!}\exp\left(-\frac{R_0^2}{2A_1^2}\right). \tag{52}$$

We now move to consider the more general model where the excited state oscillator has the frequency Ω_2 different from that of oscillator 1 (see Figure 4(b)),

$$H_2 = \Delta E + \frac{P^2}{2M} + \frac{1}{2}M\Omega_2^2(R - R_0)^2 \,. \tag{53}$$

In order to describe this situation we introduce the squeeze operator

$$S(r) = \exp\left[-\frac{r}{2}(a^{\dagger 2} - a^2)\right] \,. \tag{54}$$

This transforms the dynamical variables according to

$$\begin{aligned}
S(r)RS^{\dagger}(r) &= e^r R = sR \\
S(r)PS^{\dagger}(r) &= e^{-r}P = s^{-1}P;
\end{aligned} \tag{55}$$

where $s = e^r$.

From Equation (48) we obtain directly

$$s^2 S(r) H_1 S^{\dagger}(r) S(r)|u_\nu^{(1)}\rangle = \hbar\Omega_1 s^2 \nu S(r)|u_\nu^{(1)}\rangle \tag{56}$$

and

$$\left(\frac{P^2}{2M} + \frac{1}{2}M\Omega_1^2 s^2 R^2\right)|\tilde{u}_\nu^{(2)}\rangle = \hbar\Omega_1 s^2 \nu |\tilde{u}_\nu^{(2)}\rangle, \tag{57}$$

which has got the same oscillation frequency as the Hamiltonian (53) if

$$r = \frac{1}{2}\log\left(\frac{\Omega_2}{\Omega_1}\right) \,. \tag{58}$$

We can now displace this Hamiltonian by the distance R_0 using the transformation (46). The new eigenfunctions are then given by

$$|u_\nu^{(2)}\rangle = D(\xi)S(r)|u_\nu^{(1)}\rangle \,. \tag{59}$$

Just as in Equation (51), this can be used to evaluate the Franck-Condon factor. For example, from the ground state of the excited level

$$\langle R|u_0^{(2)}\rangle = \left(\frac{s^2}{\pi A_2^2}\right)\exp\left[-\frac{s^2}{2}\left(\frac{R - R_0}{A_2}\right)^2\right] \tag{60}$$

the coupling to the various ground state levels is proportional to

$$\begin{aligned}
|\langle u_\nu^{(1)}||u_0^{(2)}\rangle|^2 &= |\langle\nu|D(\xi)S(r)|0\rangle|^2 \\
&= \frac{2s}{s^2+1}\left(\frac{s^2-1}{s^2+1}\right)^2\frac{1}{2^\nu\nu!}H_\nu^2\left[\frac{s^2}{s^4-1}\left(\frac{R_0}{A_2}\right)\right]\exp\left[\frac{-s^2}{s^2+1}\left(\frac{R_0}{A_2}\right)^2\right] \tag{61}
\end{aligned}$$

where $H_\nu(x)$ is the νth Hermite polynomial, see Reference [12].

Using the results like (52) and (61) many conclusions about the coupling between the two levels can be drawn. In many cases, on the other hand, it is more advantageous to integrate the coupled channel Schrödinger equations (22) directly.

3   Wigner representation of quantum dynamics

3.1   Representing the density matrix

In this section we are going to look at the dynamics of a quantum system evolving in a simple potential function $V(x)$, where as before x stands for the position variable in a $3N$ dimensional configuration space. The dimensionality of the space will not be indicated explicitly, but it is understood to be implied by the notation. The Schrödinger equation of the system is given by

$$i\hbar\frac{\partial}{dt}\psi(x) = H\,\psi(x),\tag{62}$$

where the Hamiltonian is of the type

$$H = \frac{p^2}{2M} + V(x)\,.\tag{63}$$

For an ensemble of state vectors $\{\psi^i(x)\}$ we can define the density matrix

$$\rho = \sum_i |\psi^i\rangle\langle\psi^i|\,.\tag{64}$$

The states making up the ensemble need not be orthogonal; consequently it is not possible to extract the ensemble uniquely from the density matrix.

For a pure state, i.e. one consisting of one single state vector only, the density matrix is a projector

$$\rho^2 = \rho,\tag{65}$$

and its trace is unity. It is easy to see from Equation (64) that for mixed states

$$\mathrm{Tr}\left(\rho^2\right) \le 1\,.\tag{66}$$

From Equation (62) it follows that the density matrix evolves in the Schrödinger picture according to the equation

$$i\hbar\frac{\partial}{dt}\rho = [H,\rho]\,.\tag{67}$$

The density matrix is useful to describe the evolution of a full ensemble, it is bilinear in the quantum states and relates directly to the observable expectation values of the system, and it allows the description of an open system with dissipation due to interactions with the environment.

For a free particle the time evolution equation becomes in the position representation

$$\frac{\partial}{\partial t}\langle x_1|\rho|x_2\rangle = \frac{i\hbar}{2M}\left(\frac{\partial^2}{\partial x_1^2} - \frac{\partial^2}{\partial x_2^2}\right)\langle x_1|\rho|x_2\rangle = \frac{i\hbar}{2M}\frac{\partial^2}{\partial r\partial R}\rho(R,r),\tag{68}$$

where we have introduced the two new variables

$$r = x_1 - x_2,\qquad R = \frac{1}{2}\left(x_1 + x_2\right)\,.\tag{69}$$

The form of Equation (68) suggests a simplification by introducing the Fourier transform

$$
\begin{aligned}
W(R, P) &= \frac{1}{2\pi\hbar} \int dr \, \exp\left(-\frac{iPr}{\hbar}\right) \left\langle R+\frac{r}{2} \left| \rho \right| R-\frac{r}{2} \right\rangle \\
&= \frac{1}{2\pi\hbar} \int dr \, \exp\left(-\frac{iPr}{\hbar}\right) \rho(R, r) .
\end{aligned}
\tag{70}
$$

This is the function introduced by Wigner [13] in 1932 to extract the classical limit from quantum dynamics. From the definition one simply proves that the marginal distributions give the diagonal elements of the density matrix

$$
W_P(R) = \int W(R, P) \, dP = \langle R|\rho|R\rangle
\tag{71}
$$

$$
W_R(P) = \int W(R, P) \, dR = \langle P|\rho|P\rangle .
\tag{72}
$$

The Wigner function can thus be used as a probability function to calculate the expectation value of any function of position or momentum only. For functions of both, the expectation value evaluated from the Wigner function corresponds to the operator in symmetric order with respect to position and momentum. The variables $\{R, P\}$ do not consist of conjugate dynamical variables but constitute the coordinates of a Wigner analogue of classical phase space. The Wigner function is defined on this space, but in spite of its usefulness to derive expectation values, it is regarded as a quasi-distribution function. It is, for example, not guaranteed to be positive everywhere, and only distribution of Wigner functions compatible with the quantum mechanical uncertainty relation in R and P can be accepted. In classical phase space all positive normalisable functions are acceptable distributions.

With the definition (70) the free particle equation of motion (68) becomes

$$
\left(\frac{\partial}{\partial t} + \frac{P}{M}\frac{\partial}{\partial R}\right) W(R, P) = 0 .
\tag{73}
$$

This is clearly the classical Liouville equation for a free particle with the solution

$$
W(R, P, t) = W_0\left(R - R_0 - \frac{Pt}{M}, P\right) ,
\tag{74}
$$

which describes free translation of the initial distribution W_0 as in the classical case.

The Equations (73) and (74) show that the Wigner function obeys the free dynamics of a classical system. In Reference [14] it is shown that it is the only quasi-distribution function with the correct marginal distributions (71) and (72) which also gives the classical equation of motion for the free particle.

In some sense the Wigner function extracts the diagonal dependence in both the momentum and position representations. The variable R in (69) is the center of the two position variables in $\langle x_1|\rho|x_2\rangle$. Instead of the definition (70) we can introduce the Wigner function in the momentum representation by

$$
W(R, P) = \frac{1}{2\pi\hbar} \int d\pi \, \exp\left(\frac{i\pi R}{\hbar}\right) \left\langle P+\frac{\pi}{2} \left| \rho \right| P-\frac{\pi}{2} \right\rangle .
\tag{75}
$$

The momentum representation density matrix $\langle p_1 | \rho | p_2 \rangle$ defines the central momentum

$$P = \frac{1}{2}(p_1 + p_2) \tag{76}$$

and the off-diagonality

$$\pi = p_1 - p_2 \tag{77}$$

analogous to the position off-diagonality r in Equation (69). Fourier transforming $W(R, P)$ further with respect to R gives the pure momentum representation in terms of P and π.

It is also possible to introduce a pure off-diagonal representation which, to the best of my knowledge, was first utilised by Shirley [15]. We define the Shirley function

$$
\begin{aligned}
S(r, \pi) &= \frac{1}{2\pi\hbar} \int dR \, \exp\left(\frac{i\pi R}{\hbar}\right) \left\langle R + \frac{r}{2} \middle| \rho \middle| R - \frac{r}{2} \right\rangle \\
&= \frac{1}{2\pi\hbar} \int dP \, \exp\left(-\frac{iPr}{\hbar}\right) \left\langle P + \frac{\pi}{2} \middle| \rho \middle| P - \frac{\pi}{2} \right\rangle .
\end{aligned}
\tag{78}
$$

The compatibility of all these representations is easily proved. The Shirley representation has sometimes been found useful in applications, see _e.g._ [16].

3.2   Properties of the Wigner function

We are now going to enumerate some of the important features of the Wigner representation of the density matrix. Many interesting properties are derived by Moyal [17]. The properties of the Wigner representation are reviewed by Tatarski [18] and Hillary _et al._ [19].

First we assume that we have two Wigner functions representing the two pure states ψ_1 and ψ_2 . Then we have

$$\iint dR \, dP \, W_1(R, P) W_2(R, P) = \frac{|\langle \psi_1 | \psi_2 \rangle|^2}{2\pi\hbar} . \tag{79}$$

It is useful to show the validity of this theorem explicitly. We have

$$
\begin{aligned}
2\pi\hbar \iint dR \, dP \, W_1(R, P) W_2(R, P) = \iint dR \, dP \iint \frac{dr_1 dr_2}{2\pi\hbar} \, \exp\left(-i\frac{P(r_1 + r_2)}{\hbar}\right) \\
\times \psi_1\left(R + \frac{r_1}{2}\right) \psi_1^*\left(R - \frac{r_1}{2}\right) \psi_2\left(R + \frac{r_2}{2}\right) \psi_2^*\left(R - \frac{r_2}{2}\right)
\end{aligned}
\tag{80}
$$

The integral over P can be performed giving a delta-function, eliminating one r-integration. Then introducing the variables

$$x = R + \frac{r_1}{2}, \quad y = R - \frac{r_1}{2} \tag{81}$$

we find

$$2\pi\hbar \iint dR \, dP \, W_1(R, P) W_2(R, P) = \int dy \, \langle \psi_1 | y \rangle \langle y | \psi_2 \rangle \int dx \, \langle \psi_2 | x \rangle \langle x | \psi_1 \rangle \tag{82}$$

from which Equation (79) follows directly. If $\psi_1 = \psi_2$, the scalar product equals unity, which is a necessary condition for any pure state Wigner function to satisfy.

For $W_1 = W_2$, the result (79) is related to the purity of the quantum state. We express the trace (66) using the inversion of the Fourier transform (70) and obtain

$$
\begin{aligned}
\mathrm{Tr}\,\rho^2 &= \iint dx_1 dx_2 \langle x_1|\rho|x_2\rangle\langle x_2|\rho|x_1\rangle \\
&= \iint dx_1 dx_2 \iint dP_1 dP_2 \, e^{i(P_1-P_2)(x_1-x_2)} \, W\left(\frac{x_1+x_2}{2}, P_1\right) W\left(\frac{x_1+x_2}{2}, P_2\right) \\
&= 2\pi\hbar \iint dR dP \, W^2(R,P) .
\end{aligned}
\tag{83}
$$

Thus, when the density matrix is not in a pure state, we have to replace the equality in Equation (79) by

$$
\iint dR dP \, W^2(R,P) < \frac{1}{2\pi\hbar} .
\tag{84}
$$

If we choose ψ_2 in Equation (79) to have a positive Wigner function, and ψ_1 to be any function orthogonal to it, we can see that the Wigner function W_1 cannot be positive everywhere; the Wigner function is not positive definite. Smeared with suitable functions, it does often lead to distributions over the Wigner phase space which are positive everywhere. For instance the convolution of two Wigner functions

$$
\Phi(R,P) = \iint dr\, dp\, W_1(R-r, P-p) W_2(r,p) \geq 0
\tag{85}
$$

can be given a probabilistic interpretation.

We carry through the proof first for pure state Wigner functions

$$
\begin{aligned}
\Phi(R,P) &= \iint dr\, dp \int \frac{dr_1}{2\pi\hbar} \exp\left[-i\frac{(P-p)r_1}{\hbar}\right] \psi_1\left(R-q+\frac{r_1}{2}\right) \psi_1^*\left(R-q-\frac{r_1}{2}\right) \\
&\quad \times \int \frac{dr_2}{2\pi\hbar} \exp\left(-i\frac{pr_2}{\hbar}\right) \psi_2\left(q+\frac{r_2}{2}\right) \psi_2^*\left(q-\frac{r_2}{2}\right) .
\end{aligned}
\tag{86}
$$

The p integration can be performed giving a delta-function, which allows one of the r-integrations to be carried out. Then introducing the new variables

$$
\xi = q - \frac{r_2}{2}, \qquad \eta = q + \frac{r_1}{2}
\tag{87}
$$

we obtain the result

$$
\begin{aligned}
\Phi(R,P) &= \int \frac{d\eta}{\sqrt{2\pi\hbar}} \exp\left(-i\frac{P\eta}{\hbar}\right) \psi_2(\eta)\psi_1^*(R-\eta) \\
&\quad \times \int \frac{d\xi}{\sqrt{2\pi\hbar}} \exp\left(i\frac{P\xi}{\hbar}\right) \psi_2^*(\xi)\psi_1(R-\xi)
\end{aligned}
\tag{88}
$$

from which the positivity follows. Because the relation (88) is bilinear in both states, we can take an ensemble average over each one giving the Wigner functions in Equation (85).

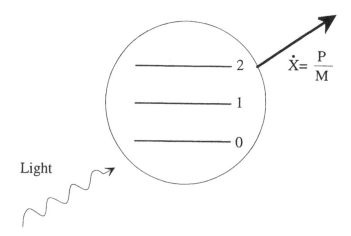

Figure 5. *Pictorial representation of the system under consideration.*

If we in particular choose as the state ψ_2 a minimum uncertainty state

$$\psi_2(x) = \frac{1}{(2\pi a^2)^{1/4}} \exp\left(i\frac{\bar{p}x}{\hbar}\right) \exp\left(-\frac{(x-\bar{x})^2}{4a^2}\right) \tag{89}$$

we find the result

$$
\begin{aligned}
\Phi(R, P) &= \frac{1}{2\pi\hbar} \left| \int \frac{d\xi}{(2\pi a^2)^{1/4}} \exp\left[i\frac{\xi(P-\bar{p})}{\hbar}\right] \exp\left(-\frac{(\xi-\bar{x})^2}{4a^2}\right) \psi_1(R-\xi) \right|^2 \\
&= \frac{1}{2\pi\hbar} \left| \int \frac{d\zeta}{(2\pi a^2)^{1/4}} \exp\left[i\frac{\zeta(P-\bar{p})}{\hbar}\right] \exp\left(-\frac{(\zeta-R+\bar{x})^2}{4a^2}\right) \psi_1(\zeta) \right|^2 . \tag{90}
\end{aligned}
$$

This is a probability filtering with space resolution a around the point $\{\bar{x}, \bar{p}\}$. Such a function was introduced by Husimi in 1940 [20], see also Reference [21].

In addition to the Wigner phase space variables $\{R, P\}$, we assume that the quantum system under consideration has a (finite) set of internal discrete quantum states (see Figure 5). Thus the Wigner function retains its matrix character in these internal indices, and it becomes a Wigner matrix W_{ij}. The potential energy $V(x)$ in Equation (63) is a Hamiltonian matrix in these indices, and the commutator character of the equation of motion (67) is retained for the Wigner function. Consequently the potential term in the Hamiltonian becomes in the Wigner representation

$$
\begin{aligned}
\frac{1}{2\pi\hbar} \int dr\, & \exp\left(-i\frac{Pr}{\hbar}\right) \left\langle R+\frac{r}{2} \left| [V(x), \rho] \right| R-\frac{r}{2} \right\rangle \\
&= \frac{1}{2\pi\hbar} \int dr\, \exp\left(-i\frac{Pr}{\hbar}\right) \left[V\left(R+\frac{r}{2}\right)\rho(R, r) - \rho(R, r)V\left(R-\frac{r}{2}\right) \right] \\
&= \frac{1}{2\pi\hbar} \int dr\, \exp\left(-i\frac{Pr}{\hbar}\right) \\
&\quad \times \left\{ [V(R), \rho] + \frac{r}{2}\left[\frac{\partial V}{\partial R}, \rho\right]_+ + \frac{r^2}{8}\left[\frac{\partial^2 V}{\partial R^2}, \rho\right] + \frac{r^3}{48}\left[\frac{\partial^3 V}{\partial R^3}, \rho\right]_+ + \cdots \right\}. \tag{91}
\end{aligned}
$$

Combining this with Equation (73) from the kinetic energy and replacing the variable r in the integral by $i\hbar\partial/\partial P$ we obtain the full equation of motion

$$\left(\frac{\partial}{\partial t}+\frac{P}{M}\frac{\partial}{\partial R}\right)W(R,P) = \frac{1}{i\hbar}[V(R),W(R,P)]+\frac{1}{2}\left[\frac{\partial V}{\partial R},\frac{\partial W(R,P)}{\partial P}\right]_+ \qquad (92)$$
$$+\frac{i\hbar}{8}\left[\frac{\partial^2 V}{\partial R^2},\frac{\partial^2 W(R,P)}{\partial P^2}\right]-\frac{\hbar^2}{48}\left[\frac{\partial^3 V}{\partial R^3},\frac{\partial^3 W(R,P)}{\partial P^3}\right]_+ +\cdots.$$

This is clearly an expansion of the Wigner function in powers of the quantum parameter $\hbar$. If the system has no internal states, all commutator terms vanish, and the anticommutator terms ($[\cdot,\cdot]_+$) become simple products. In the following sections we will consider some simple applications of this equation of motion.

3.3 Potential motion

When the system under consideration has no internal states, its dynamical time evolution is entirely determined by a potential function $V(x)$. The equation becomes up to quantum corrections of order $O(\hbar^2)$

$$\left(\frac{\partial}{\partial t}+\frac{P}{M}\frac{\partial}{\partial R}-\frac{\partial V}{\partial R}\frac{\partial}{\partial P}\right)W(R,P) = -\frac{\hbar^2}{24}\frac{\partial^3 V}{\partial R^3}\frac{\partial^3 W(R,P)}{\partial P^3}. \qquad (93)$$

The left hand side of this equation is clearly the classical Liouville equation in phase space. Quantum aspects enter, however, through the initial conditions that all Wigner functions derive from state vectors and their ensembles. The quantum correction to the dynamics is proportional to the third order derivative of $W(R,P)$; this tells us that wave packets tend to diffract in phase space. If classical motion were combined with randomness, the result would be diffusive spreading, which would add a term proportional to a second order derivative of $W(R,P)$; see *e.g.* Reference [22]. This can be seen as an indication that no hidden variable theory can lead to quantum dynamics, *i.e.* the quantum diffractive motion cannot be caused by interaction with random unobserved degrees of freedom.

The second observation is that potentials with vanishing third derivatives give no quantum corrections to the motion. Thus potentials of the form

$$V_n(R) = \kappa\,R^n, \qquad (n \leq 2) \qquad (94)$$

have a Wigner function obeying the classical Liouville equation.

The time evolution in the potentials (94) can be solved analytically in terms of the characteristics deriving from the classical equations of motion

$$\dot{R} = \frac{P}{M}; \qquad \dot{P} = -\frac{\partial V}{\partial R}. \qquad (95)$$

We look for a fundamental system of solutions $\{A(t),B(t),C(t)\}$ satisfying the initial conditions

$$\begin{aligned}
A(0) &= \dot{B}(0) = 1 \\
\dot{A}(0) &= B(0) = 0 \\
C(0) &= \dot{C}(0) = 0.
\end{aligned} \qquad (96)$$

With these we can write the solution for an arbitrary trajectory starting at $\{R_0, P_0\}$ in the form

$$R(t) = R_0\, A(t) + \frac{P_0}{M}\, B(t) + C(t).\tag{97}$$

Defining the determinant

$$D[X, Y] = X(t)\dot{Y}(t) - \dot{X}(t)Y(t),\tag{98}$$

we can write the initial conditions in the form

$$
\begin{aligned}
R_0 &= \frac{1}{D[A, B]}\left(\dot{B}R - B\frac{P}{M}\right) + \frac{D[B, C]}{D[A, B]} \\
P_0 &= \frac{1}{D[A, B]}\left(AP - M\dot{A}R\right) - M\frac{D[A, C]}{D[A, B]}.
\end{aligned}\tag{99}
$$

If we have the initial Wigner function $W_0(R_0, P_0)$, we can obtain the general solution to this equation by substituting $\{R_0, P_0\}$ from (99); this is the methods of characteristics for partial differential equations.

Acceptable initial distributions are only those which can be derived from state vectors. Thus all classically allowed distributions are not acceptable. For a pure state Wigner function, it has been shown [23], [24], that the only state leading to no negative values in Wigner phase space is a Gaussian distribution. Thus most Wigner functions take nonclassical negative values.

The simple minimum uncertainty state (89) is a Gaussian distribution, and its Wigner function is

$$W(R, P) = C \exp\left[-\frac{(R-\overline{x})^2}{2a^2}\right]\exp\left[-\frac{2\,(P-\overline{p})^2\,a^2}{\hbar^2}\right].\tag{100}$$

From this distribution function the minimum uncertainty product $\Delta R \times \Delta P = \hbar/2$ is seen to follow. When we insert $\{R_0, P_0\}$ from Equation (99) into Equation (100) we obtain the solution $W(R, P, t)$. This is no longer of the product form and the dependences on R and P have become entangled.

If we integrate over P to obtain the marginal distribution (71), we find a Gaussian function of the type

$$W_P(R) = N \exp\left[-\frac{(R-X(t))^2}{2\sigma^2(t)}\right],\tag{101}$$

where we have

$$
\begin{aligned}
X(t) &= \overline{x}\, A(t) + \frac{\overline{p}}{M}\, B(t) + C(t) \\
\sigma^2(t) &= a^2 A^2(t) + \frac{\hbar^2}{4a^2 M^2}B^2(t).
\end{aligned}\tag{102}
$$

The wave packet is seen to follow the classical path (97) starting at the center of the wave packet (100). The width is a weighted average between the initial width in position space and the spread introduced by the initial momentum uncertainty.

For the three cases solved exactly by the present method (94), we obtain

$$V_0 = 0: \qquad A = 1, \qquad B = t, \qquad C = 0$$

$$V_1 = -MgR: \qquad A = 1, \qquad B = t, \qquad C = gt^2/2 \qquad (103)$$

$$V_2 = M\Omega^2 R^2/2: \quad A = \cos(\Omega t), \quad B = \Omega^{-1}\sin(\Omega t), \quad C = 0.$$

From these results and Equations (101) and (102) we can obtain the motion of a minimum uncertainty wave packet in any of the potentials (94). The same method allows us to derive the solution starting from any other initial state, pure or mixed.

Even more general cases can be solved. As long as the R−dependence is of the general form (94) the parameters can contain arbitrary time dependences. Finding the fundamental set of solutions (96) provides a solution of the form derived above even in this case. In Reference [25] the quantum motion of an ion in a Paul trap was described using the potential function

$$V(R, t) = \frac{1}{2}\left(K_0 + K_1 \cos(\omega t)\right) R^2. \qquad (104)$$

The general solution is obtained from numerical integration of the time dependent Mathieu equation. Also other time dependent problems can be solved.

3.4 The predetermined motion approximation

The Equation of motion (92) provides a time evolution operator expanded in the quantum parameter $\hbar$. Unfortunately this cannot be used directly to generate a numerical solution, even if we regard $\hbar$ as a numerically small parameter. When the derivatives $\partial/\partial P$ are applied to an initial Wigner function, we see from the example (100) that terms in $\hbar^{-2}$ are generated. This mixes terms in the expansion as was pointed out a long time ago by Heller [26]. Because $\hbar$ is a dimensional constant it is far from clear how to extract uniquely all dependences on it from quantum dynamics.

However, the time evolution propagator or the Green function of Equation (92) can be represented as a unique series in $\hbar$; what happens to an initial state can be divided into classical and quantum contributions according to the dependence on $\hbar$. The Wigner function at time t can be written

$$W(R, P, t) = \int G(R, P, t | R_0, P_0) W_0(R_0, P_0) dR_0 dP_0. \qquad (105)$$

For classical initial distributions

$$\Delta R \, \Delta P \gg \hbar \qquad (106)$$

we can calculate the propagator $G(R, P, t | R_0, P_0)$ in perturbation theory with respect to $\hbar$. This is, however, a singular perturbation problem, the solution containing negative powers of $\hbar$ and special techniques have to be devised. In Reference [27] the method of multiple time scales was used to generate a systematic approach to the problem. Here we derive the lowest order result using more heuristic arguments.

We assume that the quantum time evolution is fast, the internal levels exchange population much faster than the particle center of mass moves. On the other hand, the drift of the particle changes the rate of coupling between these levels only slowly, and hence we assume that the potential can be written in the form

$$V(R) = \overline{U}(R) + H_{\text{eff}}(R),\tag{107}$$

where the mean potential $\overline{U}(R)$ is a scalar in the internal state indices and the effective Hamiltonian $H_{\text{eff}}(R)$ depends only weakly on position.

In this situation we can attempt an adiabatic approximation (c.f. Equation (10)) of the type

$$W(R, P, t) \approx w^0(R, P, t)\, \sigma(R, P, t),\tag{108}$$

where w^0 is a scalar distribution function and σ is a density matrix in the internal state indices, which depends only parametrically on the variables $\{R, P\}$. This is assumed to be normalised according to

$$\text{Tr}\,[\sigma(R, P, t)] = 1.\tag{109}$$

To lowest order, the Equation (92) now becomes

$$\left(\frac{\partial}{\partial t} + \frac{P}{M}\frac{\partial}{\partial R}\right)(w^0\sigma) = \frac{1}{i\hbar}\,[H_{\text{eff}}, \sigma]\,w^0 + \frac{\partial \overline{U}}{\partial R}\frac{\partial}{\partial P}\left(w^0\sigma\right).\tag{110}$$

Taking the trace of this equation and using the result (109) we obtain

$$\left(\frac{\partial}{\partial t} + \frac{P}{M}\frac{\partial}{\partial R} - \frac{\partial \overline{U}}{\partial R}\frac{\partial}{\partial P}\right)w^0(R, P, t) = 0.\tag{111}$$

This shows that the distribution function w^0 describes the flow of the probability distribution as in the case of classical phase space; for positive initial distributions this reduces to the classical problem.

Next, neglecting the $\{R, P\}$ dependence of σ, and using (111) in Equation (110) we obtain the equation

$$\left(i\hbar\frac{\partial \sigma}{\partial t} - [H_{\text{eff}}, \sigma]\right)w^0 = 0.\tag{112}$$

If we simply cancel the distribution function, we obtain an equation describing quantum evolution at each local point R. This is not too useful.

To proceed we assume that we solve Equation (111) for a trajectory starting at $\{r_0, p_0\}$ and travelling according to the classical equations of motion (95). These solutions are called $r(t|r_0, p_0)$ and $p(t|r_0, p_0)$. The distribution function then becomes

$$w^0(R, P, t) = \delta\left(R - r(t|r_0, p_0)\right)\delta\left(P - p(t|r_0, p_0)\right).\tag{113}$$

It is easily seen that this satisfies Equation (111).

The delta functions appearing in semiclassical calculations are not of the symmetric type normally encountered in physics. They derive from the Airy function in the sense that

$$\frac{1}{a}Ai\left(\frac{x}{a}\right) = \frac{1}{2\pi}\int\limits_{-\infty}^{\infty}\exp\left(itx + \frac{i}{3}t^3a^3\right)dt\tag{114}$$

reduces to

$$\frac{1}{2\pi} \int\limits_{-\infty}^{\infty} \exp(itx)\, dt = \delta(x)$$

in the limit $a \to 0$. In the semiclassical applications $a \propto \hbar^{2/3}$.

With the solution (113), we can integrate Equation (112) over R and P obtaining the result

$$i\hbar \frac{\partial \sigma}{\partial t} = [H_{\text{eff}}\left(r(t|r_0, p_0)\right), \sigma]. \tag{115}$$

This is the time evolution equation we would write down directly for a system with a finite number of internal levels travelling along the classical trajectory determined by the initial values $\{r_0, p_0\}$ and the potential $\overline{U}(R)$. This semiclassical method of predetermined trajectories has found its applications in scattering theory, magnetic resonance phenomena, laser cooling and molecular reaction theory. It is a natural way to convert a dynamical problem in phase space into a Hamiltonian problem with explicit time dependence generated by the motion of the particles.

It is, however, easy to see when the approximation introduced above breaks down. When the different initial states see different potentials, no mean trajectory can be defined. Then quantum dynamics must be treated according to the models set out in Section 2. In higher dimensions, the potential function $V(x)$ may easily lead to chaotic motion. Then the corresponding quantum problem cannot be expanded in powers of $\hbar$, and the approach developed here becomes irrelevant, see *e.g.* Reference [28]. In general, an expansion in the dimensional parameter $\hbar$ cannot be of universal validity. Additional parameters characterising the particular situation under consideration must be included to give a dimensionless quantity. Only this can be regarded as numerically small, when the validity of the semiclassical expansion is to be assessed.

3.5   Connection with classical microcanonical ensemble

We consider the equation of motion for the Wigner function (93). In the classical limit the right hand side can be set equal to zero, and then the solution with given initial conditions is (113)

$$W(R, P, t|r_0, p_0) = \delta\left(R - r(t|r_0, p_0)\right) \delta\left(P - p(t|r_0, p_0)\right). \tag{116}$$

If we choose $p_0 = 0$, we can obtain the momentum dependence from the energy conservation condition

$$E = \frac{p^2}{2M} + V(r) = V(r_0). \tag{117}$$

Thus we write

$$p(t|r_0, p_0 = 0) = p(r(t), E). \tag{118}$$

We can use the delta function in R to replace the time dependent variable $r(t)$ by R.

In order to obtain the Wigner function for a constant energy ensemble, a microcanonical ensemble, we collect an assembly of all positions along the trajectory for half a period $T_0/2$, because on this the connection between time and position is single valued (see Figure 6). We find

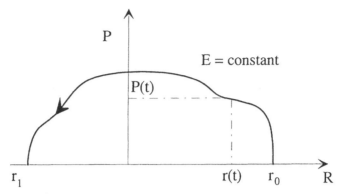

Figure 6. *Periodic 1/2 orbit in the phase space.*

$$\frac{2}{T_0} \int_0^{T_0/2} \delta\left(R - r(t|r_0, p_0)\right) dt = \frac{2}{T_0} \int_0^{T_0/2} dt \frac{\delta(t - t(R|r_0, p_0))}{v(t|r_0, p_0)} = \frac{2M}{T_0\, p(R, E)}. \tag{119}$$

The time variable $t(R|r_0, p_0)$ is now determined uniquely by R and E. From Equation (116) we now have the constant energy Wigner function

$$W(R, P|E) = \frac{2M}{T_0\, p(R, E)} \delta\left(P - p(R, E)\right). \tag{120}$$

As usual the period can be determined from the integral

$$T_0 = 2M \int_{\text{half period}} \frac{dr}{p(r(t), E)}. \tag{121}$$

We can rewrite the delta function according to

$$\delta\left(P - p(R, E)\right) = \frac{1}{2}\left(\frac{\partial E}{\partial P}\right) \delta(E - E(R, P)). \tag{122}$$

The inversion is single valued for each half period, and hence the factor $1/2$ is to be included. Inserted into Equation (120) this gives

$$W(R, P|E) = \frac{1}{T_0} \delta\left(E - E(R, P)\right). \tag{123}$$

This result becomes particularly simple if we introduce the canonical action-angle variables $\{\varphi, I\}$. Because we have

$$\frac{2\pi}{T_0} = \frac{\partial E}{\partial I}, \tag{124}$$

we find from Equation (123)

$$W(R, P|E) = \frac{1}{2\pi} \delta(I - I(R, P)). \tag{125}$$

This is the classical microcanonical ensemble in action-angle phase space; see Reference [28]. It is manifestly normalised with respect to integration over $\varphi \in [0, 2\pi]$ and I. In the present derivation we have obtained it from summing in time over a single constant energy trajectory. It is physically clear that the arbitrarily chosen starting point cannot affect the result.

4 Adiabatic approach to time dependence

4.1 Time dependent 2-level Hamiltonians

According to the theory of the previous section, a Hamiltonian may become time dependent either because of particle motion or the variation of some external parameter. The former is the case in collision and reaction problems, the latter occurs in pulsed laser experiment. In this section, we assume that the Hamiltonian is explicitly time dependent, and for simplicity only two discrete levels will be discussed.

The general Hamiltonian is of the type

$$H(t) = \begin{bmatrix} E_1(t) & U(t)e^{i\varphi(t)} \\ U(t)e^{-i\varphi(t)} & E_2(t) \end{bmatrix}, \tag{126}$$

where all functions are real. Introducing for the state vector the ansatz

$$|\psi\rangle = \exp\left[-\frac{i}{\hbar}\int^t (E_1(t')+E_2(t'))\,dt'\right]\left[\exp\left(i\frac{\varphi}{2\hbar}\right)C_1|1\rangle + \exp\left(-i\frac{\varphi}{2\hbar}\right)C_2|2\rangle\right] \tag{127}$$

we obtain the Schrödinger equation

$$i\hbar\frac{\partial}{\partial t}\begin{bmatrix} C_1 \\ C_2 \end{bmatrix} = \frac{1}{2}\begin{bmatrix} (E_1 - E_2 + \dot{\varphi}) & 2U \\ 2U & -(E_1 - E_2 + \dot{\varphi}) \end{bmatrix}\begin{bmatrix} C_1 \\ C_2 \end{bmatrix}. \tag{128}$$

This shows that, without loss of universality, we can take the Hamiltonian to be of the form

$$H(t) = \hbar\begin{bmatrix} \alpha(t) & V(t) \\ V(t) & -\alpha(t) \end{bmatrix} \tag{129}$$

with α and V real. The Schrödinger equation is now

$$\begin{aligned} i\dot{C}_1 &= \alpha C_1 + V C_2 \\ i\dot{C}_2 &= -\alpha C_2 + V C_1 . \end{aligned} \tag{130}$$

An alternative way of presenting the information about the 2-level system is in terms of the Bloch vector

$$\mathbf{R} = r_1\mathbf{e}_1 + r_2\mathbf{e}_2 + r_3\mathbf{e}_3, \tag{131}$$

which lives in an abstract (energy) space; see Reference [29]. The vector components are

$$\begin{aligned} r_1 &= \langle\sigma_1\rangle = C_1C_2^* + C_1^*C_2 \\ r_2 &= \langle\sigma_2\rangle = i\left(C_1C_2^* - C_1^*C_2\right) . \\ r_3 &= \langle\sigma_3\rangle = C_1C_1^* - C_2^*C_2 \end{aligned} \tag{132}$$

where the σ are the Pauli matrices. The equation of motion (130) for the Bloch vector is

$$\frac{d}{dt}\mathbf{R} = \mathbf{\Omega} \times \mathbf{R}, \tag{133}$$

with the effective external field vector

$$\mathbf{\Omega} = 2V\mathbf{e}_1 - 2\alpha\mathbf{e}_3 . \tag{134}$$

From Equation (133) follows that the magnitude of **R** is conserved, the dynamics is represented by a rotation of the Bloch vector on a sphere, the Bloch sphere.

All the above results are derived for purely Hamiltonian time evolution. With dissipative processes present, the equations are modified; see Reference [30]. Then a density matrix treatment or the equivalent Bloch vector representation is useful.

4.2   The adiabatic approach

Time independent quantum problems are often solved by diagonalising the Hamiltonian with a unitary transformation. If its time dependence is slow enough, we can carry out an instantaneous diagonalisation by the time dependent transformation

$$U = \begin{bmatrix} \cos(\theta/2) & -\sin(\theta/2) \\ \sin(\theta/2) & \cos(\theta/2) \end{bmatrix} \tag{135}$$

which effects the diagonalisation

$$U H U^\dagger = \begin{bmatrix} \mathcal{E}_- & 0 \\ 0 & \mathcal{E}_+ \end{bmatrix}. \tag{136}$$

Choosing the angle θ according to

$$\tan\theta = -\frac{V}{\alpha} \tag{137}$$

we obtain the diagonal elements

$$\mathcal{E}_\pm = \pm\sqrt{\alpha^2 + V^2} \equiv \pm\mathcal{E} . \tag{138}$$

The time dependence of the adiabatic state

$$|\varphi\rangle = U|\psi\rangle \tag{139}$$

is determined by the transformed Hamiltonian

$$H_{\text{ad}} = U H U^\dagger - i\hbar U\frac{dU^\dagger}{dt} = \begin{bmatrix} -\mathcal{E} & -i\gamma \\ i\gamma & +\mathcal{E} \end{bmatrix}, \tag{140}$$

where the coupling parameter is

$$\gamma = \frac{\dot\theta}{2} = \frac{V\dot\alpha - \alpha\dot V}{2\,(\alpha^2 + V^2)}. \tag{141}$$

The form of the Hamiltonian implies fixing the phases of the eigenstates in a specific way.

From Equation (137), the transformation angle θ is also defined by the expression

$$e^{2i\theta} = \frac{\alpha - iV}{\alpha + iV}. \tag{142}$$

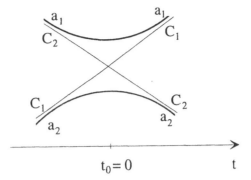

Figure 7. *Level crossings and adiabatic states.*

The amplitudes of the adiabatic state vector $|\varphi\rangle$ evolve according to the coupled equations

$$\dot{a}_1 = -\gamma e^{-i\Delta} a_2$$
$$\dot{a}_2 = \gamma e^{i\Delta} a_1,$$

(143)

where we have defined

$$\Delta(t) = 2 \int^t \mathcal{E}(t')dt'.$$

(144)

At a point where $\alpha = 0$, the bare energy levels cross, and the adiabatic levels repel each other; we have an avoided crossing as shown in Figure 7. This means that the roles of the bare and the adiabatic amplitudes are exchanged between $t = -\infty$ and $t = +\infty$. Thus we have

$$
\begin{aligned}
C_1(-\infty) &= a_2(-\infty), & C_2(-\infty) &= a_1(-\infty), \\
C_1(\infty) &= a_1(\infty), & C_2(\infty) &= a_2(\infty).
\end{aligned}
$$

(145)

We want to solve the quantum evolution with the initial conditions

$$C_2(-\infty) = a_1(-\infty) = 1,$$

(146)

and calculate the probability of transfer between the bare states

$$C_1(\infty) = a_1(\infty);$$

(147)

in the adiabatic state basis, this is the probability of no transfer. For ideal adiabatic conditions this quantity is expected to approach unity.

In perturbation theory, we can insert the initial conditions into Equation (143) and integrate

$$a_2(\infty) = \int_{-\infty}^{\infty} \gamma(\tau)e^{i\Delta(\tau)}d\tau.$$

(148)

This integral goes along the real τ-axis. Following Dykhne [31] we distort the contour into the upper complex plane in the way shown in Figure 8. This does not change the value of the transfer before we encounter a singularity of the integrand. We assume this to be a singularity of the adiabatic energy $\mathcal{E}(t)$ as a function of the variable t continued

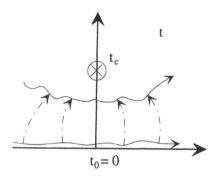

Figure 8. *Shifting of integration path.*

out into the complex plane. We deform the integration contour to pass through this point.

Davis and Pechukas [32] have carried out a detailed analysis of the validity of this method. They assume that the parameters of the Hamiltonian (129) are analytic functions of the time parameter and that the first zero encountered when leaving the real axis is at t_c. We expand

$$\mathcal{E}^2 = \alpha^2 + V^2 = (\alpha - iV)(\alpha + iV) = A^2(t - t_c) . \tag{149}$$

The singularity in $\mathcal{E}(t)$ then becomes a simple branch point.

Because a zero of a product is also a zero of one of the factors, we obtain from (149) and (142) the result

$$e^{2i\theta} \propto (t - t_c)^{\pm 1} \tag{150}$$

from which follows that

$$\gamma = \frac{\dot{\theta}}{2} = \frac{1}{4i(t - t_c)} . \tag{151}$$

This surprisingly universal result implies a certain fixing of the signs in the adiabatic eigenvectors. Expanding the function $\Delta(t)$ around the singular point t_c we find to lowest order

$$\Delta(t) = \Delta(t_c) + \frac{4A}{3}(t - t_c)^{3/2} . \tag{152}$$

We choose the path of integration along the level curves

$$\text{Im}\,\Delta(t) = \text{constant} \tag{153}$$

(see Figure 9). Along these lines, the main change in the occupation of the quantum state occurs near the singularity point t_c and we can evaluate the integral (148) near this point; see discussion in Reference [32]. Choosing the variable

$$x = \frac{4A}{3}(t - t_c)^{3/2} \tag{154}$$

we obtain

$$a_2(\infty) = \frac{\exp[-\text{Im}\,\Delta(t_c)]}{6i} \int_{-\infty}^{\infty} \frac{\exp(ix)dx}{x} = \frac{\pi}{3}\exp[-\text{Im}\,\Delta(t_c)] . \tag{155}$$

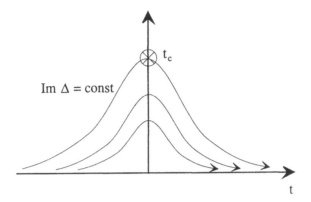

Figure 9. *Paths in the pole approximation.*

The irrelevant phase factor is omitted. The square of this is the perturbative result for the adiabatic transition probability. The prefactor is, however, incorrect. The reason is that in a full calculation of the solution of Equations (143) there are terms of the form $\exp(-\operatorname{Im}\Delta(t_c))$ in each order. In Reference [32] it is shown that they sum up to replace the prefactor $\pi/3$ by unity, a surprisingly small but essential correction.

Consequently, when the transition rate is dominated by the nearest singularity in the complex plane, the transition probability is given by the simple result

$$|a_2(\infty)|^2 = \exp[-2\operatorname{Im}\Delta(t_c)]. \tag{156}$$

The smaller this is, the more adiabatic is the transition. The correction terms derive from the other singularities in the complex plane; when their distance from the real axis greatly exceeds that of t_c, the result (156) is the dominant contribution.

4.3 Solvable models

4.3.1 The Landau-Zener model

In 1932 both Landau [33] and Zener [34] provided the solution to the simplest possible case of a level crossing. In the Hamiltonian (129) they chose

$$\alpha(t) = \lambda t, \qquad V(t) = V_0. \tag{157}$$

The adiabatic energy eigenvalue is

$$\mathcal{E} = \sqrt{V_0^2 + \lambda^2 t^2}. \tag{158}$$

The branch point singularity is then at the position

$$t_c = i\frac{V_0}{\lambda} \tag{159}$$

and the adiabatic result (156) requires the integral

$$\Delta(t_c) = 2\int^{iV_0/\lambda} \sqrt{V_0^2 + \lambda^2 t^2}\, dt = \frac{V_0^2}{\lambda}\log(i). \tag{160}$$

Taking the imaginary part of this and inserting into the Equation (156) we obtain

$$|a_2(\infty)|^2 = e^{-\pi\Lambda}, \tag{161}$$

where the adiabaticity parameter is

$$\Lambda = \frac{V_0^2}{\lambda}. \tag{162}$$

For adiabatically slow motion, $\lambda \to 0$, the right-hand side of (161) goes to zero, the population is adiabatically transferred to the other bare level.

Because there is only one singularity in each half-plane, we expect no correction to the result (161). Indeed, it is exact and can be used also in the perturbative limit, $V_0 \to 0$. We have then

$$|a_1(\infty)|^2 = 1 - e^{-\pi\Lambda} = \pi\frac{V_0^2}{\lambda}, \tag{163}$$

which follows from second order time dependent perturbation theory too. In fact, Landau used this requirement to fix the prefactor in the result (155). For the Landau-Zener model the adiabatic result is not asymptotic only but exact. Zener [34] expresses the general solution in terms of Weber functions and their asymptotic properties. This has the advantage that also the phase change imposed at a crossing can be obtained. In situations with repeated crossings this may become a measurable quantity; see References [35] and [36].

4.3.2   The Rosen-Zener model

A simple model for pulsed excitation was treated by Rosen and Zener in 1932 [37]. They introduced the Hamiltonian

$$H_{RZ} = \begin{bmatrix} -\Omega & -\frac{i\omega_0}{2}\operatorname{sech}\left(\frac{\pi t}{\tau}\right) \\ \frac{i\omega_0}{2}\operatorname{sech}\left(\frac{\pi t}{\tau}\right) & \Omega \end{bmatrix}. \tag{164}$$

The model contains a time scale τ, a detuning parameter Ω and a coupling strength ω_0. The time evolution can be solved using the method to be explained in Section 4.4, and the solution follows in terms of hypergeometric functions. The exact transition probability is given by

$$|a_2(\infty)|^2 = \sin^2\left(\frac{\omega_0\tau}{2}\right)\operatorname{sech}^2(\Omega\tau). \tag{165}$$

This result contains both a periodic variation with ω_0 similar to the Area Theorem, and an exponential decrease of probability in the adiabatically slow limit ($\tau \to \infty$).

4.3.3   The hyperbolic model

One may think that the only parameters determining the population transfer would be the values of α and V near $t = 0$. In order to investigate this question we look at

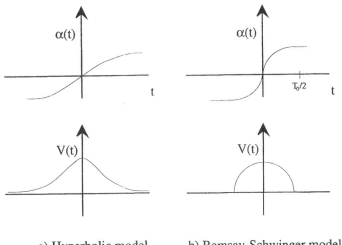

a) Hyperbolic model   b) Ramsay-Schwinger model

Figure 10. *Solvable models: (a) the Hyperbolic model, (b) the Ramsey-Schwinger model.*

another model [38] with hyperbolic functions

$$\alpha(t) = A \tanh\left(\frac{t}{T}\right)$$
$$V(t) = \left(A^2 + B^2\right)^{1/2} \operatorname{sech}\left(\frac{t}{T}\right) \tag{166}$$

(see Figure 10(a)). The model has two strength parameters, A and B, and one time scale T. The adiabatic energy is

$$\mathcal{E} = \left(A^2 + B^2 \operatorname{sech}^2\left(\frac{t}{T}\right)\right)^{1/2}, \tag{167}$$

and the coupling strength is

$$\gamma(t) = \frac{1}{2T} \operatorname{sech}\left(\frac{t}{T}\right). \tag{168}$$

For the special case $B = 0$, the energy $\mathcal{E}$ becomes constant, and the model reduces to the Rosen-Zener case, but in the adiabatic basis. Identifying the parameters with those in Equation (164) according to

$$\omega_0 = \frac{1}{T}, \quad \Omega = A, \quad \tau = \pi T, \tag{169}$$

we obtain from Equation (165) the solution

$$|a_2(\infty)|^2 = \operatorname{sech}^2(\pi A T) \to 4e^{-2\pi A T}. \tag{170}$$

Linearising the model (166) around $t = 0$, we find that the adiabatic parameter $\Lambda = AT$. Thus the result (170) goes to zero twice as fast as it should according to its Landau-Zener approximation. We conclude that the adiabatic result does depend on the exact form of the time dependent Hamiltonian.

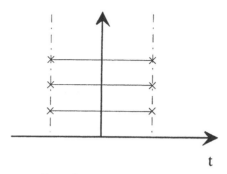

Figure 11. *Branch points in the hyperbolic model.*

We can treat the model (166) using the adiabatic approximation of Section 4.2. We look for the zeroes of the adiabatic energy (167) and find a series

$$t_c = \pm T \, ar\sinh\left(\frac{B}{A}\right) + iT\left(\frac{\pi}{2} + n\pi\right) \tag{171}$$

(see Figure 11). The zeroes closest to the real axis follow with $n = 0$, and there are two of these. At these points the integral (144) can be evaluated to give (see Reference [38])

$$\Delta(t_c) = \pi T\left(\pm B + iA\right) . \tag{172}$$

When several singularities are at equal distance from the real axis, Davis and Pechukas [32] suggest that their contributions are added coherently. Consequently we keep the phase from the real part of $\Delta(t_c)$ and find from (156)

$$|a_2(\infty)|^2 = 4\cos^2\left(\pi BT\right) e^{-2\pi AT} . \tag{173}$$

When $B = 0$, this result agrees with the asymptotic limit of Equation (170), but for finite values of B it displays periodic oscillations like the Rosen-Zener model. For such values that BT equals an odd half-integer, no transfer is observed in Equation (173).

The time dependent model defined by (166) is, in fact, exactly solvable [39],[40]. In Section 4.4, we will see that its solution is

$$|a_2(\infty)|^2 = \cos^2\left(\pi BT\right) \operatorname{sech}^2\left(\pi AT\right) . \tag{174}$$

This result contains both the results (170) and (173) as limiting cases. It is, however, instructive to see how these results can be derived directly. An amusing observation is that the complex integration method gives the correct asymptotic value (173) for $B = 0$, even if there exist no singularities of the adiabatic energy (167) for this special case. The influence of the full series of singularities in (171) is discussed by Suominen in Reference [41].

Another interesting limit can be obtained from the analytic result (174). We let the parameter B become imaginary

$$B = iC, \tag{175}$$

but so that $C < A$. The singularities of Equation (171) then lie along the imaginary axis, and when $C \to A$, one of them approaches the origin (see Figure 12). Then we

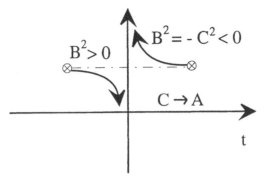

Figure 12. *Motion of the branch points; $B \to iC \to iA$.*

expect this singularity to dominate the time evolution, and the Landau-Zener situation may be expected to emerge. We linearise around $t = 0$ in this limit and find from (166)

$$V_0 = \sqrt{A^2 - C^2} \approx \sqrt{(2A)(A - C)}$$
$$\lambda = \frac{A}{T}. \tag{176}$$

From Equation (174) we find the result

$$|a_2(\infty)|^2 = \cosh^2(\pi CT) \operatorname{sech}^2(\pi AT) \approx \exp[-2\pi(A - C)T]. \tag{177}$$

Because the adiabaticity parameter from Equation (176) is

$$\Lambda = \frac{V_0^2}{\lambda} = 2(A - C)T, \tag{178}$$

the result (177) agrees with the Landau-Zener limit as we expected.

4.3.4 The Ramsey-Schwinger model

Another simple solvable model can be obtained by setting

$$
\begin{aligned}
V &= A \cos\left[\frac{\pi t}{T}\right], & \alpha &= A \sin\left[\frac{\pi t}{T}\right], & \text{for } t &\in \left[-\frac{T}{2}, \frac{T}{2}\right] \\
V &= 0, & \alpha &= -A, & \text{for } t &< -\frac{T}{2}, \\
V &= 0, & \alpha &= A, & \text{for } t &> \frac{T}{2}
\end{aligned} \tag{179}
$$

(see Figure 10(b) and Reference [42]). For this model, the adiabatic Hamiltonian (140) becomes very simple

$$H_{\text{ad}} = \begin{bmatrix} -A & i\pi/2T \\ -i\pi/2T & A \end{bmatrix}. \tag{180}$$

No time dependence remains, and the Rabi solution gives the result at time $T/2$

$$|a_2(\infty)|^2 = \frac{1}{\Omega^2}\left(\frac{\pi}{2T}\right)^2 \sin^2(\Omega T), \tag{181}$$

where

$$\Omega^2 = A^2 + \left(\frac{\pi}{2T}\right)^2 . \tag{182}$$

This result does not have any relationship to the Landau-Zener result obtained by linearising around $t = 0$; the asymptotic limit is approached in an oscillatory way when $T \to \infty$, and the adiabatic result is obtained as a power law not exponentially

$$|a_2(\infty)|^2 \sim T^{-2} . \tag{183}$$

For this model no asymptotic method is seen to work.

4.4   The analytic approach

From the Schrödinger equation (130) we can derive one single second order differential equation by eliminating C_1. We obtain the equation

$$\ddot{C}_2 - \frac{\dot{V}}{V}\dot{C}_2 + \left(\alpha^2 + V^2 - i\dot{\alpha} + i\alpha\frac{\dot{V}}{V}\right) C_2 = 0 . \tag{184}$$

The time parameter t is defined on the interval $[-\infty, \infty]$. It is often advantageous to map this to the bounded interval $[0, 1]$ by a suitable function $z(t)$. Rosen and Zener [37] introduced the mapping

$$z(t) = \frac{1}{2}\left[\tanh\left(\frac{t}{2T}\right) + 1\right], \tag{185}$$

which works for models with hyperbolic dependence on time. For the special case

$$\begin{aligned}
\alpha(t) &= D + A \tanh\left(\frac{t}{T}\right) \\
V(t) &= \left(A^2 + B^2\right)^{1/2} \operatorname{sech}\left(\frac{t}{T}\right)
\end{aligned} \tag{186}$$

Demkov and Kunike [39] and later Hioe [43] used the transformation (185) to transform Equation (184) into a special case of the hypergeometric differential equation

$$\frac{d^2 f}{dz^2} + \left[\frac{c}{z} + \frac{c-a-b-1}{1-z}\right]\frac{df}{dz} - \frac{ab}{z(1-z)}f = 0 . \tag{187}$$

This equation has the hypergeometric function as its solution between its singularities $z = 0$ and $z = 1$. This is usually written as

$$F(a, b, c; z) = (1 - z)^{c-a-b} F(c - a, c - b, c; z) . \tag{188}$$

Using this function to construct two independent solutions near $z = 0$ (see Reference [39]) and imposing the initial condition, one can then use the special result

$$F(a, b, c; 1) = \frac{\Gamma(c)\Gamma(c - a - b)}{\Gamma(c - a)\Gamma(c - b)} \tag{189}$$

to obtain the asymptotic result at $t = \infty$. The details of the calculation are messy, but in this way, one can obtain the solution to the dynamic problem of Equation (186) in the form

$$|C_2(\infty)|^2 = |a_2(\infty)|^2 = \frac{\cosh\left[\pi T\left(D + iB\right)\right]\cosh\left[\pi T\left(D - iB\right)\right]}{\cosh\left[\pi T\left(A + D\right)\right]\cosh\left[\pi T\left(A - D\right)\right]}. \tag{190}$$

This solution assumes $D < A$. When $D = 0$, this reproduced the result (174) which we referred to in Section 4.3.3. For more details of this very useful solvable model, see Reference [40].

Using similar transformations as (185) one can solve other pulse problems, see *e.g.* References [44] and [45]

5 Markovian quantum master equations

5.1 The generic master equation

The density matrix of a quantum system acquires its time evolution directly from the Schrödinger equation as shown in Equation (67), but this includes only the unitary part deriving from time reversible Hamiltonian dynamics. If we want to include irreversible phenomena, these add terms to the equation of motion; it is then usually called a master equation.

Physically the dissipative parts of the time evolution are caused by an interaction with an environment, which is called a bath if it is in thermal equilibrium. If the environment is large and responds very fast to the perturbation caused by the system under consideration, we can often take the dissipative terms in the master equation to depend only on the state at the time considered, no memory effects derive from the environment. In physics this is called the Markovian limit, and it is used to represent dissipation in many quantum systems of physical interest. However, the disentanglement of the bath degrees of freedom from those of the system of interest is an intricate procedure, which makes it hard to justify the Markovian approximation except in the simplest situations. In this lecture, I will not discuss the derivation of master equations, but I will discuss their applications and main properties. Further applications can be found in the lectures by Peter Knight at this school.

In many cases the Markovian master equation can be written in the form

$$\dot{\rho} = -\frac{i}{\hbar}\left[H, \rho\right] - \sum_q \left(C_q^\dagger C_q \rho + \rho C_q^\dagger C_q - 2C_q \rho C_q^\dagger\right), \tag{191}$$

where the operators C_q are arbitrary and they are defined over a range $\{q\}$. The equation is easily seen to conserve probability, *i.e.* the trace of the density matrix is not changing. It is also easy to see that the master equation evolves pure quantum states into mixtures. Start with a pure state density matrix at time $t = 0$

$$\rho(t = 0) = |\Psi\rangle\langle\Psi|. \tag{192}$$

We calculate the time derivative

$$\frac{d}{dt}\text{Tr}\left(\rho^2\right) = -4\sum_q\left[\text{Tr}\left(\rho^2 C_q^\dagger C_q\right) - \text{Tr}\left(\rho C_q \rho C_q^\dagger\right)\right]. \tag{193}$$

Using Equation (192) and the projection operator

$$\Pi = 1 - |\Psi\rangle\langle\Psi| = \Pi^2, \tag{194}$$

we obtain the result

$$
\begin{aligned}
\frac{d}{dt}\text{Tr}\left(\rho^2\right) &= -4\sum_q \left[\langle\Psi|C_q^\dagger C_q|\Psi\rangle - \langle\Psi|C_q^\dagger|\Psi\rangle\langle\Psi|C_q|\Psi\rangle\right] \\
&= -4\sum_q ||\Pi C_q|\Psi\rangle||^2 \le 0.
\end{aligned}
\tag{195}
$$

This says that the pure state distributes its probability over several states, a mixed state ensues.

A Markovian master equation of the type (191) was derived by Lindblad [46] from a mathematical structure relating to semigroup theory. The main assumption, in addition to the Markovian requirement, was that probability is conserved and that density operators with positive eigenvalues are evolved into operators with positive eigenvalues. These conditions are necessary to preserve the probabilistic interpretation of the theory. The proof is formal, and it gives no insight into the physical applicability of the theory. However, any master equation like (191) is referred to as being of the Lindblad form.

In practical applications, the Lindblad form has turned out to be very popular, because it allows one to extract the density matrix time evolution as the evolution of an ensemble of state vectors. This forms the basis of a computational algorithm, which allows one to simulate the time evolution of the ensemble on a computer. Because one only needs to evolve one member of the ensemble at any one time, some restrictions posed by computer memory size can be overcome. Each state vector history can be stored separately and used to evaluate the ensemble averages in the end. One trades memory requirements for computation time.

The stochastic ensemble theory was developed separately by Carmichael [47], who calls it the Quantum Trajectory method and by Mølmer *et al.* [48], who call it the Wave Function Monte Carlo method. It has been used widely in applications concerned with the exchange of momentum between photons and atoms.

The main features of the Monte Carlo approach are easily seen. We rewrite Equation (191) in the form

$$\dot{\rho} = -\frac{i}{\hbar}\left[\left(H - i\hbar\sum_q C_q^\dagger C_q\right)\rho - \rho\left(H + i\hbar\sum_q C_q^\dagger C_q\right)\right] + 2\sum_q C_q\rho C_q^\dagger. \tag{196}$$

The first term can be interpreted as a time evolution with the effective Hamiltonian

$$H_{\text{eff}} = H - i\hbar\sum_q C_q^\dagger C_q. \tag{197}$$

Starting with a state vector $|\Psi(t)\rangle$ at time t, we can obtain the time evolution with this Hamiltonian over an infinitesimal interval Δt in the form

$$|\Phi\rangle = \exp\left(-i\frac{H_{\text{eff}}\Delta t}{\hbar}\right)|\Psi(t)\rangle = \left(1 - \frac{i}{\hbar}H\Delta t - \sum_q C_q^\dagger C_q\Delta t\right)|\Psi(t)\rangle. \tag{198}$$

Owing to the anti-Hermitian part of the effective Hamiltonian, the norm of the state has been decreased. This we interpret to mean that the state has suffered no irreversible quantum jump in the interval Δt with the probability

$$\langle \Phi | \Phi \rangle = \langle \Psi(t) | \Psi(t) \rangle - 2 \sum_q \langle \Psi(t) | C_q^\dagger C_q | \Psi(t) \rangle \Delta t = 1 - \sum_q \Delta p_q, \qquad (199)$$

where

$$\Delta p_q = 2 ||C_q | \Psi(t) \rangle||^2 \Delta t \qquad (200)$$

is interpreted as the probability of decay into the channel denoted by the index q.

Now we make a Monte Carlo step. We generate a random number $r \in [0, 1]$ and perform one of the following operations.

(a) If $r > \sum_q \Delta p_q$, no quantum jump is supposed to have taken place, and the state vector at the end of the interval has to be renormalised to occur with unit probability. We set

$$|\Psi(t + \Delta t)\rangle_a = \frac{\exp\left(-i H_{\text{eff}} \Delta t / \hbar\right) |\Psi(t)\rangle}{\sqrt{1 - \sum_q \Delta p_q}}. \qquad (201)$$

(b) If $r < \sum_q \Delta p_q$, a quantum jump is assumed to have taken place. Now we have to perform a second Monte Carlo step to determine the branching ratios into the different channels $\{q\}$ according to the probabilities given by Δp_q. The normalised state jumping into the channel q is given by the expression

$$|\Psi(t + \Delta t)\rangle_{b,q} = \frac{C_q |\Psi(t)\rangle}{\sqrt{||C_q |\Psi(t)\rangle||^2}}. \qquad (202)$$

We have now evolved the state vector over the interval Δt; the result is either (201) or one of the results (202). Taking this as a new initial value, we can proceed with the time integration and generate a history of the time evolution of the state vector. When we have proceeded far enough, we start from the initial state again. This generates an ensemble of histories of the state vector evolution. If the initial density matrix is a mixed state, we have to include this ensemble into the simulation procedure. If the quantum jumps are taken to signify actual observations, each sequence of recorded events must eventually be found as a member of the ensemble generated. Carmichael chooses to interpret the approach as a consequence of continuous measurements; it is, however, possible to regard the procedure as a numerical algorithm only.

If the time step Δt is made short enough, it is straightforward to show that the ensemble of state vectors generated gives the same ensemble averages as the solution of the original density matrix equation of motion (191). We write the average density matrix at time $t + \Delta t$ in the form

$$\overline{|\Psi(t + \Delta t)\rangle\langle\Psi(t + \Delta t)|}$$

$$= \left(1 - \sum_q \Delta p_q\right) |\Psi(t + \Delta t)\rangle_{aa}\langle\Psi(t + \Delta t)| + \sum_q \Delta p_q |\Psi(t + \Delta t)\rangle_{b,q}{}_{b,q}\langle\Psi(t + \Delta t)|$$

$$= \left(1 - \frac{i}{\hbar} H \Delta t - \sum_q C_q^\dagger C_q \Delta t\right) |\Psi(t)\rangle\langle\Psi(t)| \left(1 + \frac{i}{\hbar} H \Delta t - \sum_q C_q^\dagger C_q \Delta t\right)$$

$$+ 2 \sum_q C_q |\Psi(t)\rangle\langle\Psi(t)| C_q^\dagger \Delta t, \qquad (203)$$

where we have used Equations (201), (198), (202) and (200). If we now let Δt go to zero, and replace the difference by the derivative according to

$$\overline{|\Psi(t + \Delta t)\rangle\langle\Psi(t + \Delta t)|} - |\Psi(t)\rangle\langle\Psi(t)| = \dot{\rho}\,\Delta t \tag{204}$$

we obtain the original equation of motion (191). This proves the equivalence between the Monte Carlo procedure and the solution of the density matrix.

Another approach to the simulation is possible. According to Equation (198) we have the probability

$$P(T) = \langle\Psi(t)|\exp\left(-2\sum_q C_q^\dagger C_q T\right)|\Psi(t)\rangle \tag{205}$$

that no quantum jump has occurred in the interval $[t, t+T]$. We can generate a sequence of instants $\{t_1, t_2, ...\}$, which realises this statistical distribution, and we assume that the quantum jumps in a single realisation take place at these times. Between them the time evolution is taken to derive only from the effective Hamiltonian (197). At each instant, when a jump occurs, we distribute the outcomes according to the results (201) and (202) as in the previous method. In this way we have generated one member of the ensemble of state vector histories. The next member is obtained by generating another random sequence of jump times. The procedure is formally equivalent, but may differ in the amount of numerical work. The former method assumes that we test for randomness after every time interval Δt, the latter may allow uninterrupted evolution for some rather long intervals. On the other hand, the latter method may spend a lot of time on a history with very small probability; such histories are efficiently eliminated in the former one. Both methods tend to accumulate inaccuracy from single histories which progress exceptionally long without a jump. According to Equation (201) they have to be divided by a very small number, which greatly amplifies the numerical error accumulated in the integration process.

5.2   Simple examples

5.2.1   The master equation for a cavity mode

In a radiation resonance cavity, the losses are due to photons absorbed in the material structures or escaped by tunneling through the end reflectors. For a small damping of the cavity eigenmodes, the losses are described by a master equation of the Lindblad type (191)

$$\begin{aligned}
\dot{\rho} &= -i\Omega\left[a^\dagger a, \rho\right] - \frac{\gamma}{2}(\bar{n} + 1)\left(a^\dagger a\,\rho + \rho\,a^\dagger a - 2a\,\rho\,a^\dagger\right)\\
&\quad -\frac{\gamma}{2}\bar{n}\left(aa^\dagger\rho + \rho\,a\,a^\dagger - 2a^\dagger\rho\,a\right),
\end{aligned} \tag{206}$$

where the thermal excitation of the fluctuating bath of vacuum modes enters through the factor

$$\bar{n} = \frac{1}{\exp[(\hbar\Omega)/(kT)] - 1}. \tag{207}$$

The master equation (206) can be used to evaluate the equations of motion for the moments of the annihilation and creation operators. Thus *e.g.* we find

$$\frac{d}{dt}\langle a \rangle = \text{Tr}\,(\dot{\rho}a) = -\left(i\Omega + \frac{\gamma}{2}\right)\langle a \rangle. \tag{208}$$

Integrating this we find the time evolution expected for a complex field amplitude. The result is the same as in the classical case. The quantum fluctuations are found in the second moments of the operators. Thus the population is determined by

$$\frac{d}{dt}\langle a^\dagger a \rangle = -\gamma(\bar{n}+1)\langle a^\dagger a \rangle + \gamma\bar{n}\left(\langle a^\dagger a \rangle + 1\right) = \gamma(\bar{n} - \langle a^\dagger a \rangle). \tag{209}$$

From the first form of the equation, we note that the term proportional to $(\bar{n}+1)$ denotes losses and the term proportional to $\bar{n}$ denotes thermally induced gain. Together these force the population of the mode towards the thermal equilibrium distribution (207). Equations of motion for higher moments can be obtained in the same manner.

5.2.2 The radiatively damped two-level system

We consider a two-level system with the levels $\{|1\rangle, |2\rangle\}$ with the energy separation $\hbar\omega$. Introducing the Pauli operator

$$\sigma^+ = |2\rangle\langle 1| = \left(\sigma^-\right)^\dagger \tag{210}$$

we obtain the master equation

$$\dot{\rho} = -\frac{i}{\hbar}[H,\rho] - \frac{\Gamma}{2}\left(\sigma^+\sigma^-\rho + \rho\sigma^+\sigma^- - 2\sigma^-\rho\sigma^+\right). \tag{211}$$

The projector

$$\sigma^+\sigma^- = |2\rangle\langle 2| \tag{212}$$

restricts the decay to the upper level, and the term

$$\sigma^-\rho\,\sigma^+ = \langle 2|\rho|2\rangle|1\rangle\langle 1| \tag{213}$$

transfers the upper state population to the lower state in a quantum jump event. Deriving the equation of motion for the expectation values of the 3 independent Pauli operators from the master equation (211) we obtain the ordinary optical Bloch equations.

In order to see the working of the Monte Carlo simulation method discussed above, we look at the two-level system when no coupling between the levels exists. The only transfer from one level to the other is then through spontaneous decay. The initial state at time $t = 0$ is given by

$$|\Psi(0)\rangle = c_1|1\rangle + c_2|2\rangle. \tag{214}$$

As long as no quantum jump event has taken place up to the time t, the normalised state vector is to be taken in the form

$$|\Psi(t)\rangle = \frac{c_1|1\rangle + c_2\,e^{-i\omega t}e^{-\Gamma t/2}|2\rangle}{\sqrt{|c_1|^2 + e^{-\Gamma t}|c_2|^2}}. \tag{215}$$

The probability of decaying from the upper state at time t is given by

$$P_2(t) = \frac{\Gamma e^{-\Gamma t}|c_2|^2}{|c_1|^2 + e^{-\Gamma t}|c_2|^2},\qquad(216)$$

which goes to zero for long times. However, if no photon is emitted, no decay is observed with an absolutely efficient detector, the state (215) goes to the lower level at infinite time. This happens with the probability $|c_1|^2$, and thus even when no photon is emitted the state is for long times reduced to the ground state with the correct probability given by the initial state. The non-observation of a photon acts as a measurement.

If a quantum jump takes place at time t, a photon is emitted, and according to the formalism, we have to reduce the state to

$$|\Psi(t+)\rangle = \frac{\Gamma\sigma^-|\Psi(t)\rangle}{\sqrt{\Gamma^2\langle\Psi(t)|\sigma^+\sigma^-|\Psi(t)\rangle}} = e^{-i\omega t}|1\rangle.\qquad(217)$$

The phase factor depends on the random jump time t, and no physical significance can be assigned to it. The phase of a state is undetermined after spontaneous emission. With the free Hamiltonian evolution only, no second photon will be seen because the ground state is not re-excited.

The form of Equation (215) opens the question of the reality of the states generated by the Monte Carlo simulation. Even if we take the quantum jump to signify the observation of a photon by a hypothetical perfect detector, the state (215) is to be assigned to the system when no observation has been made. The amount of reality belonging to such an entity is hard to grasp; may be we can choose its interpretation according to our preferences. The only certain thing is that it gives the correct ensemble averages. In addition, if the quantum jumps are recorded photon counts, the simulated state history is a possible sequence of observed photons. There seems to be a lesson about quantum reality in the procedure, but in the time honoured manner of quantum theory the formalism stubbornly resists answering too penetrating questions.

The examples treated here, the leaking cavity and the decaying atom, are both based on the emission of photons that rapidly propagate out of the space where the experiments are performed. Thus they are well described by a Markovian time evolution, and the Lindblad form is a natural representation of the evolution equation. In the next section we are going to treat a different type of dissipation, where the situation is much more complicated.

5.3   Quantum theory of friction

5.3.1   The Lindblad form

We are now going to consider quantum evolution in a simple one-dimensional dynamic system with the Hamiltonian

$$H = \frac{p^2}{2M} + U(x),\qquad(218)$$

where $\{p, x\}$ form the canonical pair of dynamical variables. As a special example we will often take the harmonic oscillator

$$H = \frac{p^2}{2M} + \frac{1}{2} M \Omega^2 x^2 = \hbar \Omega b^\dagger b. \tag{219}$$

Such motion is expected to suffer friction in a thermally excited environment. Classically the well developed theory of Brownian motion describes the situation in a physically and mathematically satisfactory way. The problem we are going to address here is the possibility of constructing an equivalent quantum theory.

Gallis [49] has recently shown how the theory of friction can be put into a form compatible with the Lindblad theory. I am going to present the main results of his formulation in a simplified manner, which makes it easier to see the emergence of the physically interesting features.

Firstly we notice, that the theory should be valid for low energy, and hence we assume that an expansion in the momentum operator p is allowed; the exact expansion parameter will be given shortly. We write for the operators in the equation of motion (191) the expressions

$$C_q = \frac{1}{\sqrt{2}} e^{iqx} A_q \left(1 - \frac{\kappa q p}{2} \right), \tag{220}$$

where the parameter κ is chosen to be

$$\kappa = \frac{\hbar \beta}{2M} = \frac{\hbar}{2MkT}. \tag{221}$$

The exponent $\exp(iqx)$ shifts the momentum by the amount $\hbar q$ which thus is the kick given to the system by the environment. The probability for this kick is determined by the (real) parameter A_q which is defined over a symmetric range of q values

$$A_q = A_{-q}. \tag{222}$$

Using the relation (220) in Equation (191) we find

$$\dot{\rho} = -\frac{i}{\hbar} [H, \rho] - \sum_q A_q^2 \left(\rho - e^{iqx} \rho e^{-iqx} \right) - \frac{\kappa}{2} \sum_q A_q^2 q \, e^{iqx} [p, \rho]_+ \, e^{-iqx}$$
$$- \frac{\kappa^2}{8} \sum_q A_q^2 q^2 \left(p^2 \rho + \rho p^2 - 2 e^{iqx} p \rho p \, e^{-iqx} \right). \tag{223}$$

This equation is our main result, by construction it is compatible with the Lindblad form.

In order to see the physical consequences of Equation (223) we look at the equations of motion for the moments. Using the relation

$$e^{-iqx} p e^{iqx} = p + \hbar q \tag{224}$$

we obtain the results

$$\frac{d}{dt} \langle x \rangle = \frac{\langle p \rangle}{M}$$
$$\frac{d}{dt} \langle p \rangle = -\left\langle \frac{\partial U}{\partial x} \right\rangle - \gamma \langle p \rangle, \tag{225}$$

where the damping constant for the momentum is given by

$$\gamma = \frac{1}{kT} \sum_q A_q^2 \frac{\hbar^2 q^2}{2M}. \tag{226}$$

If we define the relaxation time constant τ by setting

$$\sum_q A_q^2 = \frac{1}{\tau} \tag{227}$$

and introduce a mass m for the particles constituting the environment, we can write the damping constant in the form

$$\gamma = \left(\frac{\bar{\varepsilon}}{kT}\right) \left(\frac{m}{M}\right) \frac{1}{\tau}. \tag{228}$$

The first factor contains the average kinetic energy of the particles in the bath

$$\bar{\varepsilon} = \left(\sum_q A_q^2\right)^{-1} \sum_q A_q^2 \frac{\hbar^2 q^2}{2m}; \tag{229}$$

this factor is expected to be of the order unity. The second factor is the ratio between the mass of the particles in the bath to that of the particle under consideration; this is the familiar small expansion parameter of the classical Brownian motion theory. The final factor is the relaxation time constant.

An important fact about Equations (225) is that the coordinate is not damped. For a physical particle, this is correct, in contrast to the situation for the damped cavity mode of Section 5.2.1. There the two quadrature components are physically equivalent and must be damped equally. The other significant fact is that there are no terms of the form $\langle p^2 \rangle$ and $\langle p^3 \rangle$. This is essential, because the equation is supposed to be used down to $\langle p \rangle \to 0$, and if present, the higher order terms would eventually dominate. Thus the Gallis operator gives the expected first moments.

For the second moments we find from the master equation (223) the results

$$\begin{aligned}
\frac{d}{dt}\langle p^2 \rangle &= \left\langle \left(p\frac{\partial U}{\partial x} + \frac{\partial U}{\partial x}p \right) \right\rangle - \Gamma \langle p^2 \rangle + 2D_p \\
\frac{d}{dt}\langle (xp + px) \rangle &= -\gamma (xp + px) + \frac{2\langle p^2 \rangle}{M} - 2\left\langle x\frac{\partial U}{\partial x} \right\rangle \\
\frac{d}{dt}\langle x^2 \rangle &= \frac{\langle (xp + px) \rangle}{M} + 2D_x.
\end{aligned} \tag{230}$$

The parameters appearing in these equations are

$$\begin{aligned}
\Gamma &= \frac{2m}{M} \sum_q A_q^2 \left(\frac{\hbar^2 q^2}{2mkT}\right) \left(1 - \frac{m}{4M}\left(\frac{\hbar^2 q^2}{2mkT}\right)\right) \equiv 2\gamma(1 - \eta) \\
D_p &= \frac{1}{2} \sum_q A_q^2 \hbar^2 q^2 = \gamma M kT \\
D_x &= \frac{\hbar^2 m}{8M^2 kT} \sum_q A_q^2 \left(\frac{\hbar^2 q^2}{2mkT}\right) = \frac{\hbar^2}{8M^2 (kT)^2} D_p.
\end{aligned} \tag{231}$$

The relation between the momentum diffusion coefficient D_p and the linear damping coefficient γ is seen to follow the classical Einstein relation. The parameter η is a small expansion parameter of the order of the mass ratio.

For the harmonic oscillator potential (219) we have

$$\frac{\partial U}{\partial x} = M\Omega^2 x, \tag{232}$$

and the second moment equations (230) close. It is easy to obtain the steady state solutions

$$\langle (xp + px) \rangle = -2MD_x \sim O(\hbar^2). \tag{233}$$

The correlations between position and momentum acquires a quantum correction which disappears in the classical limit. We also obtain

$$\Gamma \langle p^2 \rangle = 2 \left(D_p - M^2 \Omega^2 D_x \right), \tag{234}$$

from which we find

$$\frac{\langle p^2 \rangle}{2M} = \frac{kT}{2(1-\eta)} \left(1 - \frac{1}{8} \left(\frac{\hbar\Omega}{kT} \right)^2 \right). \tag{235}$$

Finally we have

$$\frac{1}{2} M\Omega^2 \langle x^2 \rangle = \frac{\langle p^2 \rangle}{2M} + \frac{1}{2} \gamma M D_x = \frac{\langle p^2 \rangle}{2M} + \frac{\hbar\gamma}{4} \left(\frac{\hbar\gamma}{kT} \right). \tag{236}$$

From Equations (235) and (236) we see the quantum corrections to the virial result; when they are neglected, and we put η equal to zero, the classical equipartition emerges.

The small parameter in this approach is proportional to the momentum. In Equation (220) the expansion parameter is κqp, which we estimate according to

$$(\kappa qp)^2 = \left(\frac{m}{M} \right) \left(\frac{p^2}{2MkT} \right) \left(\frac{\hbar^2 q^2}{2mkT} \right). \tag{237}$$

The last factor is expected to be unity; near equilibrium the second factor is of the same order. The remaining first factor is small because of the basic assumption of a Brownian motion situation, the environment mass is much smaller than that of the particle of interest. In this limit we can also replace the momentum factor in Equation (220) by an exponential

$$\left(1 - \frac{\kappa qp}{2} \right) \approx \exp\left(-\frac{\kappa qp}{2} \right). \tag{238}$$

The two factors in (220) do not commute; it is, however, easy to show that the corrections coming from choosing a different order are of higher order in the small expansion parameter (237).

The results derived in this section are exact consequences of the Gallis ansatz (220). Similar results have been given by Diosi [50] and also by Dalibard and Castin [51]. They are of the Lindblad form, but it is far more difficult to base a Monte Carlo algorithm on these results than it was in the cases with photon emission, treated in the previous section.

5.3.2 The weak kick limit

We are now going to expand the master equation in a limit we call the "weak kick" limit. That assumes a very narrow range of values in the summation over q, so that we can set

$$\sum_q A_q^2 q^{2n} \approx 0, \quad \text{for } n > 1. \tag{239}$$

In addition we neglect the quantum correction D_x. Using the expansion

$$e^{iqx} B e^{-iqx} = B + iq\,[x, B] - \frac{q^2}{2}\,[x, [x, B]] + O(q^3) \tag{240}$$

we find from Equation (223) the master equation [52], [53]

$$\dot{\rho} = -\frac{i}{\hbar}\,[H, \rho] - \frac{i\gamma}{2\hbar}\,\big[x, [p, \rho]_+\big] - \frac{D_p}{\hbar^2}\,[x, [x, \rho]] . \tag{241}$$

This equation is easily seen to give the classical values for the second moments in Equation (230) without any quantum corrections.

If we write the equation of motion for the Wigner function with the harmonic oscillator Hamiltonian, we find from (241)

$$\frac{\partial W}{\partial t} + \frac{p}{M}\frac{\partial W}{\partial x} - \frac{\partial U}{\partial x}\frac{\partial W}{\partial p} = \frac{\partial}{\partial p}\left(\gamma p W\right) + D_p \frac{\partial^2 W}{\partial p^2} . \tag{242}$$

This is exactly the classical Liouville equation for Brownian motion, and quantum considerations enter only through the restrictions on the initial conditions for the Wigner function. Thus Equation (241) appears to be a satisfactory result, which should be used to describe quantum damping. However, a more detailed consideration shows that it is seriously flawed and provides a questionable starting point for quantum calculations. For classical initial states, it is perfectly acceptable.

As a preliminary consideration of the behaviour of the master equation, we rewrite it using the creation and annihilation operators defined by

$$x = \sqrt{\frac{\hbar}{2M\Omega}}\left(b + b^\dagger\right) \qquad p = \sqrt{\frac{\hbar\Omega M}{2}}\left(\frac{b - b^\dagger}{i}\right) . \tag{243}$$

Using the expansions

$$
\begin{aligned}
x\rho p - p\rho x &= \frac{2\hbar}{i}\left(b^\dagger \rho b - b\rho b^\dagger\right) \\
px &= \frac{1}{2}\left(xp + px\right) - \frac{i}{2}\left[b, b^\dagger\right] \\
xp &= \frac{1}{2}\left(xp + px\right) + \frac{i}{2}\left[b, b^\dagger\right]
\end{aligned}
\tag{244}
$$

we can rewrite the master equation in the form

$$
\begin{aligned}
\dot{\rho} =\ & -i\Omega\left[b^\dagger b, \rho\right] - \frac{i\gamma}{4\hbar}\left[\left(xp + px\right), \rho\right] \\
& -\frac{D_p}{\hbar^2}[x, [x, \rho]] - \frac{\gamma}{4}\left(b^\dagger b\rho + \rho b^\dagger b - 2b\rho b^\dagger\right) \\
& +\frac{\gamma}{4}\left(bb^\dagger \rho + \rho bb^\dagger - 2b^\dagger \rho b\right) .
\end{aligned}
\tag{245}
$$

The terms on the first line clearly denote Hamiltonian evolution; the dissipative term adds a renormalisation to this. The terms on the second line are clearly of the Lindblad form, but the final term has got the wrong sign. In a simulation, it denotes a process occurring with negative probability, and as such it cannot represent any physical process. Something is wrong!

One more calculation can be carried out before we continue our discussion of the problems with the master equation. The renormalised Hamiltonian is given by

$$H_{\text{ren}} = \hbar\Omega\, b^\dagger b + \frac{\gamma}{4}(xp + px) = \hbar\Omega\, b^\dagger b + i\frac{\gamma}{4}\left(b^\dagger b^\dagger - bb\right). \tag{246}$$

This Hamiltonian can be diagonalised by the unitary transformation

$$b = \sqrt{i}\left[\cosh\left(\frac{\theta}{2}\right)A - \sinh\left(\frac{\theta}{2}\right)A^\dagger\right], \tag{247}$$

when the angle θ is determined by

$$\tanh\theta = \frac{\gamma}{2\Omega}. \tag{248}$$

The result is the Hamiltonian

$$H_{\text{ren}} = \sqrt{\Omega^2 - \frac{\gamma^2}{4}}\, A^\dagger A. \tag{249}$$

The renormalised oscillation frequency thus agrees with the well known result for the damped classical oscillator. The unitary part of the master equation is found to give the correct frequency shift.

With a harmonic oscillator, it is often permissible to carry out a rotating wave approximation on the double x-commutator in Equation (245). We set

$$\frac{D_p}{\hbar^2}[x,[x,\rho]] = \frac{D_p}{2M\Omega\hbar}\left([b,[b^\dagger,\rho]] + [b^\dagger,[b,\rho]]\right). \tag{250}$$

This gives the master equation

$$\dot\rho = -i\Omega\left[b^\dagger b,\rho\right] - D_+\left(b^\dagger b\rho + \rho b^\dagger b - 2b\rho b^\dagger\right) - D_-\left(bb^\dagger\rho + \rho bb^\dagger - 2b^\dagger\rho b\right), \tag{251}$$

where

$$D_\pm = \frac{D_p}{2M\hbar\Omega} \pm \frac{\gamma}{4} = \frac{D_p}{2M\hbar\Omega}\left(1 \pm \frac{\hbar\Omega}{2kT}\right). \tag{252}$$

The master equation (251) is exactly of the form (209) with the ratio between the coefficients given by

$$\frac{D_+}{D_-} = \left(1 - \frac{\hbar\Omega}{2kT}\right)^{-1}\left(1 + \frac{\hbar\Omega}{2kT}\right) \approx \frac{\overline{n}+1}{\overline{n}} = \exp\left(\frac{\hbar\Omega}{kT}\right). \tag{253}$$

Thus, in the rotating wave limit, the master equation is of the acceptable type, compatible with the Lindblad form, and well tested in numerous applications.

We still need to see exactly where the original master equation (241) goes wrong; this will be considered in detail in the following section.

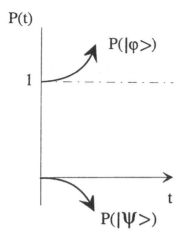

Figure 13. *Breakdown of the probability interpretation.*

5.3.3 Nonphysical features of the master equation

I am now going to show in some detail what goes wrong with the application of the master equation (241). To this end I introduce an initial quantum state

$$\rho(t = 0) = |\varphi\rangle\langle\varphi|, \tag{254}$$

where the initial wave function is assumed real and symmetric

$$\varphi(x) = \langle x|\varphi\rangle = \varphi(-x). \tag{255}$$

We are now going to calculate the probability that the system remains in the original state over a short time interval near the origin. The Hamiltonian part of the evolution is a rotation in the Hilbert space and gives only a small contribution initially; hence only the dissipative part needs to be considered.

We obtain

$$
\begin{aligned}
\frac{d}{dt}\langle\varphi|\rho|\varphi\rangle &= -\frac{i\gamma}{2\hbar}\left(\langle\varphi|xp|\varphi\rangle - \langle\varphi|px|\varphi\rangle\right) - \frac{2D_p}{\hbar^2}\left(\langle\varphi|x^2|\varphi\rangle - \langle\varphi|x|\varphi\rangle^2\right) \\
&= \frac{\gamma}{2} - \frac{2D_p}{\hbar^2}\overline{\Delta x^2}.
\end{aligned}
\tag{256}
$$

For initial states narrow enough,

$$\overline{\Delta x^2} < \frac{\hbar^2\gamma}{4D_p}, \tag{257}$$

the probability of finding the system in the initial state $|\varphi\rangle$ grows, but having been initially unity it must exceed this value, at least for some time (see Figure 13). This breakdown of the probability interpretation has been found by Gardiner and Munro [54].

However, it is easy to see that the master equation (241) preserves the trace of the density matrix. This being unity originally, the appearance of a probability larger than one must be accompanied by some negative probability. Otherwise the probabilities could not sum to unity. The state which acquires this negative probability was published by Ambegaokar already in 1991 [55].

In addition to the initial state $|\varphi\rangle$ we now define another state

$$|\psi\rangle = p|\varphi\rangle. \tag{258}$$

We have already chosen $\varphi(x)$ to be even, and hence the state $\psi(x)$ is odd and orthogonal to $\varphi(x)$. Thus it occurs with zero probability originally. We calculate the rate of change of its occurrence probability for a short time after the initial instant. Since we have

$$\rho|\psi\rangle = 0, \qquad \langle\psi|\rho = 0 \tag{259}$$

we find

$$\frac{d}{dt}\langle\psi|\rho|\psi\rangle = -\left(\frac{i\gamma}{2\hbar}\right)\left(\langle\psi|x\rho p|\psi\rangle - \langle\psi|p\rho x|\psi\rangle\right) + \frac{2D_p}{\hbar^2}\langle\psi|x\rho x|\psi\rangle. \tag{260}$$

Because of the relation (258), we have

$$\begin{aligned}
\langle\psi|x\rho p|\psi\rangle - \langle\psi|p\rho x|\psi\rangle &= \langle\varphi|p^2|\varphi\rangle\left(\langle\varphi|px|\varphi\rangle\right) - \langle\varphi|xp|\varphi\rangle\right) \\
&= -i\hbar\langle\varphi|p^2|\varphi\rangle.
\end{aligned} \tag{261}$$

Further, we obtain

$$\langle\psi|x\rho x|\psi\rangle = |\langle\varphi|xp|\varphi\rangle|^2. \tag{262}$$

This matrix element can be calculated from a partial integration

$$\begin{aligned}
\langle\varphi|xp|\varphi\rangle &= -i\hbar\int\varphi(x)x\frac{d}{dx}\varphi(x)dx = i\hbar\int\varphi(x)\frac{d}{dx}x\varphi(x)dx \\
&= i\hbar\int\varphi(x)^2dx + i\hbar\int\varphi(x)x\frac{d}{dx}\varphi(x)dx.
\end{aligned} \tag{263}$$

From this we find that the matrix element in Equation (262) equals $i\hbar/2$. Inserting into Equation (260) we find

$$\frac{d}{dt}\langle\psi|\rho|\psi\rangle = \frac{1}{2}\left(D_p - \gamma\langle p^2\rangle\right). \tag{264}$$

Because the wave function is real, there is no average momentum and $\langle p^2\rangle$ is directly the spread. Thus we find that the probability of finding the system in the state $|\psi\rangle$ decreases initially if

$$\overline{\Delta p^2} > \frac{D_p}{\gamma}. \tag{265}$$

Because the probability is initially zero, it must take negative values at least for some short period of time (see Figure 13). This is what allows the probabilities to sum to zero even if they separately violate the physical interpretation. If we assume the initial state to be a minimum uncertainty state

$$\Delta x\,\Delta p = \frac{\hbar}{2}, \tag{266}$$

the two conditions (257) and (265) are simultaneously satisfied.

We have found that the master equation (241), in spite of all its attractive features, cannot consistently describe dissipative quantum evolution except in the region where it can be approximated by its rotating wave equivalent. It leads to nonphysical probabilities and the conclusion is that even starting with an allowed Wigner function, Equation (242) may evolve it into a function unacceptable as a Wigner function. This is in stark contrast to the classical phase space evolution; an allowed distribution function (*i.e.* positive and normalisable) is always developed into another allowed distribution by Equation (242). The equation itself is, however, exactly the same.

Thus we end on a curious note. The master equation we have considered gives acceptable first and second moments, but it does not give acceptable density matrices for future times. Thus the approach is not safe to use except in the rotating wave approximation. This, however, is only useful for the harmonic oscillator or potentials very similar to it. For a general potential, the treatment of linear friction in a correct and practically useful way still poses an open problem.

References

[1] M Gruebele and A H Zewail, *Physics Today*, May 1990, page 24; A H Zewail, *Scientific American*, **263**, December 1990, page 40

[2] T Baumert, V Engel, C Röttgermann, W T Strunz, and G Gerber, *Chem Phys Lett*, **191**, 639 (1992); T Baumert, V Engel, C Meier, and G Gerber, *Chem Phys Lett*, **200**, 488 (1992)

[3] T J Dunn, J N Sweetser, and I A Walmsley, *Phys Rev Lett*, **70**, 3388, (1993)

[4] P S Julienne, A M Smith, and K Burnett, *Adv At Mol Opt Phys*, **30**, 141 (1993)

[5] E O Göbel, *Optics and Photonics News*, **3**, 33, (1992)

[6] B M Garraway, S Stenholm, and K-A Suominen, *Physics World*, **6**, April 1993, page 46

[7] G Baym, *Lectures on Quantum Mechanics* (W A Benjamin, Inc , New York, 1969)

[8] C Alden Mead, *Rev Mod Phys*, **64**, 51 (1992)

[9] S Stenholm, *Rev Mod Phys*, **58**, 699 (1986)

[10] C S Adams, M Sigel, and J Mlynek, *Phys Rep*, **240**, 143 (1994)

[11] J Janszky and Y Y Yusin, *Opt Comm*, **59**, 151 (1986) and A V Vinogradov and J Janszky, *Phys Rev Lett*, **64**, 2771 (1990)

[12] H P Yuen, *Phys Rev A*, **13**, 2226 (1976)

[13] E P Wigner, *Phys Rev*, **40**, 749 (1932)

[14] E P Wigner, in *Perspective in Quantum Theory*, edited by W Yourgrau and A van der Merwe (Dover, New York, 1971), page 25

[15] J H Shirley and S Stenholm, *J Phys A: Math Gen*, **10**, 613 (1977)

[16] W G Unruh and W H Zurek, *Phys Rev D*, **40**, 1071 (1989)

[17] J E Moyal, *Proc Cambridge Philos Soc*, **45**, 99 (1949)

[18] V I Tatarski, *Sov Phys -Usp*, **26**, 311 (1983)

[19] M Hillary, R F O'Connell, M O Scully, and E P Wigner, *Phys Rep*, **106**, 121 (1984)

[20] K Husimi, *Proc Phys Math Soc Japan*, **22**, 264 (1940)

[21] S Stenholm, *Ann Phys (N Y)*, **218**, 233 (1992)

[22] S Chandrasekhar, *Rev Mod Phys*, **15**, 1 (1943)

[23] R L Hudson, *Rep Math Phys*, **6**, 249 (1974)

[24] F Soto and P Claverie, *J Math Phys*, **24**, 97 (1983)

[25] S Stenholm, *J Mod Opt*, **39**, 279 (1992)

[26] E J Heller, *J Chem Phys*, **65**, 1289 (1976)

[27] S Stenholm, *Phys Rev A*, **47**, 2523 (1993)

[28] A M Ozorio de Almeida, *Hamiltonian Systems: Chaos and Quantization* (Cambridge Univ Press, Cambridge, 1988)

[29] R P Feynman, F L Vernon, and R W Hellwarth, *J Appl Phys*, **28**, 49 (1957)

[30] S Stenholm, *Foundations of Laser Spectroscopy* (J Wiley, New York, 1984)

[31] A M Dykne, *Sov Phys JETP*, **11**, 411 (1960) and *ibid* **14**, 941 (1962)

[32] J P Davis and P Pechukas, *J Chem Phys*, **64**, 3129 (1976)

[33] L D Landau, *Phys Z Soviet Union*, **2**, 46 (1932)

[34] C Zener, *Proc Roy Soc (London) A*, **137**, 696 (1932)

[35] B M Garraway and S Stenholm, *Phys Rev A*, **46**, 1413 (1992)

[36] K-A Suominen, *Opt Comm*, **93**, 126 (1992)

[37] N Rosen and C Zener, *Phys Rev*, **40**, 502 (1932)

[38] K-A Suominen, B M Garraway, and S Stenholm, *Opt Comm*, **82**, 260 (1991)

[39] Yu Demkov and M Kunike, *Vestn Leningr Univ Ser Fiz Khim*, **16**, 39 (1969)

[40] K-A Suominen and B M Garraway, *Phys Rev A*, **45**, 374 (1992)

[41] K-A Suominen, *Time dependent two-state models and wave packet dynamics*, Report Series in Theoretical Physics, University of Helsinki, HU-TFT-IR-92-1, Helsinki 1992

[42] N F Ramsey, *Molecular Beams* (Oxford Univ Press, Oxford, 1990)

[43] F T Hioe, *Phys Rev A*, **30**, 2100 (1984)

[44] F T Hioe and C E Carroll, *J Opt Soc Am B*, **2**, 497 (1985) and *Phys Rev A*, **41**, 2835 (1990)

[45] A Bambini and P Berman, *Phys Rev A*, **23**, 2496 (1981)

[46] G S Lindblad, *Rep Math Phys*, **10**, 393 (1976)

[47] H Carmichael *An Open Systems Approach to Quantum Optics* (Springer-Verlag, Heidelberg, 1993)

[48] K Mølmer, Y Castin, and J Dalibard, *J Opt Soc Am B*, **10**, 527 (1993)

[49] M R Gallis, *Phys Rev A*, **48**, 1028 (1993)

[50] L Diosi, *Physica A*, **194**, 517 (1993), and to be published

[51] J Dalibard and Y Castin, to be published

[52] G Agarwal, *Phys Rev*, **4**, 739 (1971)

[53] A O Caldeira and A J Leggett, *Physica A*, **121**, 587 (1983)

[54] C W Gardiner and W J Munro, to be published

[55] V Ambegaokar, *Ber Bunsenges Phys Chem*, **95**, 400 (1991)

Atomic hydrogen in magnetostatic traps

J T M Walraven

University of Amsterdam

1   Introduction

This course discusses the properties of atomic quantum gases at ultra-low temperatures with atomic hydrogen as a primary example. As hydrogen is known, to the public at large, mostly for the various ways in which it participates in violent processes, it must be emphasised right from the start that we will meet in this course atomic hydrogen as a delicate substance that behaves much alike an inert gas. In the lectures presented in Stirling three topics where emphasised: (a) the position of hydrogen among the quantum gases, (b) experiments with magnetically trapped hydrogen and (c) quantum adsorption of atomic hydrogen on the surface of liquid helium. In these written notes the attention is concentrated on the first two topics. The third topic has been described in pedagogical detail by the present author in the lecture notes of the Les Houches summer school of 1990 [1].

Rather than outlining the peculiarities of atomic hydrogen, emphasis will be put on general properties of quantum gases. Building on the extensive experience of various research groups with the low temperature gas phase of atomic hydrogen, many interesting aspects could be illustrated with well established (and sometimes very familiar) experimental and theoretical results. The main topics of the course are described in two sections: properties of weakly interacting Bose gases in external potential fields (Section 2) and evaporative cooling (Section 3) as a prominent method of achieving ultra-low temperatures in trapped atomic gases. The text starts with an introductory summary of basic properties of spin-polarised hydrogen. Comprehensive reviews that may serve for further introduction to atomic hydrogen *as a quantum gas* were written by Greytak and Kleppner [2] and by Silvera and Walraven [3]. Also several articles in the book *Bose-Einstein Condensation* [4] may be valuable to meet the dilute Bose gases in a broader context.

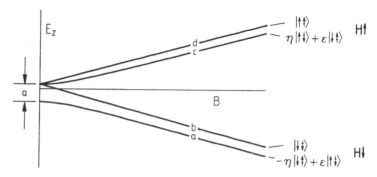

Figure 1. *The hyperfine structure of the 1S manifold of the hydrogen atom. E_z is the Zeeman energy, $\eta = \cos\theta$ and $\varepsilon = \sin\theta$; the other symbols are defined in the text.*

In the final stages of preparation of the manuscript the observation of Bose-Einstein condensation in ultra-cold atomic rubidium vapour was announced by Cornell and Wieman at JILA in Boulder [5]. This exciting result adds a very timely aspect to the article. It is hoped that the emphasis on general properties will make the paper valuable to a wide audience as an introduction to a fascinating field.

1.1 Single atom properties

Atomic hydrogen (H) is a bosonic atom composed of an electron tightly bound to a proton and having a $^2S_{1/2}$ electronic ground state. Due to the presence of nonzero nuclear spin the ground state has hyperfine structure. The electronic and nuclear magnetic moment follow from $\boldsymbol{\mu}_e = -g_e\mu_B\mathbf{s}$ and $\boldsymbol{\mu}_p = g_p\mu_N\mathbf{i}$, respectively, where μ_B is the Bohr magneton, μ_N the nuclear magneton and g_e, g_p are the associated g factors (defined as positive numbers). In terms of these quantities the gyromagnetic ratios are defined by $\gamma_e = g_e\mu_B/\hbar \approx 1.76 \times 10^{11}\text{s}^{-1}\text{T}^{-1}$ and $\gamma_p = g_p\mu_N/\hbar \approx 2.68 \times 10^8\text{s}^{-1}\text{T}^{-1}$, where $2\pi\hbar$ is the Planck constant. The Hamiltonian describing the ground state hyperfine structure is given by the expression

$$H = (g_e\mu_B\mathbf{s} - g_n\mu_N\mathbf{i})\cdot\mathbf{B} + a_h\mathbf{i}\cdot\mathbf{s} \,, \tag{1}$$

where $a_h/(2\pi\hbar) \approx 1420\text{MHz}$ is the hyperfine interaction constant. For $s = 1/2$ and $i = 1/2$ the Hamiltonian (1) leads to the well known energy level diagram shown in Figure 1 with a zero-field hyperfine splitting equal to a_h [6]. In zero field $\mathbf{f} = \mathbf{i} + \mathbf{s}$ and m_f are good quantum numbers, whereas in high fields ($B \gg a_h/\mu_e \simeq 50.7\text{mT}$) this holds for m_s, m_i and m_f.

By convention the ground state hyperfine levels are labeled a, b, c and d in order of increasing energy in small magnetic field. The c and d levels cross at $B = 16.7\text{T}$. The b and d states are pure spin states, the a and c states are hyperfine-mixed linear combinations of the high field basis states $|m_s, m_i\rangle$:

$$\begin{aligned}
|a\rangle &= \sin\theta \,|\uparrow\downarrow\rangle - \cos\theta\,|\downarrow\uparrow\rangle\,, & |b\rangle &= |\downarrow\downarrow\rangle \\
|c\rangle &= \cos\theta\,|\uparrow\downarrow\rangle + \sin\theta\,|\downarrow\uparrow\rangle\,, & |d\rangle &= |\uparrow\uparrow\rangle
\end{aligned} \tag{2}$$

where $\tan 2\theta \equiv a_{\text{h}}/[\hbar(\gamma_e + \gamma_p)B]$. The simple arrows $\uparrow$ and $\downarrow$ refer to the magnetic quantum number of the electron spins and crossed arrows $\Uparrow$ and $\Downarrow$ to that of the proton spins.

The gas phase is characterised by the spin-polarisation of the constituent atoms. Unpolarised gas is referred to as H, up or down electron-spin-polarised gases as H↑ and H↓, respectively. Further, one distinguishes the doubly (both electron and proton spin) polarised gases, consisting predominantly of b-state (H↓⇓) or d-state (H↑⇑) atoms. Sometimes it is convenient to describe the atoms by the direction of the force caused by a magnetic field gradient. For this purpose we introduce the terminology 'high-field-seekers' (for H↓) and 'low-field-seekers' (for H↑) . The first experiments with spin-polarised hydrogen were done with H↓ [7]. In these lectures hydrogen is used as an example to introduce the properties of ultra-cold atomic gases confined near the field minimum of magnetostatic traps. Therefore, the discussion will be restricted primarily to the low-field-seeking gas, H↑.

1.2 Interatomic interactions

The interaction between two H atoms depends on their spin states. Thus, the four $1^2S_{1/2}$ hyperfine states give rise to 16 potential energy curves. However, for the collisional motion a description in terms of two potential curves is often sufficient. The dominant interaction is the Coulomb interaction, which is usually written in a spin-dependent form as the sum of a direct (V_D) and an exchange (J) contribution

$$H_{\text{int}} = V_{\text{D}}(r_{ij}) + J(r_{ij})\,\mathbf{s}_i\cdot\mathbf{s}_j \,, \tag{3}$$

where $r_{ij} = |\mathbf{r}_i - \mathbf{r}_j|$ is the internuclear distance between atoms i and j at position $\mathbf{r}_i$ and $\mathbf{r}_j$. The Hamiltonian (3) conserves the total spin $\mathbf{S} = \mathbf{s}_i + \mathbf{s}_j$ and gives rise to a singlet ($S = 0$) potential, $V_s = V_D - J/4$, and a triplet ($S = 1$) potential, $V_t = V_D - 3J/4$, which correspond, respectively, to the X-$^1\Sigma_g^+$ and the b-$^3\Sigma_u^+$ electronic states of the quasi-molecule. The exchange interaction $J = V_t - V_s$ vanishes exponentially with the interatomic distance. The X-$^1\Sigma_g^+$ and the b-$^3\Sigma_u^+$ states are known to high accuracy [8, 9]. Both $V_t(r)$ and $V_s(r)$ are shown schematically in Figure 2. The triplet potential has a very shallow attractive minimum, only 6.5K deep, located at $r_{\text{min}} \approx 4.16$Å. The zero crossing of this potential occurs at an interatomic distance $r_0 \approx 3.68$Å and is thus rather large for a small atom like H. A convenient fitting function for the triplet potential is presented in the literature [10].

For two H atoms in the 'pure' hyperfine d state (or similarly for two atoms in the b state) we have full spin polarisation and the atoms interact purely via the triplet potential. Collisions between two c state atoms cannot be described in terms of a pure triplet or singlet potential but involve a mixed state due to the asymptotic presence of the hyperfine interaction. Neglecting some triplet-singlet crossover near $r_{ij} = 5.8$Å (where $J(r_{ij}) \approx a_{\text{h}}$) the interaction (3) enables a fairly accurate description of hyperfine transitions due to 'spin-exchange'. A full account using a coupled-channel approach has been given [11].

The next interaction to consider is the magnetic dipole-dipole interaction

$$H_{\text{dip}} = \frac{\mu_0}{4\pi\, r_{ij}^3}\left[\boldsymbol{\mu}_i\cdot\boldsymbol{\mu}_j - 3(\boldsymbol{\mu}_i\cdot\hat{\mathbf{r}}_{ij})(\boldsymbol{\mu}_j\cdot\hat{\mathbf{r}}_{ij})\right] \,, \tag{4}$$

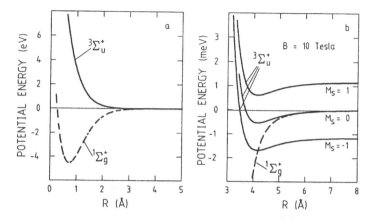

Figure 2. *The singlet and triplet interatomic potentials as calculated by Kołos and Wolniewicz. Figure (b) shows the lifting of electron-spin-degeneracy by a magnetic field on a magnified vertical scale with respect to Figure (a).*

where μ_i and μ_j are the electronic spin-dipole moments of atoms i and j at relative position $r_{ij} = r_i - r_j$, with modulus r_{ij} and direction unit vector $\hat{r}_{ij}$. This interaction is responsible for the magnetic relaxation in d-d collisions and ultimately limits the stability of H↑ in a magnetic trap. Note that the operator (4) does not induce triplet to singlet transitions as it conserves the total spin of the two interacting electrons (property of $s = 1/2$ system only!). The relaxation is associated with a change in total spin projection M_S which is possible because H_{dip} is a second rank tensor (in contrast to Equation (3), which is isotropic) and therefore allows transfer of spin angular momentum to the orbital angular momentum of the atomic motion. For all interatomic distances relevant at low temperature the electron spin-dipole interaction is even weaker than the hyperfine interaction. To very good approximation the relaxation may be calculated in first order perturbation theory, using $V_t(r)$ to describe the relative atomic motion (distorted wave Born approximation). Spin-dipole interactions between an electron spin on one atom and a proton spin on the other atom are weaker by a factor γ_p/γ_e and unimportant in magnetostatic traps. The nuclear spin-dipole interaction is weaker by another factor γ_p/γ_e and is negligible in H for any practical purpose.

1.3 Stability considerations

In the context of the present paper we are mainly interested in the stability of the trappable, *i.e.* low-field-seeking gas, H↑. A general treatment of the decay kinetics of atomic hydrogen, including two-body and three-body phenomena, was first given by Kagan *et al.* [12]. An introductory discussion is given in the review by Silvera and Walraven [3]. For magnetically trapped H the stability was analysed and calculated in specialised papers (see [13, 11]). Here the discussion will be restricted to a brief summary.

The main causes for decay of H↑ are spin exchange and magnetic dipolar interactions between the atoms. Spin exchange is most efficient (provided the hyperfine mixing angle is not too small) but only leads to relaxation in collisions between two c state atoms. For c-d collisions spin-exchange only proceeds via odd partial waves which are not populated at temperatures below 1K (see Section 2.2). Hence, in the case of d-d and c-d collisions, relaxation proceeds more slowly because it is induced by the relatively weak dipolar interaction. This implies a preferential depletion of the c-state component and ultimately leads to a pure d-state gas [13]. For a pure d-state sample the decay rate is given by the expression

$$\tau_{\text{dip}}^{-1} = n_0 \langle G_{dd} \rangle, \tag{5}$$

where $\langle G_{dd} \rangle$ is the loss rate constant for dipolar relaxation averaged over the trap. In the notation of Reference [11] the rate constant may be expressed as $G_{dd} = 2G_{ddaa}^d + G_{ddac}^d + G_{ddad}^d$. This notation shows the various contributions to the rate in terms of the detailed initial and final hyperfine states. Note that the relaxation events can produce atoms both in trapped and untrapped hyperfine states. Importantly, in the low temperature limit, $kT \ll a_h$, the dipolar rate constant (and thus also τ_{dip}^{-1}) is temperature independent. For the trap used in Amsterdam [14] with $n_0 = 10^{12} \text{cm}^{-3}$ and $T = 10\text{mK}$ the average yields $\langle G_{dd} \rangle \approx 2 \times 10^{-15} \text{cm}^3/\text{s}$ and $\tau_{\text{dip}} \approx 1000\text{s}$. For spin-exchange in a pure c-state gas under the same conditions one calculates $\langle G_{cc} \rangle \approx 10^{-13} \text{cm}^3/\text{s}$ and $\tau_{\text{ex}} \approx 20\text{s}$. Here the rate constant may be expressed as $G_{cc} = 2G_{ccaa}^c + G_{ccac}^c + G_{ccbd}^c$. The dipolar relaxation rate has been measured both in Amsterdam [15] and at MIT [16]. The results are in agreement with theory. Thus far, the spin exchange rates have not been established experimentally in magnetic traps.

Unlike in other systems [17], such as the cold alkali vapours, three-body recombination is completely negligible in a pure d-state gas. This finds its origin in the extremely weak elastic potential that does not allow dimerisation. Three-body recombination in d-state hydrogen can only occur in combination with an electronic spin flip induced by the spin-dipole interaction [12].

2 Quantum gases

Let us now turn to a more general level and consider a thermal gas of cold atoms at number density n and temperature T. Assuming the atoms to interact pairwise via an isotropic central potential U the Hamiltonian has the following form

$$H = -\frac{\hbar^2}{2m} \sum_i \nabla_i^2 + \sum_{i<j} U(r_{ij}). \tag{6}$$

In such a gas the characteristic thermal momentum $\hbar k$ is given by $k \approx \Lambda^{-1}$, where $\Lambda \equiv [2\pi\hbar^2/(mk_BT)]^{1/2}$ is the thermal de Broglie wavelength ($\Lambda T^{1/2} \approx 17.4\text{Å K}^{1/2}$ for H). If the temperature is sufficiently low, quantum mechanical effects determine the behaviour of the gas phase and the gas is referred to as a quantum gas. We raise the question 'what makes a gas into a quantum gas?' and analyse the difference between H and other experimentally investigated quantum gases, such as the optically cooled alkali systems or the metastable inert gases. Three quantum indicators will be analysed for this purpose. These are the zero-point motion of atoms (Section 2.1), nonclassical scattering in binary collisions (Sections 2.2 and 2.3), and quantum degeneracy effects (Section 2.4).

Before entering into this analysis, it should be emphasised that stable quantum gases do not exist. All stable substances condense into solid or liquid phases at temperatures approaching absolute zero. The quantum gases are therefore at best metastable, heavily oversaturated, vapours. Spin-polarised hydrogen is the only exception to this rule: it behaves at all temperatures as a fluid above its critical point. In this sense hydrogen is the only 'true' quantum gas. The metastability of spin-polarised hydrogen is therefore also of a different nature than that of other quantum gases. Because the spin-dipole interaction is so much weaker than the Van der Waals interaction, the spin-dipole induced recombination rate is very slow [12], 10 orders of magnitude slower than the dimerisation rate in caesium [17]. It is interesting to note here that, unlike the three-body processes such as dimerisation, two-body processes such as spin-dipole relaxation do not impose a fundamental limit on the densities that can be achieved in ultra-cold gases. Although the gas may survive longer at lower densities, it also takes longer to establish thermal (quasi-)equilibrium. The highest density that can be studied experimentally with hydrogen is therefore determined mostly by practical considerations such as the minimum time required to do a meaningful measurement. In other quantum gases this density is limited by the dimerisation rate. In the following sections we take a practical attitude and address the properties of quantum gases as if these were stable. This means that we consider quantum gases on a time scale short as compared to their lifetime.

2.1   True quantum gases

The first quantum indicator that we consider is the importance of kinetic energy in substances at absolute zero, *i.e.* the importance of zero-point motion. This is conveniently done within the quantum theory of corresponding states (QTCS) in which the Hamiltonian (6) is rewritten in the following dimensionless form [18, 19]

$$H = -\frac{\eta}{2}\sum_i \nabla_i^{*2} + \sum_{i<j} u^*(r_{ij}^*),\tag{7}$$

where $\eta \equiv \hbar^2/m\varepsilon r_0^2$ and

$$u^*(r_{ij}^*) \equiv U(r_{ij})/\varepsilon \quad \text{with} \quad r_{ij}^* \equiv r_{ij}/r_0.\tag{8}$$

The QTCS applies to any class of systems that share an interaction potential of a given shape characterised by two parameters, one which sets the energy scale (ε) and the other the length scale (r_0). This holds, for instance, for all systems that may be described by a Lennard-Jones potential. Interestingly, the difference between the various systems in a given class (characterised by a potential of a given shape) appears exclusively through the quantum parameter η in front of the kinetic energy term of the Hamiltonian (7). The kinetic energy term gives rise to a positive contribution to the energy of the ground state, the potential energy contribution is negative except for very high densities where the repulsive cores dominate the interaction. For $\eta \to 0$ the kinetic energy is negligible and the ground state energy is minimal for conditions corresponding to a classical solid at its equilibrium density. With increasing η this solid will melt to form a quantum liquid. Further increasing η the ground-state energy will become positive at any density. This means that the liquid will become unbound: there is no many-body bound state.

We are dealing with a true quantum gas that has to be confined by walls or in a trap to maintain constant density.

The ground-state energy of a many-body system described by Hamiltonian (6) is given by

$$E_0 = \frac{\langle \Psi | H | \Psi \rangle}{\langle \Psi | \Psi \rangle} , \tag{9}$$

where $|\Psi\rangle$ is the ground state. For the quantum liquids and gases E_0 has been calculated with variational wave-functions of the Jastrow type, approximating the ground state of an N-body system by a product of $N(N-1)/2$ pair wave-functions $f(r_{ij})$

$$\Psi = F(r_1, \cdots, r_N) \equiv \prod_{i<j} f(r_{ij}) . \tag{10}$$

Then we have

$$\langle \Psi | H | \Psi \rangle = \sum_{i<j} \int \cdots \int F^2 \left[-\frac{\hbar^2}{2m} \nabla_i^2 \ln f(r_{ij}) + U(r_{ij}) \right] dr_1 \cdots dr_N . \tag{11}$$

This expression is obtained after integration by parts and repetitive use of the relation $\nabla_i F = F \sum_{j \neq i} \ln f(r_{ij})$. Introducing the pair correlation function

$$g(r) = \frac{N(N-1)}{n^2} \frac{\int \cdots \int \Psi^2 dr_3 \cdots dr_N}{\int \cdots \int \Psi^2 dr_1 \cdots dr_N} . \tag{12}$$

one obtains for the ground state energy *per atom*

$$\frac{E_0}{N} = \frac{n}{2} \int g(r) \left[-\frac{\hbar^2}{2m} \nabla^2 \ln f(r) + U(r) \right] dr . \tag{13}$$

For a given choice for the Jastrow function $f(r)$, for example the form [20]

$$f(r) = \exp \left[-\frac{1}{2} \left(\frac{br_0}{r} \right)^5 \right] \tag{14}$$

(with only a single variational parameter b), one can calculate $g(r)$ by a cluster expansion method [21] or a Monte Carlo procedure [22] and then find E_0/N by integration of Equation (13) and variation of the parameter b.

It was established [21] that Lennard-Jones Bose systems have a positive ground state energy for $\eta > 0.46$. Comparing the Lennard-Jones quantum parameters for various substances

	H	D	^{4}He	H_2	Li	Na	Cs
η	0.55	0.275	0.18	8×10^{-2}	2×10^{-3}	5×10^{-4}	3×10^{-5}

shows that only spin-polarised hydrogen satisfies this condition. H is therefore the only substance that can remain gaseous under equilibrium conditions, *i.e.* the only true quantum gas. All other substances tend to form many-body bound states, usually crystalline solids. This may seem paradoxical in view of many practical experiments which have been done with the optically cooled alkali systems. At this point the metastability of all ultra-cold gases enters as an essential ingredient in the discussion and allows us to put the various systems in relative perspective. Although spin-polarised H has an intrinsically gaseous ground state, due to the presence of depolarisation mechanisms spin polarisation will never be complete and H is at best a metastable gas. Expanding the discussion to *metastable* systems in quasi-equilibrium it will follow from the next section that also systems with $\eta \ll 0.46$ can show gaseous behaviour as long as the formation of the many-body bound state can be excluded kinetically.

2.2 Quantum aspects in binary collisions

2.2.1 *s*-wave scattering regime

Let us leave many-body behaviour for a while and turn to quantum mechanical aspects of the relative motion of a pair of atoms under the influence of the interaction potential U (V_t for spin-polarised H). The standard treatment involves partial waves of different angular momentum l (see for instance Section 132 in [23])

$$\psi = \Sigma_l A_l P_l(\cos\theta) R_{kl}(r) \,, \tag{15}$$

where the A_l are constants and the wave-functions R_{kl} satisfy the radial Schrödinger equation

$$\left[-\frac{d^2}{dr^2} + \frac{l(l+1)}{r^2} + \mathcal{V}(r) - k^2 \right] r R_{kl}(r) = 0 \,. \tag{16}$$

Here $\mathcal{V}(r) \equiv (m/\hbar^2)U(r)$, $k \equiv (mE/\hbar^2)^{1/2}$ is the wavenumber corresponding to the relative energy E and m (1.6735×10^{-27} kg for H) the mass of the atoms. For sufficiently large values of r the $\mathcal{V}$ term may be neglected in Equation (16), which implies that the solutions always have the same asymptotic form $r R_{kl}(r) \sim \sin(kr - (l\pi/2) + \eta_l)$ for $r \to \infty$. For $\mathcal{V}(r) \equiv 0$ the phase shifts η_l are identically zero, in any other case the η_l have to be evaluated explicitly. The amplitude $f(\theta)$ for scattering over an angle θ by the potential U is fully determined by the asymptotic behaviour of the wave-function ψ and may therefore be expressed in terms of the η_l. Using the expansion

$$f(\theta) = \Sigma_l (2l+1) f_l P_l(\cos\theta) \,, \tag{17}$$

the partial amplitudes f_l are related to the phases η_l by $f_l = (\exp(2i\eta_l) - 1)/2ik$.

For $l > 0$ the third term of Equation (16) dominates over the (repulsive) second term only for distances $r < R_0$, where R_0 (~ 4.7Å for spin-polarised H) is the range of the interaction (radius of action) defined by $R_0^2 = \mathcal{V}(R_0)$. This means that for

$$kR_0 \ll 1 \tag{18}$$

the $\mathcal{V}$ term may be neglected in Equation (16) (for these k values the classical turning point R_{cl} for the radial motion is much larger than R_0). Therefore, all scattering has to result from the $l = 0$ channel. In the gas phase, where $k \approx \Lambda^{-1}$, the condition (18) is satisfied for

$$\Lambda \gg R_0 \,. \tag{19}$$

The range of temperatures corresponding to condition (19) defines the *s*-wave scattering regime. For hydrogen this regimes covers temperatures $T \ll 1$K. Under these conditions we are dealing with a pure quantum gas with only isotropic scattering through the s-wave channel.

For s-waves and for $k \to 0$ the phase shift is given by the following asymptotic expression

$$k \cot(\eta_0(k)) \simeq -\frac{1}{a} + \frac{r_e k^2}{2} \quad \text{for } k \to 0 \tag{20}$$

where a is the s-wave scattering length and r_e the effective range of interaction. For the V_t potential of hydrogen we have $a = 1.33a_0$ and $r_e = 323a_0$ (see [24, 25]). For $k \ll [ar_e/2]^{-1/2}$ Equation (20) reduces to $\eta_0 \simeq -ka$. Note that this inequality hardly differs from condition (18) so that in the s-wave scattering regime the amplitude of scattering in any direction Ω becomes

$$f(\Omega) \approx f_0 = \frac{e^{i\eta_0}}{k} \sin \eta_0 \approx -a , \tag{21}$$

independent of both k and Ω. For bosons the differential cross section is given by the expression

$$d\sigma(\Omega) = |f(\Omega) + f(-\Omega)|^2 \, d\Omega = 4 |f_0|^2 \, d\Omega . \tag{22}$$

Integrating over a hemisphere yields the total elastic cross section $\sigma = 8\pi a^2$ ($\sim 1.3 \times 10^{-15} \text{cm}^2$ for spin-polarised H).

The formal expression for the l-wave amplitude of scattering in the direction Ω is given by

$$f_l(k, \Omega) = -\int j_l(kr) \, \mathcal{V}(r) R_{kl}(r) r^2 dr \tag{23}$$

where j_l is a spherical Bessel function (see for instance [26]). The asymptotic expression for R_{k0} coincides with the corresponding expression for s-wave scattering of hard spheres of diameter a:

$$R_{k0}(r) \simeq (kr)^{-1} \sin k(r - a) \approx 1 - a/r , \quad (\text{for } kr \to 0, \ r \to \infty) . \tag{24}$$

Hence the wave-function equals unity everywhere except in a small region of radius R_0 around the scattering centre.

2.2.2 Binary collision approximation

As long as the density is sufficiently low, the properties of gases may be described in terms of binary collisions with asymptotic wave-functions of the type discussed above. Such gases are known as 'nearly ideal gases'. This pair approximation is valid when collisions within the range of interaction with a third atom are of negligible importance, *i.e.* when the mean particle separation $n^{-1/3}$ is much larger than the range of the interaction

$$n^{1/3} R_0 \ll 1 \tag{25}$$

($n \ll 10^{20} \text{cm}^{-3}$ for spin-polarised H). More generally, if no other length scales of order R_0 or smaller have are considered, one may replace U by a point interaction, or pseudo potential

$$U(r) = v_0 \delta(\mathbf{r}) \quad \text{with} \quad v_0 \equiv (4\pi\hbar^2/m)a . \tag{26}$$

Note that by substituting Equation (26) into Equation (23), observing that $R_{k0}(r) = j_0(kr)$ in the absence of a phase shift, the scattering amplitude is found to be $-a$ as in Equation (21). The microscopic theory for nearly ideal Bose gases has been developed under the condition (25) using the quantity na^3 as a small parameter, the gas parameter [27]. For spin-polarised hydrogen we calculate $v_0 n/k_B \approx 40\text{nK}$ with $n = 10^{14}\,\text{cm}^{-3}$. Interestingly, as a tends to grow with growing mass, v_0 has, usually, more or less the same value.

2.3 Quantum statistical aspects

Large and heavy atoms, such as rubidium and caesium, can also form quantum gases, even though they fail to satisfy the corresponding states criterion mentioned in Section 2.1. To understand why, one has to analyse how the quantum mechanical aspects of binary collisions translate themselves in the properties of a gas. In the coming sections this will be done for s-wave Bose gases. It will be shown that the ground-state energy per atom, E_0/N, is (to leading order) linearly dependent on both the density of the atoms and their scattering length. Thus, in contrast to the corresponding states prediction for quantum gas behaviour, heavy atoms *with positive scattering length* can also satisfy the $E_0 > 0$ condition, *i.e.* behave as a quantum gas. The case of negative scattering length is special. Such systems want to lower their energy by contracting. This adds another level of metastability to the discussion which is not addressed further in this introductory context. Preference is given to showing that the interactions are experienced differently by ground-state atoms than by atoms in excited states of motion. All this assumes, of course, that the density fluctuations that give rise to triple, quadruple or higher order collisions can be neglected. This assumption is certainly true on time scales short as compared to the lifetime of the gas clouds.

2.3.1 Bose symmetrisation

Equation (26) represents a very powerful approximation for calculating many-body properties of low-density gases. Before turning for this purpose to a formal many-body description we first have a look how effects related to Bose symmetrisation come about.

For a pair of atoms the wave-function is written as

$$\psi_{1,2} = \frac{1}{V}\,e^{-i\mathbf{k}_1\cdot\mathbf{r}_1}e^{-i\mathbf{k}_2\cdot\mathbf{r}_2} = \frac{1}{V}\,e^{-i\mathbf{K}\cdot\mathbf{R}}e^{-i\mathbf{k}\cdot\mathbf{r}}, \tag{27}$$

where $\mathbf{K} = \mathbf{k}_1 + \mathbf{k}_2$, and $\mathbf{R} = (\mathbf{r}_1 + \mathbf{r}_2)/2$ are the centre of mass wavevector and position vector, respectively. Similarly $\mathbf{k} = (\mathbf{k}_1 - \mathbf{k}_2)/2$, and $\mathbf{r} = \mathbf{r}_1 - \mathbf{r}_2$ are the relative wavevector and position vector corresponding to the reduced mass. V is the normalisation volume. The expression (27) is ideal for mathematical manipulation but does not include the proper correlations at short distances. However, as long as we are not interested in properties that vary over a length scale of order R_0, this objection may be circumvented by turning to the point interaction (26). Note further that the wave-function (27) is not yet symmetrised. Both aspects will be considered here explicitly in calculating, to leading order in the density, the energy associated with the interaction potential V_t that gives rise to the elastic scattering in the gas. In the individual scattering events the direction of the relative momentum changes from $\mathbf{k}$ to $\mathbf{k}'$ ($|\mathbf{k}| = |\mathbf{k}'|$). The momentum transfer is $\mathbf{q} = \mathbf{k} - \mathbf{k}'$.

(a) We first consider the case $\mathbf{k}_1 = \mathbf{k}_2$. Then Equation (27) has the proper symmetrisation for bosons and with Equation (26) the perturbation matrix element of elastic interaction with momentum transfer $\mathbf{q}$ is given by the following expression

$$\tilde{U}(q) = \frac{1}{V} \int^V v_0 \delta(\mathbf{r}) e^{i\mathbf{q}\cdot\mathbf{r}} d\mathbf{r}, \tag{28}$$

where the $\mathbf{R}$ dependence has been integrated out. Because the relative momentum is zero ($\mathbf{k} = 0$), we also have $\mathbf{q} = 0$ (elastic scattering) and Equation (28) is easily integrated using to yield $\tilde{U}(0) = v_0/V$. We see that the elastic interaction energy is positive (*i.e.* effectively repulsive) for $a > 0$.

A particularly interesting special case is a gas of N bosons in a volume V with $\mathbf{k}_1 = \mathbf{k}_2 = \cdots = \mathbf{k}_N = 0$ which corresponds to Bose condensed ideal gas at $T = 0$. With $N(N-1)/2 \approx N^2/2$ pairs ($N \gg 1$) and assuming pair-wise interaction the total energy is $E_0 = N^2 \tilde{U}(0)/2$. The energy per atom is found to be

$$\frac{E_0}{N} = \frac{nv_0}{2}. \tag{29}$$

(b) We now turn to the case $\mathbf{k}_1 \neq \mathbf{k}_2$. Then, Equation (27) has to be symmetrised while conserving normalisation. This leads to

$$\psi_{1,2} = \frac{1}{V} e^{-i\mathbf{K}\cdot\mathbf{R}} \frac{1}{\sqrt{2}} \left(e^{i\mathbf{k}\cdot\mathbf{r}} + e^{-i\mathbf{k}\cdot\mathbf{r}} \right) \tag{30}$$

and we obtain instead of Equation (28)

$$\tilde{u}(q, Q) = \frac{1}{2V} \int^V v_0 \delta(\mathbf{r}) \left[e^{i\mathbf{q}\cdot\mathbf{r}} + e^{i\mathbf{Q}\cdot\mathbf{r}} + e^{-i\mathbf{Q}\cdot\mathbf{r}} + e^{-i\mathbf{q}\cdot\mathbf{r}} \right] d\mathbf{r}, \tag{31}$$

where $\mathbf{Q} = \mathbf{k} + \mathbf{k}'$. Hence, we have $\tilde{u}(q, Q) \approx 2v_0/V$ provided the approximation (26) is valid (kR_0, $QR_0 \ll 1$). For a Bose condensed gas of N atoms with all except $N' \ll N$ in the condensate we find for the interaction energy per non-condensate atom

$$\frac{E'}{N'} = nv_0, \tag{32}$$

twice as big as Equation (29).

2.3.2 Many-body formalism

The differences in treatment between singly and multiply occupied states in many-body systems, as illustrated in the above example, are conveniently handled in the number representation (second quantisation). The peculiarities of Bose systems become here explicit through the definition of the number states, the construction operators and the commutations rules. The symmetrised many-body state $|\Psi\rangle = |n_1, n_2, \cdots\rangle$, with n_i bosons in state $|i\rangle$ and $\Sigma_i n_i = N$, is given by

$$|n_1, n_2, \cdots\rangle = \left(\frac{1}{N! n_1! n_2! \cdots} \right)^{1/2} \sum_P |n_1, n_2, \cdots\rangle, \tag{33}$$

where the sum runs over all possible permutations P of the unsymmetrised product states $|n_1, n_2, \cdots)$. In second quantisation, using field operators $\hat{\psi}^\dagger(\mathbf{r})$ and $\hat{\psi}(\mathbf{r})$, the potential energy operator has the following form (see for instance Section 64 of [23])

$$\hat{U} = \frac{1}{2} \iint d\mathbf{r}_1 \, d\mathbf{r}_2 \, \hat{\psi}^\dagger(\mathbf{r}_1)\hat{\psi}^\dagger(\mathbf{r}_2)U(\mathbf{r}_1 - \mathbf{r}_2)\hat{\psi}(\mathbf{r}_2)\hat{\psi}(\mathbf{r}_1) \tag{34}$$

which may be rewritten as

$$\hat{U} = \frac{1}{2} \iint d\mathbf{r}_1 \, d\mathbf{r}_2 \, U(\mathbf{r}_1 - \mathbf{r}_2) \, \hat{g}(\mathbf{r}_1, \mathbf{r}_2), \tag{35}$$

where $\hat{g}(\mathbf{r}_1, \mathbf{r}_2) = \hat{\psi}^\dagger(\mathbf{r}_1)\hat{\psi}^\dagger(\mathbf{r}_2)\hat{\psi}(\mathbf{r}_2)\hat{\psi}(\mathbf{r}_1)$ is the correlation operator. Choosing a particular basis $\hat{\psi}^\dagger(\mathbf{r}) = \sum_k \varphi_k^*(\mathbf{r})a_k^\dagger$ the expectation value of $\hat{g}$ becomes

$$g(\mathbf{r}_1, \mathbf{r}_2) = \sum_{klpq} \varphi_k^*(\mathbf{r}_1)\varphi_l^*(\mathbf{r}_2)\varphi_p(\mathbf{r}_2)\varphi_q(\mathbf{r}_1) \langle \Psi| \, a_k^\dagger a_l^\dagger a_p a_q \, |\Psi\rangle \,. \tag{36}$$

The only non-zero terms in the summation correspond to the combinations $k = p$, $l = q$ and $k = q, l = p$. Using the Bose commutation relations $[a_k, a_{k'}^\dagger] = \delta_{k,k'}$ we find for plane wave basis functions

$$g(\mathbf{r}_1 - \mathbf{r}_2) = \frac{1}{V^2} \sum_{k \neq l} \left(1 + e^{i(l-k)\cdot(\mathbf{r}_1 - \mathbf{r}_2)}\right) n_k n_l + \frac{1}{V^2} \sum_k n_k (n_k - 1) \,. \tag{37}$$

In a thermal gas typical values of wavevectors k, l are of order Λ^{-1}. Hence, for Equation (37) two regimes should be distinguished. Within the quantum correlation range, $|\mathbf{r}_1 - \mathbf{r}_2| \ll \Lambda$, the two-point correlation function g is twice as big as outside this range where the effect of the oscillating terms averages out. For the operator $\hat{U}$ of (35), representing the elastic interactions, this implies that at (low) temperatures where $\Lambda \gg R_0$, Bose correlations should be taken into account. Note that we arrived at the same conclusion in Section 2.2, see Equation (19), in a different context. Equation (37) shows that the Bose correlations are present independent of the existence of any interaction.

Replacing U by a point interaction of the type (26) the operator $\hat{U}$ can be expressed in terms of the operator density $\hat{\mathcal{Z}}(\mathbf{r}) = \hat{\psi}^\dagger(\mathbf{r})\hat{\psi}^\dagger(\mathbf{r})\hat{\psi}(\mathbf{r})\hat{\psi}(\mathbf{r})$ that enables the determination of the interaction energy as a function of position, for example in inhomogeneous systems. By integrating over position the full operator $\hat{U}$ is obtained,

$$\hat{U} = \frac{v_0}{2} \int d\mathbf{r} \, \hat{\mathcal{Z}}(\mathbf{r}). \tag{38}$$

Using the Bose commutation relations we find with Equation (36)

$$\hat{\mathcal{Z}}(\mathbf{r}) = \sum_k |\varphi_k(\mathbf{r})|^4 \, \hat{n}_k (\hat{n}_k - 1) + 2 \sum_{k \neq l} |\varphi_k(\mathbf{r})|^2 \, |\varphi_l(\mathbf{r})|^2 \, \hat{n}_k \hat{n}_l \,. \tag{39}$$

Splitting-off the ground state explicitly

$$\hat{\psi}^\dagger(\mathbf{r}) = \hat{\psi}_0^\dagger(\mathbf{r}) + \hat{\psi}'^\dagger(\mathbf{r}) = \varphi_0^*(\mathbf{r})a_0^\dagger + \sum_{k \neq 0} \varphi_k^*(\mathbf{r})a_k^\dagger \tag{40}$$

we have

$$\hat{\mathcal{Z}}(\mathbf{r}) = |\varphi_0(\mathbf{r})|^4 \, \hat{n}_0 (\hat{n}_0 - 1) + 4\hat{n}_0(\mathbf{r}) \sum_{k \neq 0} \hat{n}_k(\mathbf{r}) + \hat{\mathcal{Z}}_{(k,l) \neq 0}(\mathbf{r}) \tag{41}$$

where $\hat{n}_k(\mathbf{r}) = |\varphi_k(\mathbf{r})|^2 \hat{n}_k$, with $\hat{n}_k = a_k^\dagger a_k$. Using $\sum_{k \neq 0} \hat{n}_k = \hat{n}' = \widehat{N} - \hat{n}_0$ with $N_0 \gg n_k$ and retaining only terms of order n'^2 or larger we find

$$\langle U \rangle \approx v_0 \int d\mathbf{r} \left[\frac{n_0^2(\mathbf{r})}{2} + 2n_0(\mathbf{r})n'(\mathbf{r}) + n'^2(\mathbf{r}) \right] . \tag{42}$$

Note by comparing the special cases $n_0 = 0$ and $n' = 0$ the difference of a factor 2 (at equal density) for the interaction energy per unit volume. This difference has the same origin as the difference between Equations (29) and (32). It embodies a stabilising effect for the Bose condensate $n_0(\mathbf{r})$ at positive values of v_0. For $v_0 < 0$ it points to a destabilising factor.

2.4 Quantum degeneracy in inhomogeneous Bose gases

The current interest in Bose gases mostly concerns the nearly ideal gas in the presence of an externally applied confining potential $\mathcal{U}(\mathbf{r})$. In relation to atomic hydrogen this problem was first addressed at the beginning of the last decade by Walraven and Silvera [28] and further investigated by Goldman *et al.* [29] and Huse and Siggia [30]. Later, after the first trapping proposals for hydrogen were published [31, 32] the topic was discussed by Lovelace and Tommila [33], Bagnato *et al.* [34] and in relation to stability considerations by Hijmans *et al.* [35]. Most recently dynamical aspects were investigated using time-dependent mean field solutions for the condensate [36]. These authors mostly considered quasi-homogeneous conditions as defined by the inequality

$$kT \gg nv_0 \gg \hbar\omega . \tag{43}$$

The low density case $nv_0 \ll \hbar\omega$, investigated in relation to the stability of gases with negative scattering length [37, 36], is not considered in the present text.

It is the purpose of the coming sections to show that at temperatures above the critical temperature for Bose-Einstein condensation, T_c, the elastic interactions do not affect the density distribution of the gas to any appreciable extent (*i.e.* it can be neglected), but that even slightly below T_c they are of major importance to determine the shape of the Bose condensate.

2.4.1 Ideal Bose gas in confining potential

To introduce the notation, first the case of an ideal Bose gas in an external potential is briefly summarised. The single particle eigenstates $\{\varphi_k(\mathbf{r})\}$, labeled by index k, with eigenvalues ε_k satisfy the following Schrödinger equation

$$\left[-\frac{\hbar^2}{2m}\Delta + \mathcal{U}(\mathbf{r}) \right] \varphi_k(\mathbf{r}) = \varepsilon_k \varphi_k(\mathbf{r}) . \tag{44}$$

With the grand ensemble the thermodynamic properties of the gas follow from the statistical operator

$$\hat{\rho}_0 = \exp\left[\left(-\Omega_0 + \mu\hat{N} - \hat{H}_0 \right)/kT \right] = \frac{1}{Z_{\text{gr}}} \exp\left[\left(\mu\hat{N} - \hat{H}_0 \right)/kT \right] \tag{45}$$

with normalisation $\text{Tr}\langle \hat{\rho}_0 \rangle = 1$. Here Ω_0 is the thermodynamic potential (that normalises the trace of the density matrix), $\hat{H}_0$ is the Hamiltonian operator of the ideal gas, μ is the chemical potential and $\hat{N}$ is the number operator. The grand canonical partition function is defined by

$$Z_{\text{gr}} \equiv \exp[\Omega_0/kT] = \text{Tr} \left\langle \exp\left[\left(\mu\hat{N} - \hat{H}_0\right)/kT\right]\right\rangle . \tag{46}$$

The Hamiltonian $\hat{H}_0$ is diagonal in the number representation of the basis $\{\varphi_k(\mathbf{r})\}$

$$H_0(\mathbf{r}) = \sum_k \varepsilon_k \left|\varphi_k(\mathbf{r})\right|^2 \text{Tr}\langle\hat{\rho}_0\hat{n}_k\rangle . \tag{47}$$

Evaluating the statistical average we find $\text{Tr}\langle\hat{\rho}_0\hat{n}_k\rangle = \bar{n}_k$, where

$$\bar{n}_k = \frac{1}{\exp[(\varepsilon_k - \mu)/kT] - 1} . \tag{48}$$

The $\bar{n}_k$ are the mean occupation numbers for the ideal Bose gas. By choosing the chemical potential such that $\sum_k \bar{n}_k = N$, the ensemble describes a trapped gas of N atoms at temperature T. At high temperatures μ has to be chosen large and negative. With decreasing temperature the chemical potential has to be chosen larger and larger until, at a finite temperature (T_c), it approaches ε_0. Because expression (48) diverges for $\mu = \varepsilon_0$, the value of μ will never grow beyond ε_0. An arbitrarily large ground state occupation is obtained by choosing μ sufficiently close to ε_0. Therefore, below T_c, the value of μ is fixed and mean occupation numbers $\bar{n}_k$ no longer depend on N but only on T. Their sum $\sum_{k\neq0} \bar{n}_k = N'$ represents a finite number, vanishing with decreasing temperature. Hence, at T_c, the ground state occupation N_0 starts to become a macroscopic fraction of the total number of trapped atoms:

$$N_0 = N - \sum_{k\neq0} \bar{n}_k . \tag{49}$$

This phenomenon, in which, at finite temperature, the statistics favours a macroscopic occupation of the ground state, is known as Bose-Einstein condensation (BEC).

Density distributions and BEC.   For future reference it is useful to have a closer look at the density distribution $n(\mathbf{r})$ of a trapped Bose gas. With a large number of oscillator states occupied (quasi-homogeneous case) the ideal gas expression (48) may be approximated by a continuum expression for the distribution function

$$f(\varepsilon) = \frac{1}{z^{-1}\exp[-\varepsilon/kT] - 1} = \sum_{\ell=1}^{\infty} z^\ell \exp[-\ell\varepsilon/kT], \tag{50}$$

here also written as an expansion in terms of the fugacity $z = \exp[\mu/kT]$ (with $z < 1$ in the absence of a condensate), and normalised on the total number of trapped atoms

$$N = \int d\varepsilon \rho(\varepsilon) f(\varepsilon). \tag{51}$$

Here

$$\rho(\varepsilon) \equiv (2\pi\hbar)^{-3} \int d\mathbf{r} d\mathbf{p}\, \delta\left[\varepsilon - \mathcal{U}(\mathbf{r}) - p^2/2m\right] \tag{52}$$

is the density of states in the quasi-classical phase space $(\mathbf{r}, \mathbf{p})$. By changing to the continuum distribution we have implicitly set $\varepsilon_0 = 0$. Substituting Equations (50) and (52) into Equation (51) and integrating over the momentum space we obtain

$$N = \frac{1}{\Lambda^3} \int d\mathbf{r} \sum_{\ell=1}^{\infty} \frac{z^\ell}{\ell^{3/2}} \exp[-\ell \mathcal{U}(\mathbf{r})/kT], \tag{53}$$

where Λ is the thermal wavelength. This expression fixes the value of the chemical potential for any given temperature. Writing $N = \int d\mathbf{r}\, n(\mathbf{r})$, we find for the density distribution

$$n(\mathbf{r}) = \frac{g_{3/2}}{\Lambda^3} \{ z \exp[-\mathcal{U}(\mathbf{r})/kT] \}. \tag{54}$$

Throughout this paper the usual notation $g_\alpha(x) \equiv \sum_{\ell=1}^{\infty} x^\ell/\ell^\alpha$ is used. For the centre of a (deep) trap $(\mathbf{r} = 0)$ Equation (54) can be written in the following form

$$n(0)\Lambda^3 = g_{3/2}(z). \tag{55}$$

This quantity is known as the degeneracy parameter of the trapped gas. Well above T_c we have the non-degenerate regime in which the degeneracy parameter is small, the fugacity expansion may be approximated by its first term, $g_{3/2}(z) \approx z$, and the chemical potential can be expressed as

$$\mu = kT \ln[n(0)\Lambda^3] \tag{56}$$

(note that μ is large and negative). At the onset of BEC the degeneracy parameter reaches its critical value, $n(0)\Lambda^3 = g_{3/2}(1) = 2.612$, and $\mu = 0$. Clearly, the critical density for BEC is first reached in the centre of the trap. The degeneracy parameter is the same for any type of trap as long as the continuum approximation is valid and $\varepsilon_0 = 0$ [34].

The shape of the density distribution (54) changes with temperature. For $T \leq T_c$ it has a fixed form which, in the absence of an adjustable chemical potential μ, can only accommodate a fixed number of atoms

$$N' = \frac{1}{\Lambda^3} \int d\mathbf{r}\, g_{3/2}\{\exp[-\mathcal{U}(\mathbf{r})/kT]\} \tag{57}$$

(for any given temperature). Above T_c the distribution changes form until Equation (54) reaches its classical form,

$$n(\mathbf{r}) = n(0) \exp[-\mathcal{U}(\mathbf{r})/kT], \tag{58}$$

in the non-degenerate regime (*i.e.* for $z \lesssim 0.1$). For H at density $n = 10^{14}$ cm^{-3}, BEC occurs at $T_c \approx 30$ μK. Note that for these numbers the conditions (19), (25) and the l.h.s. of Equation (43) are extremely well satisfied. Note further that at this density the mean free path, given by $\lambda \sim 1/n\sigma$, is of order 10cm, much larger than the typical size of a trapped gas cloud.

Example: BEC in harmonic traps.   It is instructive to consider BEC for the example a specific type of trap. For this purpose we first rewrite Equation (52) in the form

$$\rho(\varepsilon) = \frac{2\pi(2m)^{3/2}}{(2\pi\hbar)^3} \int_{\mathcal{U}(\mathbf{r}) \leq \varepsilon} d\mathbf{r} \sqrt{\varepsilon - \mathcal{U}(\mathbf{r})} \tag{59}$$

which implies that the density of states may be expressed as

$$\rho(\varepsilon) = A_{\mathrm{PL}}\varepsilon^{1/2+\delta} \tag{60}$$

for square ($\delta = 0$), harmonic ($\delta = 3/2$) and spherical-quadrupole ($\delta = 3$) traps, for spherically symmetric power-law traps with $\mathcal{U}(\mathbf{r}) \propto r^{3/\delta}$ and all power-law traps that may be written as $\mathcal{U}(x, y, z) \sim |x|^{1/\delta_1} + |y|^{1/\delta_2} + |z|^{1/\delta_3}$ with $\delta = \sum_i \delta_i$. In Equation (60) A_{PL} is a trap dependent constant.

For the case of an isotropic harmonic potential

$$\mathcal{U}(\mathbf{r}) = \frac{1}{2}m\omega^2 r^2, \tag{61}$$

where $r = |\mathbf{r}|$ and ω is the single atom oscillator frequency, we calculate

$$\rho(\varepsilon) = \frac{1}{2}(1/\hbar\omega)^3 \varepsilon^2. \tag{62}$$

For $T \leq T_c$ the fugacity equals unity and the number of above condensate particles is found by integration of Equation (51) using Equation (50) and substituting Equation (62) to yield $N' = (kT/\hbar\omega)^3 g_3(1)$. Thus, condensate fractions $N_0/N = 1 - (N'/N)$ close to unity are to be expected at temperatures even mildly below T_c. At T_c we have $N' = N_c$ and, therefore,

$$kT_c = \left[N^{1/3}/g_3(1)\right]^{1/3} \hbar\omega \approx 0.941\, N^{1/3}\hbar\omega. \tag{63}$$

Hence, to satisfy the inequality (43) appropriately at BEC, typically $N \gtrsim 10^6$ atoms are needed. With fewer atoms the quasi-homogeneous approximation starts to break down.

2.4.2   Nearly ideal Bose gas in a confining potential

In a nearly ideal Bose gas, *i.e.* if the gas parameter is small (see Section 2.2.2), the analysis is done with a statistical variational method that leads to a mean field expression for the motion of the trapped atoms. Under condition (43) it follows with thermodynamic perturbation theory (see for instance [38]) that the thermodynamic potential Ω satisfies the following inequality

$$\Omega \leq \tilde{\Omega} = \Omega_0 + \mathrm{Tr}\left\langle \hat{\rho}_0 \left(\hat{H} - \hat{H}_0\right)\right\rangle. \tag{64}$$

Here Ω_0 is the thermodynamic potential of the ideal Bose gas described by the Hamiltonian $\hat{H}_0$ which is used here as a trial Hamiltonian and $\hat{\rho}_0$ is the statistical operator of the non-interacting gas defined by Equation (45). Choosing the number representation of the single particle basis $\{\varphi_k(\mathbf{r})\}$ we can derive using Equation (46)

$$\Omega_0(\mathbf{r}) = -kT \sum_k |\varphi_k(\mathbf{r})|^2 \ln\left[1 - e^{(\mu-\varepsilon_k)/kT}\right]. \tag{65}$$

Using operator densities, the full Hamiltonian is written as

$$\hat{H}(\mathbf{r}) = \hat{\psi}^\dagger(\mathbf{r})\left[-\frac{\hbar^2}{2m}\Delta + \mathcal{U}(r)\right]\hat{\psi}(\mathbf{r}) + \frac{v_0}{2}\hat{\mathcal{Z}}(\mathbf{r}). \tag{66}$$

Here approximation (26) was assumed to be valid. For inhomogeneous systems this requires the additional condition that interaction range should be much smaller than size of the oscillator ground state.

(a) **No condensate** $(T > T_c)$.   We first consider the case of a weakly interacting gas without Bose condensate. In this case the energy density of the thermodynamic potential $\tilde{\Omega}(\mathbf{r})$ can be written as

$$\tilde{\Omega}(\mathbf{r}) = \Omega_0(\mathbf{r}) \; + \; \sum_k \varphi_k^*(\mathbf{r}) \left[-\frac{\hbar^2}{2m}\Delta + \mathcal{U}(\mathbf{r}) - \varepsilon_k \right] \varphi_k(\mathbf{r}) \, \mathrm{Tr}\langle \hat{\rho}_0 \hat{n}_k \rangle$$

$$+ \; v_0 \sum_k |\varphi_k(\mathbf{r})|^4 \, \frac{1}{2} \mathrm{Tr}\langle \hat{\rho}_0 \hat{n}_k [\hat{n}_k - 1]\rangle$$

$$+ \; v_0 \sum_{k \neq l} |\varphi_k(\mathbf{r})|^2 \, |\varphi_l(\mathbf{r})|^2 \, \mathrm{Tr}\langle \hat{\rho}_0 \hat{n}_k \hat{n}_l \rangle \,. \tag{67}$$

Evaluating the statistical averages we find $\mathrm{Tr}\langle \hat{\rho}_0 \hat{n}_k \rangle = \bar{n}_k$, $\mathrm{Tr}\langle \hat{\rho}_0 \hat{n}_k [\hat{n}_k - 1]\rangle = 2\bar{n}_k^2$ and $\mathrm{Tr}\langle \hat{\rho}_0 \hat{n}_k \hat{n}_l \rangle = \bar{n}_k \bar{n}_l$, where the $\bar{n}_k$ are the mean occupation numbers for the ideal Bose gas. For the interacting gas a better distribution function can be found by variation of the $\bar{n}_k$ in order to minimize $\tilde{\Omega}(\mathbf{r})$:

$$\frac{\partial \tilde{\Omega}(\mathbf{r})}{\partial \bar{n}_k} = \varphi_k^*(\mathbf{r}) \left[-\frac{\hbar^2}{2m}\Delta + \mathcal{U}(\mathbf{r}) + 2v_0 n(\mathbf{r}) - \varepsilon_k \right] \varphi_k(\mathbf{r}) \,. \tag{68}$$

The minimum is reached if $\partial\tilde{\Omega}(\mathbf{r})/\partial\bar{n}_k = 0$ for all values of k simultaneously, *i.e.* if the following set of Hartree-Fock equations, coupled through the condition $n(\mathbf{r}) = \sum_k |\varphi_k(\mathbf{r})|^2 \bar{n}_k$, is satisfied

$$\left[-\frac{\hbar^2}{2m}\Delta + \mathcal{U}(\mathbf{r}) + 2v_0 n(\mathbf{r}) \right] \varphi_k(\mathbf{r}) = \varepsilon_k \varphi_k(\mathbf{r}) \,. \tag{69}$$

The Equations (69) are known as the Ginzburg-Gross-Pitaevskii equations for an inhomogeneous Bose gas in the absence of a condensate [29, 30].

It is straightforward to show that the correction of $\mathcal{U}(\mathbf{r})$ by the mean field of the gas is negligible in the non-degenerate regime. For this we calculate the mean field correction to first order using the density distribution $n(\mathbf{r})$ of the ideal Bose gas given by Equation (54). Expanding the potential energy terms in the Ginzburg-Gross-Pitaevskii equation around the origin

$$\mathcal{U}(\mathbf{r}) + 2v_0 n(\mathbf{r}) = 2v_0 n(0) + \frac{1}{2}m\omega^2 r^2 \left[1 - \frac{g_{1/2}(z)}{g_{3/2}(z)} \frac{2v_0 n(0)}{kT} + \cdots \right] \tag{70}$$

we find that the mean field correction consists of an energy shift $2v_0 n(0)$ and a change of the spring constant of the oscillator of order $2v_0 n(0)/kT$ which is very small in view of condition (43). Note that in the non-degenerate regime we have $g_{1/2}(z)/g_{3/2}(z) \approx 1$. Only very close to BEC, where the ratio $g_{1/2}(z)/g_{3/2}(z)$ diverges, the mean field correction to the spring constant may not be neglected and Equation (54) should be replaced by the self-consistent expression

$$n(\mathbf{r}) = \frac{g_{3/2}}{\Lambda^3} \{\exp\left[(\mu - 2v_0 n(\mathbf{r}) - \mathcal{U}(\mathbf{r}))/kT\right]\} \,, \tag{71}$$

which is expected to be correct as long as the self-consistent potential varies sufficiently slowly that quasi-classical solutions to the Ginzburg-Gross-Pitaevskii equations are justified. The degeneracy parameter becomes in this case

$$n(0)\Lambda^3 = g_{3/2}\{\exp[(\mu - 2v_0 n(0))/kT]\} \tag{72}$$

and the onset of BEC, characterised (as in the non-interacting case) by the condition $n(0)\Lambda^3 = g_{3/2}(1)$, occurs at

$$\mu = 2v_0 n(0). \qquad (73)$$

In terms of the total atom number the BEC criterion is given by the following self-consistent expression

$$N_c = \int d\mathbf{r}\, n(\mathbf{r}) = \frac{1}{\Lambda^3} \int d\mathbf{r}\, g_{3/2}\{\exp[(2v_0 n(0) - 2v_0 n(\mathbf{r}) - \mathcal{U}(\mathbf{r}))/kT]\}, \qquad (74)$$

which shows that, in comparison to ideal gases, repulsive Bose gases require more atoms to Bose condense at the same T_c.

(b) **With condensate** $(T < T_c)$. Below T_c we use the same variational procedure for $\tilde{\Omega}(\mathbf{r})$ but, since in the presence of a condensate we have $\mu = \varepsilon_0$ and $\text{Tr}\langle\hat{\rho}_0\hat{n}_0\rangle$ is diverging, we should single out the ground state from the ensemble average. We first single out the ground state in the Hamiltonian (61) using Equation (41) and setting $\langle\hat{n}_0\rangle = N - \sum_{k\neq 0}\text{Tr}\langle\hat{\rho}_0\hat{n}_k\rangle \equiv N_0$ and $\langle\hat{n}_0(\hat{n}_0-1)\rangle = N_0(N_0-1) \approx N_0^2$ (assuming $N_0 \gg 1$) before the ensemble average is evaluated in expression (67) for the thermodynamic potential. Variation with respect to $\bar{n}_k$ yields, for $k \neq 0$, the same result for $\partial\tilde{\Omega}(\mathbf{r})/\partial\bar{n}_k$ as obtained above T_c, *i.e.* Equation (66) with $n(\mathbf{r}) = n_0(\mathbf{r}) + n'(\mathbf{r})$. Variation of N_0 gives rise to a different expression in which the exchange effects discussed in Sections 2.3.1 and 2.3.2 again show up:

$$\frac{\partial\tilde{\Omega}(\mathbf{r})}{\partial N_0} = \varphi_0^*(\mathbf{r})\left[-\frac{\hbar^2}{2m}\Delta + \mathcal{U}(\mathbf{r}) + v_0[n_0(\mathbf{r}) + 2n'(\mathbf{r})] - \mu\right]\varphi_0(\mathbf{r}). \qquad (75)$$

Consequently, below T_c we have the following set of Hartree-Fock equations

$$\left[-\frac{\hbar^2}{2m}\Delta + \mathcal{U}(\mathbf{r}) + v_0[n_0(\mathbf{r}) + 2n'(\mathbf{r})]\right]\varphi_0(\mathbf{r}) = \mu\varphi_0(\mathbf{r}) \qquad (76)$$

$$\left[-\frac{\hbar^2}{2m}\Delta + \mathcal{U}(\mathbf{r}) + 2v_0[n_0(\mathbf{r}) + n'(\mathbf{r})]\right]\varphi_k(\mathbf{r}) = \varepsilon_k\varphi_k(\mathbf{r}) \quad (k \neq 0). \qquad (77)$$

For $T \to 0$ Equation (76) reduces to the Ginzburg-Gross-Pitaevskii equation for the many-body ground state [39, 40]. To find the expression for the chemical potential in the presence of a condensate we have to solve Equation (76). This is particularly simple for the quasi-homogeneous case since the condition $nv_0 \gg \hbar\omega$ enables us to neglect the kinetic energy term in Equation (76). Thus we find

$$\mu = \mathcal{U}(\mathbf{r}) + v_0\left[n_0(\mathbf{r}) + 2n'(\mathbf{r})\right] \qquad (78)$$

and, since μ should be constant over the sample, we have

$$\mu = v_0\left[n_0(0) + 2n'(0)\right]. \qquad (79)$$

Note that for $N_0 \to 0$ this equation reduces to Equation (73). Assuming the size of the condensate to be small as compared to the thermal size of the sample we may set

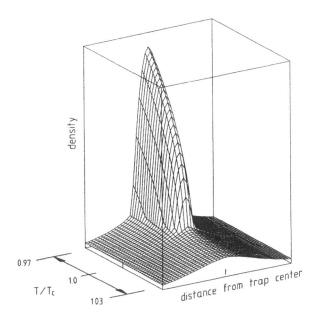

Figure 3. *The density distribution calculated for atomic hydrogen in the trap used in Amsterdam as a function of temperature near T_c. The condensate shows up as a sharp feature in the centre of the distribution.*

$n'(\mathbf{r}) \approx n'(0)$ everywhere within the condensate and combine Equations (78) and (79) to derive the following expression for the density distribution of the condensate

$$n_0(\mathbf{r}) = \frac{v_0 n_0(0) - \mathcal{U}(\mathbf{r})}{v_0} \qquad \text{(for } |\mathbf{r}| \leq l_c). \tag{80}$$

For $|\mathbf{r}| \geq l_c$ we have $n_0(\mathbf{r}) \equiv 0$. The approximation $n'(\mathbf{r}) \approx n'(0)$ is readily verified for the example of the harmonic confining potential (61) where the e^{-1} thermal Gaussian width is given by $l_{th} = [2kT/m\omega^2]^{1/2}$. With Equation (80) the condensate radius is calculated to be $l_c = [2v_0 n_0(0)/m\omega^2]^{1/2}$. Thus with condition (43) it follows that $l_c \ll l_{th}$, *i.e.* the condensate manifests itself (also in the presence of interactions) as a spatially segregated high density phase in the centre of the trapping potential. This is illustrated in Figure 3.

Equation (80) presents a very important result for inhomogeneous Bose gases under quasi-homogeneous conditions. It shows that, for a given number (N_0) of atoms in the condensate, the density in the centre of a condensate depends inversely on v_0. In the absence of kinetic energy the condensate tends to be compressed by the confining potential. This leads to the divergence of the density in the case of an ideal gas. This unphysical divergence occurs because of the continuum approximation, which neglects the finite width of the oscillator ground state. The above analysis has shown that, below T_c, it is essential to include the repulsive interaction. Above T_c the presence of the interaction is much less important, actually almost negligible, because, under condition (43), the thermal motion counteracts the compression.

Figure 3 clearly shows one of the most fascinating and novel aspects of Bose condensates in inhomogeneous traps: the condensate appears as a small high-density cloud of atoms in the centre of the trapped gas and is, therefore, almost completely spatially separated from the non-condensate part. Thus we have the unique situation of a ground state surrounded by a cloud of excitations. It was shown in very recent BEC experiments with ^{87}Rb at JILA [5] that it is possible to remove this cloud of excitations rapidly while keeping the condensate. This implies cooling speeds which are limited primarily by the time required to remove the excited atoms and not by intrinsic properties such as the thermalisation time.

3   Evaporative cooling

3.1   Introduction

Evaporative cooling is a very powerful cooling method based on the preferential removal of atoms with an energy higher than the average energy per trapped atom and on thermalisation by elastic collisions. The evaporation process leads, under suitable conditions, to an increase in phase-space density, and was suggested for this purpose as a means of attaining Bose-Einstein condensation (BEC) in atomic hydrogen [31, 41]. Presently, the method has been applied to a wider class of ultra-cold gases.

In this section the essential features of evaporative cooling will be discussed for Bose gases in the non-degenerate quantum regime. It will be assumed that the atoms are confined in a static potential well, $\mathcal{U}(\mathbf{r})$, from which they escape with unit probability as soon as their total energy, ε, exceeds a threshold value, ε_t, which is chosen to be much larger than the average energy per atom in the trap. As the most energetic atoms are removed in this way the gas will cool.

The basic assumption underlying the physical picture described above is 'sufficient ergodicity' where it is assumed that the distribution of atoms in phase space (position and momentum) depends only on their energy. This will be the case, for example, in a trap with ergodic single-particle motion, in which case all parts of the equipotential hypersurface corresponding to the total energy of the atom are sampled with equal probability. However, as for many traps the single-particle motion is (partly) non-ergodic [42], interatomic collisions are usually essential to assure 'sufficient ergodicity'.

In practical situations the escape can be arranged in various ways: by passing over a potential barrier [43, 44], by adsorbing onto a pumping surface [45, 46] or by optical [14] or RF [48, 49] pumping of the atoms to non-trapped states. If the evaporation is to be efficient it is essential that the escape occurs on a time scale short compared to the collisional time, *i.e.* that the mean free path λ is much larger than the size of the gas cloud l,

$$\lambda \gg l. \qquad (81)$$

One easily verifies that for H at density $n = 10^{14}$ cm^{-3} (close to the highest density achieved in a trap), the mean free path, given by $\lambda \sim 1/n\sigma$, is of order 10cm, indeed much larger than the typical size of a trapped sample. Spatial restrictions can limit the escape probability of energetic atoms and reduce the evaporative cooling power, in particular in the case of non-ergodic single particle motion [42]. In this respect we discuss here 'full-power evaporation', evaporation limited only by the rate at which elastic collisions promote atoms to the escape energy.

Inevitably, the cooling of the gas will cause the evaporation process to slow down as fewer and fewer atoms can acquire the threshold energy. Eventually, the cooling rate is balanced by a competing heating mechanism or becomes negligibly small. The process just described will be referred to as 'plain' evaporative cooling, as opposed to 'forced evaporative cooling' where evaporation is forced to continue by lowering the escape threshold as the gas cools.

The principle of evaporative cooling was first demonstrated experimentally at MIT in the group of Greytak and Kleppner [50] who studied plain evaporative cooling by the escape of spin-polarised hydrogen atoms across a magnetic potential energy barrier. It was established, by measuring the quantity of gas remaining in the trap after a certain holding time, that this procedure causes the gas to cool to a temperature well below the temperature of the surrounding walls. The dynamics of this type of evaporative cooling was further studied optically [45, 46].

Forced evaporative cooling was also first applied to hydrogen [16, 51, 43, 44] and enabled temperatures as low as $100 \, \mu K$ and densities as high as $8 \times 10^{13} \, cm^{-3}$. This corresponds to $n\Lambda^3 \approx 0.4$, within one order of magnitude from the degenerate quantum regime. The need for sophisticated detection methods to establish density and temperature is limiting the progress towards lower temperatures here. An optical version of forced evaporative cooling of H↑ was demonstrated in Amsterdam [14].

Recently, evaporative cooling methods were successfully applied to optically pre-cooled (and magnetically trapped) gases of sodium [48] and rubidium [49], both using RF-induced evaporation. In the final stage of preparation of the manuscript for the present paper BEC was announced for ^{87}Rb at JILA by the group of Cornell and Wieman [5] and by the group of Hulet at Rice University for ^7Li [52]. In both experiments evaporative cooling played a decisive role.

3.2   Truncated energy distributions

By its very nature (atoms in thermally accessible states are removed from the sample) evaporative cooling is an intrinsically non-equilibrium process in which a temperature, strictly speaking, cannot be defined. However, if the average energy per trapped atom is much smaller than the evaporation threshold, most interatomic collisions lead to a thermal redistribution of the energy over the trapped states. Thus, keeping in mind the discussion of Section 2.4.2 (that, if the conditions $kT > kT_c \gg nv_0$ are satisfied, interaction effects are of minor importance) it is very appealing to assume that a quasi-equilibrium distribution will be established having the form (50), given in Section 2.4.1, but with the tail of energies $\varepsilon > \varepsilon_t$ lacking:

$$f(\varepsilon) = \sum_{\ell=1}^{\infty} z^\ell \exp[-\ell\varepsilon/kT] \, \theta(\varepsilon_t - \varepsilon) \,. \tag{82}$$

Here $\theta(x - x_0)$ is Heaviside's unit step function, zero for $x < x_0$ and unity for $x \geq x_0$. Such a distribution is characterised by three independent parameters: the truncation (escape) energy, ε_t, the quasi-temperature, T, and the degeneracy parameter, z. In the limit $\varepsilon_t \to \infty$, the parameters T and z reduce to the equilibrium temperature and fugacity, respectively. From Equation (82) it is clear that the truncation mostly affects the first term in the expansion. Being further aware that the thermodynamic properties are only mildly affected by degeneracy effects above the critical temperature for BEC, it is plausible that the essential aspects of evaporative cooling down to T_c can be obtained with the expression for the energy distribution of the non-degenerate regime

$$f(\varepsilon) = z\,e^{-\varepsilon/kT}\,\theta(\varepsilon_t - \varepsilon)\,, \tag{83}$$

where z is determined by the normalisation condition (51) which becomes with (83)

$$N = z \int_0^{\varepsilon_t} d\varepsilon\,\rho(\varepsilon)\,e^{-\varepsilon/kT}\,. \tag{84}$$

Unlike (55), due to the truncation, the expression for the fugacity differs from the equilibrium expression (*i.e.* $z \neq n(0)\Lambda^3$). However, we can still formally write the degeneracy parameter as

$$z \equiv n_0\Lambda^3\,, \tag{85}$$

choosing $\Lambda \equiv [2\pi\hbar^2/mkT]^{1/2}$ to coincide in the limit $\varepsilon_t \to \infty$ with the equilibrium expression for the thermal wavelength. With this procedure n_0 is a density, determined by the normalisation condition (84) and (only) coinciding in the limit $\varepsilon_t \to \infty$ with the density $n(0)$ in the trap centre.

The truncated exponential form has been used as a starting point for several descriptions of evaporative cooling [53, 44, 47, 46]. It was shown by numerical solution of the Boltzmann equation [56] that for non-degenerate Bose gases the evaporation process, indeed rather accurately, conserves a quasi-equilibrium energy distribution of the form (82), a property noticed earlier in numerical simulations of evaporation experiments with trapped H at MIT [54].

One should be aware that the presence of the truncation edge affects all thermal properties. For example, to calculate the average thermal speed, the thermal wavelength, or the internal energy it is not sufficient to know the quasi-temperature (T), as these quantities also depend on the truncation energy (ε_t). In spite of this complication, once the truncated energy distribution (83) is adopted, a thermodynamic description of the sample follows naturally, with ε_t as one of the thermodynamic variables. In addition, the Boltzmann equation for the trapped gas can be radically simplified, allowing an explicit calculation of the rate of evaporation for different kinds of traps [56].

3.3   Thermodynamic properties

In this section we discuss a number of thermodynamic properties relevant for the description of evaporative cooling using the truncated exponential energy distribution (83) as a starting point. For the power-law traps defined in Section 2.4.1 this can be done explicitly. In more general cases the classical canonical partition function of the truncated distribution turns out to be a central quantity. This can be seen directly from Equation (84), which can be rewritten as

$$N = n_0\Lambda^3\zeta \tag{86}$$

with

$$\zeta = \int_0^{\varepsilon_t} d\varepsilon\, \rho(\varepsilon)\, e^{-\varepsilon/kT} \,. \tag{87}$$

Substituting expression (52) for the density of states and introducing the classical single particle Hamiltonian, $H(\mathbf{r}, \mathbf{p}) = \mathcal{U}(\mathbf{r}) + p^2/2m$, we obtain

$$\zeta = (2\pi\hbar)^{-3} \int d\mathbf{p}\, d\mathbf{r}\; e^{-H(\mathbf{r},\mathbf{p})/kT}\theta[\varepsilon_t - H(\mathbf{r}, \mathbf{p})] \,. \tag{88}$$

which is the classical canonical partition function with the (truncated) region of phase space, $\mathcal{U}(\mathbf{r}) + p^2/2m \geq \varepsilon_t$, excluded.

3.3.1 Density distribution and effective volume

The density distribution $n(\mathbf{r})$ of the gas over the trap is obtained by evaluating only the momentum integral of the partition function in Equation (86) and can be written as [55]

$$n(\mathbf{r}) = n_0 e^{-\mathcal{U}(\mathbf{r})/kT} P(3/2, \eta_t(\mathbf{r}))\,, \tag{89}$$

where $P(a, \eta)$ is the incomplete gamma function (see the Appendix at the end of this article) and $\eta_t(\mathbf{r}) \equiv (\varepsilon_t - \mathcal{U}(\mathbf{r}))/kT$. Thus, for the truncated distribution, the true density in the trap centre is given by

$$n(0) = n_0 P(3/2, \eta) \leq n_0 \,, \tag{90}$$

where the parameter $\eta \equiv \eta_t(0) = \varepsilon_t/kT$ will be referred to as the truncation parameter and the (true) effective volume of the gas cloud is given by the expression

$$V_{\text{eff}} = \frac{N}{n(0)} = \int \frac{dn(\mathbf{r})}{n(0)} \,. \tag{91}$$

Although the exact knowledge of V_{eff} is of practical importance, for example for the experimental determination of N, often it is more convenient to use a reference volume, V_e, defined in terms of the parameter n_0

$$V_e = \frac{N}{n_0} = \Lambda^3 \zeta = \Lambda^3 \int_0^{\varepsilon_t} d\varepsilon\, \rho(\varepsilon)\, e^{-\varepsilon/kT} \,. \tag{92}$$

The volume V_e has very nice properties which result from its simple relation to the partition function. Moreover, since the degeneracy parameter for truncated distributions involves n_0 and not $n(0)$, see Equation (85), it may be better not to introduce $n(0)$ at all into the formalism.

Power-law traps. For the special case of a power-law trap the exponent δ, introduced in Equation (60), reappears in the expression for the partition function

$$\zeta = A_{\text{PL}} \int_0^{\varepsilon_t} d\varepsilon\; \varepsilon^{1/2+\delta}\, e^{-\varepsilon/kT} = \zeta_\infty P(3/2 + \delta, \eta) \,. \tag{93}$$

Here the limiting value, $\zeta_\infty = A_{\text{PL}}[kT]^{3/2+\delta}\Gamma(3/2+\delta)$, corresponds to an infinitely deep trap (*i.e.* $\eta \to \infty$) and $\Gamma(a)$ is the Euler gamma function. In particular, for (infinitely) deep harmonic traps, the partition function is given by $\zeta_\infty = (kT/\hbar\omega)^3 /2$.

The quasi-classical reference volume follows directly with Equation (92):

$$V_e = \Lambda^3 \zeta_\infty P(3/2 + \delta, \eta). \tag{94}$$

Notice for deep power-law traps that they have a particularly simple temperature dependence of their reference volume,

$$V_e \propto T^\delta \quad \text{for} \quad \eta \to \infty. \tag{95}$$

3.3.2   Internal energy

The internal energy (total energy) of a non-degenerate trapped gas of N atoms under (quasi-)equilibrium conditions is given by

$$E = n_0 \Lambda^3 \int_0^{\varepsilon_t} d\varepsilon \, \varepsilon \, \rho(\varepsilon) \, e^{-\varepsilon/kT}. \tag{96}$$

This expression can be rewritten in terms of the partition function (87)

$$E = NkT^2 \frac{1}{\zeta} \frac{\partial \zeta}{\partial T}, \tag{97}$$

a form, very well-known from statistical mechanics, that is seen to remain valid under quasi-equilibrium conditions in the trap. Substituting $\zeta = V_e/\Lambda^3$ it follows that

$$E = (3/2 + \tilde{\gamma})NkT \tag{98}$$

where

$$\tilde{\gamma} \equiv \left(\frac{\partial \ln V_e}{\partial \ln T} \right)_{\varepsilon_t}. \tag{99}$$

The tilde symbol is used to remind us that the scaling parameter $\tilde{\gamma}$ depends on the truncation energy (ε_t) and further has a (slight) T dependence, $\tilde{\gamma} = \gamma(T, \varepsilon_t)$, due to the truncation or due to a density of states which deviates from the pure power-law density of states introduced in Section 2.4.1. In cases where the momentum and position integrals in the partition function (88) can be separated, the terms $3NkT/2$ and $\tilde{\gamma}NkT$ in Equation (98) may be interpreted as kinetic and potential energy contributions to the total energy, respectively.

Note that Equation (98) remains valid also if the separation of momentum and position integrals is not possible. Therefore, it can serve as a good starting point for a general description of the quasi-equilibrium thermodynamic properties of trapped gases in terms of three independent variables: the total number of trapped atoms (N), the quasi-temperature (T) and the truncation energy (ε_t) or, equivalently, the truncation parameter (η). This description is the subject of Section 3.5.

Power-law traps.   For power-law traps it follows with Equations (99) and (94) that the scaling parameter $\tilde{\gamma}$ depends only on the ratio of ε_t and T, *i.e.* on the truncation parameter (η),

$$3/2 + \tilde{\gamma} = (3/2 + \delta)R(3/2 + \delta, \eta). \tag{100}$$

By definition $R(a, \eta) \equiv P(a + 1, \eta)/P(a, \eta)$, see further Section 3.5.5. With Equation (100) the internal energy is found to be

$$E = E_\infty R(3/2 + \delta, \eta) , \qquad (101)$$

where $E_\infty = (3/2 + \delta)NkT$ is the internal energy of a sample of N atoms in thermal equilibrium in an infinitely deep trap. Note that the internal energy is independent of the details of the trap and depends only on the thermodynamic variables N, T and ε_t and on the scaling behaviour of the density of states (characterised for power-law traps by the δ parameter).

3.4   Kinetics of evaporation from a trap

Various questions arise from the intrinsic feature of particle loss in evaporative cooling. First of all one may have doubts about the efficiency of the cooling process. One may wonder whether there is any appreciable amount of sample left after it has been cooled to an interesting temperature or that the sample losses are excessive and disabling to the experimentalist. A related question concerns the choice of ε_t. It may be obvious that the extracted energy per evaporating atom is larger for larger ε_t but if this results in an impractically long cooling time, the evaporation is of little use. Actually, what is meant by impractically long? Is this determined by the patience of the experimentalist, by instrumental limitations or by intrinsic properties of the gas? Starting from a kinetic equation, these and other questions and their relation to the shape of different traps will be discussed in the coming sections. Several characteristic parameters appear in this discussion: (a) the truncation parameter, $\eta \equiv \varepsilon_t/kT$, expressing the escape energy in terms of the quasi-temperature, (b) the 'spilling' parameters $\tilde{\xi}$, describing the scaling of the reference volume with ε_t and a measure for the spilling of atoms when ramping-down the escape threshold to force the evaporation and (c) the thermalisation ratio expressing the ratio of elastic collisions to loss collisions and thus determining whether or not thermal (quasi-)equilibrium can be established within the lifetime of the sample.

3.4.1   Kinetic equation

For a Bose gas in the s-wave regime and under sufficiently ergodic conditions (see Section 3.1) the Boltzmann equation can be reduced to a kinetic equation for the distribution function $f(\varepsilon) = n_0\Lambda^3 \exp[-\varepsilon/kT]$ of a particularly simple form [56]

$$\rho(\varepsilon_4)\dot{f}(\varepsilon_4) = \frac{m\sigma}{\pi^2\hbar^3} \int d\varepsilon_1 d\varepsilon_2 d\varepsilon_3\, \delta[\varepsilon_1 + \varepsilon_2 - \varepsilon_3 - \varepsilon_4]$$
$$\times \rho(\min[\varepsilon_1, \varepsilon_2, \varepsilon_3, \varepsilon_4])\, \{f(\varepsilon_1)f(\varepsilon_2) - f(\varepsilon_3)f(\varepsilon_4)\} . \qquad (102)$$

This equation represents the detailed balance between collisions in which atoms with energy ε_1 and ε_2 produce atoms with energy ε_4 (under conservation of energy and momentum) and collisions in which atoms with energy ε_4 are lost from the distribution while producing atoms with energy ε_1 and ε_2 (conserving energy and momentum). Equation (102) describes the time evolution of the energy distribution function $f(\varepsilon)$ as a result of the elementary collisional processes: s-wave collisions with cross section $\sigma = 8\pi a^2$ (see Section 2.2). Notably, this expression only involves the density of states of the atom with the lowest energy involved in each collisional process. This property was first used in the BEC literature to describe the kinetics of Bose condensation in homogeneous systems in relation to excitons [57] and in spin-polarised hydrogen [58, 59].

Equation (102) was derived for inhomogeneous systems to calculate the rate of evaporation from a trap (see Section 3.4.2) and to study, by numerical solution, the time evolution of the shape of the truncated energy distribution [56]. In this way the use of the quasi-equilibrium approximation (83) could be justified.

3.4.2   Evaporation rate

By definition, the evaporation rate equals the rate of loss of atoms to non-trapped states ($\varepsilon_4 > \varepsilon_t$) as a result of collisions between trapped atoms, and is thus given by the following expression

$$\dot{N}_{\mathrm{ev}} = -\int_{\varepsilon_t}^{\infty} d\varepsilon_4 \, \rho(\varepsilon_4) \, \dot{f}(\varepsilon_4) \,. \tag{103}$$

In the truncated Boltzmann approximation we see that if $\varepsilon_4 > \varepsilon_t > \varepsilon_1, \varepsilon_2$ only the gain term in the kinetic equation (102) contributes and that $\varepsilon_3 = \varepsilon_1 + \varepsilon_2 - \varepsilon_4$ is necessarily the lowest energy involved in any collision involving escape of one of the atoms. Hence,

$$\dot{N}_{\mathrm{ev}} = -\frac{m\sigma}{\pi^2\hbar^3} \int_0^{\varepsilon_t} d\varepsilon_3 \int_{\varepsilon_3}^{\varepsilon_t} d\varepsilon_2 \int_{\varepsilon_3+\varepsilon_t-\varepsilon_2}^{\varepsilon_t} d\varepsilon_1 \, \rho(\varepsilon_3) \, f(\varepsilon_1) f(\varepsilon_2) \tag{104}$$

where the domain of integration is determined by the requirements that $\varepsilon_1, \varepsilon_2 < \varepsilon_t$ and $\varepsilon_1 + \varepsilon_2 - \varepsilon_3 = \varepsilon_4 > \varepsilon_t$. Evaluating the integral is straightforward and yields

$$\dot{N}_{\mathrm{ev}} = -n_0^2 \, \bar{v} \, \sigma \, e^{-\eta} \, V_{\mathrm{ev}} \tag{105}$$

with $\bar{v} \equiv [8kT/\pi m]^{1/2}$ defined to coincide in the limit $\eta \to \infty$ with the equilibrium expression for the mean thermal speed, and V_{ev} is the reference volume for evaporation defined by

$$V_{\mathrm{ev}} = \frac{\Lambda^3}{kT} \int_0^{\varepsilon_t} d\varepsilon \, \rho(\varepsilon) \left[(\varepsilon_t - \varepsilon - kT)e^{-\varepsilon/kT} + kTe^{-\varepsilon_t/kT} \right] \,. \tag{106}$$

Hence, the characteristic evaporational decay time τ_{ev}, defined by $-(\dot{N}_{\mathrm{ev}}/N)^{-1}$, can be written as

$$\tau_{\mathrm{ev}}^{-1} = \frac{n_0 \bar{v} \, e^{-\eta} V_{\mathrm{ev}}}{V_e} \,. \tag{107}$$

Here it was used that all atoms that acquire, as a result of an elastic interatomic collision, an energy satisfying the evaporation condition ($\varepsilon_4 > \varepsilon_t$) are lost from the trap with unit probability. The volume ratio V_{ev}/V_e can vary between zero and infinity and may be calculated numerically for any type of trap once the density of states is known. Note that the Equations (105) and (107) hold for given values of N, T and ε_t. Clearly, for a proper description of evaporative cooling of a trapped cloud of gas we have to bring into the picture how the evaporation process affects N and T for fixed (plain evaporation) or time-varying (forced evaporation) value of the truncation parameter ε_t. We return to this point in Section 3.5.

Equation (107) can be interpreted as a collision rate multiplied by an escape probability. However, the usual discussion associated with such an interpretation, concerning the influence of the thermalisation rate on the escape probability, is avoided by deriving expressions directly from the Boltzmann equation. All limitations with regard to the validity of Equation (107) are contained in the limitations to validity of the quasi-equilibrium approximation (83) and these limitations are known to be small for the quasi-homogeneous non-degenerate Bose gas discussed here [56].

Power-law traps. For these traps (introduced in Section 2.4.1) the ratio V_{ev}/V_e can be expressed analytically in terms of incomplete gamma functions

$$\frac{V_{ev}}{V_e} = \eta - (5/2 + \delta)R(3/2 + \delta, \eta). \tag{108}$$

Taking notice of the asymptotic behaviour of the volume ratio

$$\frac{V_{ev}}{V_e} \simeq \frac{\eta^2}{(5/2 + \delta)(3/2 + \delta)} \ (\eta \to 0) \quad \text{and} \quad \frac{V_{ev}}{V_e} \simeq \eta - 5/2 - \delta \ (\eta \to \infty) \tag{109}$$

one may say, somewhat handwaving, that in many practical cases the assumption $V_{ev}/V_e \approx 1$ is not a bad approximation to estimate the order of magnitude of the evaporation rate.

3.4.3 Forced evaporation

Thus far ε_t was assumed to be constant, *i.e.* we only considered plain evaporation. In forced evaporation the escape threshold is slowly lowered in order to have the evaporation proceed at the desired rate. The evaporation process is very sensitive to the value of the truncation parameter (η) because it appears in the exponent of the expressions (105) and (107). Inevitably, the lowering of ε_t gives rise to an additional loss, not related to collision induced escape (evaporation), but to the change in escape condition itself. Therefore, in forced evaporation, the loss of atoms from the trap consists of two contributions,

$$\dot{N} = \dot{N}_{ev} + \dot{N}_t. \tag{110}$$

The first term is the evaporation loss rate discussed in the previous section. The second term is the truncation loss rate, only non-zero for $d\varepsilon_t/dt < 0$,

$$\dot{N}_t = \rho(\varepsilon_t)f(\varepsilon_t)\dot{\varepsilon}_t. \tag{111}$$

Rewriting in terms of the partition function, $\rho(\varepsilon_t)e^{-\eta} = (\partial\zeta/\partial\varepsilon_t)_T$, and substituting the expression $\zeta = V_e/\Lambda^3$ we obtain

$$\frac{\dot{N}_t}{N} = \frac{\tilde{\xi}\dot{\varepsilon}_t}{\varepsilon_t}, \tag{112}$$

where the scaling parameter $\tilde{\xi}$ is defined by

$$\tilde{\xi} \equiv \left(\frac{\partial \ln V_e}{\partial \ln \varepsilon_t}\right)_T. \tag{113}$$

Note that $\tilde{\xi} \approx 0$ for deep traps. Equation (112) expresses the relation between the characteristic time for atom loss due to truncation (τ_t), defined by $-(\dot{N}_t/N)^{-1}$, and the ramp-down time (τ_{ramp}), defined by $-(\dot{\varepsilon}_t/\varepsilon_t)^{-1}$ and is easily rewritten for this purpose as $\tau_t^{-1} = \tilde{\xi}\tau_{\text{ramp}}^{-1}$. Hence, $\tilde{\xi}$ is a measure for the spilling of particles by the ramping procedure.

The overall decay time of the sample due to forced evaporation (τ) can be written as the sum of two terms,

$$\tau^{-1} = \tau_{\text{ev}}^{-1} + \tau_t^{-1}, \tag{114}$$

representing the rates of evaporation and truncation, respectively. In view of this result, we may distinguish two regimes:

- *Quasi-static ramping* ($\tau_t^{-1} \ll \tau_{\text{ev}}^{-1}$). In this regime the evaporation threshold ε_t is 'slowly' reduced and the atom loss is mainly due to evaporation. This is the regime of 'pure' forced evaporative cooling ($\dot{N} \approx \dot{N}_{\text{ev}}$) in which both N and T change. In the limit $\tau_t^{-1} \to 0$ we have plain evaporation.

- *Fast ramping (filtering)* ($\tau_t^{-1} \gg \tau_{\text{ev}}^{-1}$). In this regime essentially all particle loss is due to the truncation ($\dot{N} \approx \dot{N}_t$). The quasi-temperature remains *constant* but the internal energy, E, is reduced as the most energetic atoms are filtered from the sample (see next section).

Power-law traps.   Rewriting Equation (113) for power-law traps we obtain

$$\tilde{\xi} = (3/2 + \delta)[1 - R(3/2 + \delta, \eta)]. \tag{115}$$

It is noteworthy that, for power-law traps, the sum of the two partial derivatives of the reference volume V_e reduces to a particularly simple form,

$$\tilde{\gamma} + \tilde{\xi} = \delta. \tag{116}$$

For the special case of harmonic traps the condition for quasi-static ramping may be expressed by an inequality for the product of the trap frequency (ω) and the ramp-down time (τ_{ramp}) in terms of three characteristic ratios, the truncation parameter, the ratio of quasi-thermal wavelength to the s-wave scattering length and the ratio of the quasi-temperature and level splitting

$$\omega\tau_{\text{ramp}} \gg \frac{1}{N} \left(\frac{\Lambda}{a}\right)^2 \left(\frac{kT}{\hbar\omega}\right)^2 \frac{\eta^3}{[\eta - 4R(3,\eta)]}. \tag{117}$$

3.4.4   Loss of internal energy

The rate at which the internal energy of the trapped gas decreases during the evaporation process is composed of two contributions,

$$\dot{E} = \dot{E}_{\text{ev}} + \dot{E}_t. \tag{118}$$

The first contribution ($\dot{E}_{\text{ev}}$) is the evaporation term, resulting from the interatomic collisions and representing the energy removed with the evaporating atoms. The second term $(\dot{E}_t)$ gives the contribution due to a changing truncation energy, only present for $d\varepsilon_t/dt < 0$. In plain evaporation only the term $\dot{E}_{ev}$ contributes. If ε_t is varied the truncation losses are given the following expression

$$\dot{E}_t = \varepsilon_t \, \rho(\varepsilon_t) f(\varepsilon_t) \dot{\varepsilon}_t = \varepsilon_t \, \dot{N}_t \,. \tag{119}$$

The internal energy loss due to evaporation follows from evaluation of the following integral

$$\dot{E}_{ev} = - \int_{\varepsilon_t}^{\infty} d\varepsilon_4 \, \varepsilon_4 \, \rho(\varepsilon_4) \, \dot{f}(\varepsilon_4) \,. \tag{120}$$

With the same procedure as used in the previous section for calculating the evaporation rate $(\dot{N}_{ev})$, see Equation (105), we find

$$\dot{E}_{ev} = \dot{N}_{ev} \left[\varepsilon_t + (1 - X_{ev}/V_{ev}) \, kT \right], \tag{121}$$

where

$$X_{ev} = \frac{\Lambda^3}{kT} \int_0^{\varepsilon_t} d\varepsilon \, \rho(\varepsilon) \left[kT e^{-\varepsilon/kT} - (\varepsilon_t - \varepsilon + kT) e^{-\varepsilon_t/kT} \right]. \tag{122}$$

The volume ratio X_{ev}/V_{ev} can vary between zero and unity and is easily calculated numerically for arbitrary traps once the density of states is known. Like Equations (105) and (107) also Equation (121) is valid for given N, T and ε_t. Equation (121) expresses the physical picture that atoms escaping from the trap reduce the internal energy by an amount slightly larger than the energy of the truncation edge ε_t.

Power-law traps. For these traps (see Section 2.4.1) the ratio X_{ev}/V_{ev} can be expressed as

$$\frac{X_{ev}}{V_{ev}} = \frac{P(7/2 + \delta, \eta)}{P(3/2 + \delta, \eta)} \frac{V_e}{V_{ev}} \tag{123}$$

and the asymptotic behaviour is given by

$$\begin{aligned}
\frac{X_{ev}}{V_{ev}} &\simeq 1 - \frac{\eta}{9/2 + \delta} \quad \text{(for } \eta \to 0) \\
\frac{X_{ev}}{V_{ev}} &\simeq \frac{1}{\eta - 5/2 - \delta} \quad \text{(for } \eta \to \infty).
\end{aligned} \tag{124}$$

3.4.5 Relaxation heating and the thermalisation ratio

Just as evaporation leads to cooling because only the most energetic atoms escape, magnetic relaxation will lead to heating because in this process the low-energy atoms are preferentially removed. Because spin relaxation is a two-body process (see Section 1.3) the relaxation rate is proportional to the square of the gas density. Thus, the atoms are lost preferentially from the central region of the trap (region of highest density) where the atoms have lower-than-average energy. We assume here that *all* relaxation products leave the trapped gas: high-field-seeking atoms are ejected from the trap in accordance with their nature and the atoms in 'trapped' spin states because the recoil that accompanies (inelastic) relaxation events makes the atoms too energetic to be confined in the trap.

The rate of change of the number of trapped atoms due to spin relaxation can be written as

$$\dot{N}_{\rm rel} = -\int d^3{\rm r}\, G({\rm r})n^2({\rm r}) \tag{125}$$

where the rate constant G is assumed to be independent of temperature (low temperature limit). Neglecting a possible position dependence (*e.g.* field dependence) of the rate constant and defining a characteristic relaxation time $\tau_{\rm rel}$ by $\tau_{\rm rel}^{-1} = -(\dot{N}_{\rm rel}/N)$, Equation (125) can be reexpressed as

$$\tau_{\rm rel}^{-1} = n_0 G V_{2e}/V_e\,, \tag{126}$$

where the reference volume for binary collisions is given by

$$V_{2e} = \int d^3r \left[\frac{n({\rm r})}{n_0}\right]^2. \tag{127}$$

The associated rate of change of internal energy,

$$\dot{E}_{\rm rel} = \dot{N}_{\rm rel}(3/2 + \tilde{\gamma}_2)kT \tag{128}$$

where $\tilde{\gamma}_2 = [T/(2V_{2e})]\partial V_{2e}/\partial T$ is the scaling parameter for the collision reference volume. One may show that $\tilde{\gamma}_2 < \tilde{\gamma}$. Therefore, the relaxation effectively acts as a heating mechanism.

As both relaxation and evaporation are two-body processes their density dependence is identical. This allows us to combine Equations (121) and (128) into a single expression for the energy loss rate,

$$\dot{E}_{\rm ev} + \dot{E}_{\rm rel} = \dot{N}_{\rm ev}[\eta + (1 - X_{\rm ev}/V_{\rm ev}) + (3/2 + \tilde{\gamma}_2)/\mathcal{R}]kT\,, \tag{129}$$

where $\mathcal{R}$ is the thermalisation ratio (the 'ratio of good-to-bad collisions'), defined by the expression $\mathcal{R} = \dot{N}_{\rm ev}/\dot{N}_{\rm rel}$. Substituting Equations (107) and (126) the thermalisation ratio can be written as

$$\mathcal{R} = \frac{\bar{v}\sigma V_{\rm ev}}{G V_{2e}}e^{-\eta}, \tag{130}$$

notably independent of n_0. Similarly one can write one combined expression for the atom loss rate due to binary collisional processes

$$\dot{N}_{\rm ev} + \dot{N}_{\rm rel} = \dot{N}_{\rm ev}[1 + \mathcal{R}^{-1}]. \tag{131}$$

Since the elastic collision rate scales with $T^{1/2}$ whereas G is temperature independent (at low temperature) the thermalisation ratio decreases with decreasing temperature according to $\mathcal{R} \propto T^{1/2}$. Therefore, it is also useful to define a characteristic temperature T_* at which an atom has equal probability to experience an inelastic or an elastic collision (in a full thermal distribution):

$$kT_* = \frac{\pi m G^2}{16\sigma^2}. \tag{132}$$

For hydrogen the thermalisation ratio is anomalously small as a result of the small *s*-wave cross section. Substituting the values for H↑ in low magnetic fields one calculates $T_* \simeq 1.4\,{\rm nK}$.

In many practical cases the thermalisation ratio will be large, $\mathcal{R} \gg 3/2 + \tilde{\gamma}_2$, so that the relaxation corrections may be neglected in Equations (129) and (131). In view of this, the author has chosen not to contaminate the equations in the coming sections with these straightforward corrections, discussing the physics as if relaxation is absent.

3.5 Thermodynamics of evaporative cooling

With expressions for the particle loss and energy loss at our disposal, we can return to the thermodynamical aspects and calculate the change of density (n_0) and the rate of cooling $\dot{T}/T$ of the evaporating gas cloud. Two new scaling parameters will be introduced for this purpose. The efficiency parameter, $\tilde{\alpha}$, relates the quasi-temperature to the number of atoms remaining in the trap after a certain evaporation time, $T \propto N^{\tilde{\alpha}}$. The scaling parameter $\tilde{\beta}$ expresses how the density n_0 changes with quasi-temperature during cooling.

3.5.1 Plain evaporative cooling

We first consider the case of plain evaporative cooling. Is is seen from Equations (97) and (87) that, for a given trap, the internal energy only depends on three independent variables N, T and ε_t. Therefore, during any change of state in which these variables remain well defined, the internal energy change may be described by

$$dE = C_{\varepsilon_t}\, dT + \mu_{\varepsilon_t}\, dN + W_{\varepsilon_t}\, d\varepsilon_t \,, \tag{133}$$

where $C_{\varepsilon_t} = (\partial E/\partial T)_{N,\varepsilon_t}$ is the heat capacity at constant escape energy and for a fixed number of trapped atoms, $\mu_{\varepsilon_t} = (\partial E/\partial N)_{T,\varepsilon_t}$ is like a 'chemical potential' of evaporation at constant escape energy and $W_{\varepsilon_t} = (\partial E/\partial \varepsilon_t)_{N,T}$ is a heat of truncation at constant quasi-temperature and for a fixed number of atoms. With Equation (98) it follows immediately that

$$C_{\varepsilon_t} = (3/2 + \tilde{\gamma} + T(\partial \tilde{\gamma}/\partial T)_{\varepsilon_t})\, Nk \quad \text{and} \quad \mu_{\varepsilon_t} = (3/2 + \tilde{\gamma})\, kT \,. \tag{134}$$

This shows that μ_{ε_t} is nothing else than the average energy per remaining atom during evaporation, $\mu_{\varepsilon_t} \equiv E/N$.

For plain evaporative cooling, *i.e.* evaporative cooling at constant escape energy ε_t, the third term in Equation (133) does not contribute to the internal energy change, leaving only two independent thermodynamic variables, N and T.

Power-law traps. In this case the heat capacity at constant escape energy becomes

$$C_{\varepsilon_t} = C_\infty R(3/2 + \delta, \eta)\{(5/2 + \delta)R(5/2 + \delta, \eta) - (3/2 + \delta)R(3/2 + \delta, \eta)\} \,, \tag{135}$$

with $C_\infty = (3/2 + \delta)Nk$.

3.5.2 Forced evaporative cooling

In forced evaporative cooling the escape threshold (ε_t) is lowered with decreasing temperature, thus allowing the evaporation to proceed at a desired rate. An approximately constant evaporation rate is obtained by keeping the truncation parameter (η) constant. Therefore, rather than describing the evaporation with Equation (133), in the case of forced evaporative cooling it is more convenient to use η rather than ε_t as the third independent thermodynamic variable and to write changes of the internal energy by

$$dE = C_\eta\, dT + \mu_\eta\, dN + W_\eta\, d\eta \,. \tag{136}$$

In this equation $C_\eta = (\partial E/\partial T)_{N,\eta}$ is the heat capacity at constant truncation parameter, $\mu_\eta = (\partial E/\partial N)_{T,\eta}$ is like a 'chemical potential' of evaporation at constant truncation parameter, and $W_\eta = (\partial E/\partial \eta)_{N,T}$ is a heat of truncation at constant quasi-temperature and for a fixed number of atoms. With Equation (98) it follows immediately that

$$C_\eta = (3/2 + \tilde{\gamma} + T\,(\partial\tilde{\gamma}/\partial T)_\eta)\,Nk; \quad \text{and} \quad \mu_\eta = (3/2 + \tilde{\gamma})\,kT\,. \tag{137}$$

Note that $\mu_\eta = \mu_{\varepsilon_t} = E/N$ for the non-interacting Bose gas considered here.

For the special case of forced evaporative cooling at constant η the third term in Equation (136) does not contribute to the internal energy change, leaving only two independent thermodynamic variables, N and T. In this special case the expressions for forced evaporative cooling coincide with those of plain evaporative cooling provided C_{ε_t} is replaced by C_η.

Power-law traps. It follows with expression (100) for the $\tilde{\gamma}$ parameter that the chemical potential of evaporation and the heat capacity at constant truncation parameter are given by

$$C_\eta/Nk = (3/2 + \delta)R(3/2 + \delta, \eta)\,. \tag{138}$$

3.5.3 The scaling parameters $\tilde{\alpha}$ and $\tilde{\beta}$

From Section 3.4.4 we know the rate at which internal energy is removed from the sample by the escape of atoms. For quasi-static truncation of the distribution function the change of the internal energy may be expressed in terms of thermodynamic variables of choice, N, T and ε_t or N, T and η. Here two **special cases** will be discussed: (a) plain evaporative cooling and (b) forced evaporative cooling *at constant η*. In both cases the thermodynamic properties depend only on two independent variables, N and T, and the rate of change of internal energy can be expressed in the same form,

$$\dot{E} = C_i\,\dot{T} + E\dot{N}/N\,, \tag{139}$$

where the subscript of C_i is used to indicate plain or forced conditions, with $i \in (\varepsilon_t, \eta)$ in accordance with Equations (134) and (137), respectively. Equating Equation (139) with Equation (118) and using the property that for evaporation at constant η the quantity $\dot{\varepsilon}_t/\varepsilon_t$ may be replaced by $\dot{T}/T$ we find

$$\tilde{\alpha} = \frac{d\ln T}{d\ln N} = \frac{\eta + (1 - X_{\text{ev}}/V_{\text{ev}}) - (3/2 + \tilde{\gamma})}{(C_i/Nk) + (1 - X_{\text{ev}}/V_{\text{ev}})\tilde{\xi}_i}\,, \tag{140}$$

where the second terms in the denominator is only present in the case of forced cooling as is expressed by the Kronecker delta in the definition, $\tilde{\xi}_i \equiv \delta_{i,\eta}\tilde{\xi}$, of the parameter $\tilde{\xi}_i$. Here $\tilde{\xi}$ is the spilling parameter introduced in Section 3.4.3. The scaling parameter $\tilde{\alpha}$ is an important quantity, expressing the efficiency of the cooling process. If $\tilde{\alpha}$ is negative the evaporation gives rise to heating. For $\tilde{\alpha} > 0$ this scaling parameter tells us how the quasi-temperature drops with atom loss during the evaporation process.

A related quantity, the scaling parameter $\tilde{\beta}$, tells us how the quasi-density scales with the quasi-temperature

$$\tilde{\beta} = \frac{d\ln n_0}{d\ln T}\,. \tag{141}$$

Using the identity $\dot{n}_0/n_0 = \dot{N}/N - \dot{V}_e/V_e$ and the scaling behaviour of the reference volume V_e with quasi-temperature and escape energy, given by $\dot{V}_e/V_e = \tilde{\gamma}\dot{T}/T + \tilde{\xi}\dot{\varepsilon}_t/\varepsilon_t$, we find (noting that for forced evaporative cooling at constant η the quantity $\dot{\varepsilon}_t/\varepsilon_t$ may be replaced by $\dot{T}/T$) the following expression for $\tilde{\beta}$

$$\tilde{\beta} = \frac{1}{\tilde{\alpha}} - (\tilde{\gamma} + \tilde{\xi}_i). \tag{142}$$

To have cooling with increasing n_0 the scaling parameter $\tilde{\beta}$ should be negative, corresponding to the condition $\tilde{\alpha} > 1/(\tilde{\gamma} + \tilde{\xi}_i)$ or, equivalently, the (recursive) condition

$$\eta > (3/2 + \tilde{\gamma}) + \frac{(C_i/Nk) + \delta_{i,\eta}(1 - X_{ev}/V_{ev})\tilde{\xi}_i}{\tilde{\gamma} + \tilde{\xi}_i} - (1 - X_{ev}/V_{ev}). \tag{143}$$

The lowest η value at which this condition is satisfied will be called η_0.

3.5.4 Run-away evaporation

Even with the truncation parameter constant and equal to η_0 (*i.e.* for constant n_0) the forced evaporation rate at constant η will slow down with falling temperature due to the thermal speed dependence of the atomic collision frequency $n_0 \bar{v} \sigma$. In order to prevent this, the density n_0 should increase at least inversely proportional to $T^{1/2}$. This process in which the cooling rate speeds up is known as run-away evaporation. Using the identity $d(\ln n_0 T^{1/2})/dt = \dot{n}_0/n_0 + \dot{T}/(2T)$, the condition for run-away evaporation is seen to be given by the expression

$$\frac{d\ln n_0 T^{1/2}}{d\ln T} = \frac{1}{\tilde{\alpha}} - \left(\tilde{\gamma} + \tilde{\xi}\right) + \frac{1}{2} < 0, \tag{144}$$

i.e. $\tilde{\alpha} > 1/(\tilde{\gamma} + \tilde{\xi} - 1/2)$, so that for this case the recursive condition (143) should be replaced by the condition

$$\eta > (3/2 + \tilde{\gamma}) + \frac{C_\eta/Nk + (1 - X_{ev}/V_{ev})\tilde{\xi}}{\tilde{\gamma} + \tilde{\xi} - 1/2} - (1 - X_{ev}/V_{ev}). \tag{145}$$

The lowest η value at which this condition is satisfied will be called the critical truncation parameter η_c.

3.5.5 Examples

Plain evaporative cooling. To have cooling at initially constant n_0, the parameter $\tilde{\beta}$ for the scaling of n_0 with temperature should be equal to zero at the start of the cooling process, *i.e.* the truncation parameter η should be equal to η_0. Substituting the expressions (100), (135) and (121) for $\tilde{\gamma}$, C_{ε_t} and X_{ev}/V_{ev}, respectively, into Equation (143), and using Equation (108) for V_{ev}/V_e, we can solve for η_0.

We first discuss the case of a 3D harmonic trap ($\delta=3/2$). Here we calculate $\eta_0 = 2.9$. With this value for η_0 we can inspect values for the thermodynamic quantities: $E/NkT = 1.8$, $C_{\varepsilon_t}/Nk = 0.44$ and calculate the ratios $X_{ev}/V_{ev} = 0.6$ and $V_{ev}/V_e = 0.5$.

Then we can calculate the scaling parameters $\tilde{\gamma} = 0.29$ with Equation (100) and $\tilde{\alpha} = 3.4$ with Equation (140) to verify that indeed the scaling parameter $\tilde{\beta}$ is zero (*i.e.* $\tilde{\alpha} = 1/\tilde{\gamma}$). Then, the (initial) characteristic cooling time can be calculated with the following expression

$$-\frac{\dot{T}}{T} = \frac{V_{\mathrm{ev}}}{\tilde{\gamma} V_e} n_0 \bar{v} \, \sigma \, e^{-\eta_c} \approx 0.1 \, n_0 \bar{v} \, \sigma. \tag{146}$$

Initially n_0 remains constant as the quasi-temperature drops. However, since ε_t is constant this leads to an increase in truncation parameter and n_0 will start to increase while the cooling rate is cut-off exponentially.

For several power-law traps, characterised by their δ value, the results for plain evaporation (constant ε_t) are summarised below:

δ	3	2.5	2	1.5	1	0.5
η_0	2.25	2.39	2.59	2.91	3.50	5.05
$\tilde{\alpha}$	5.0	4.5	3.9	3.4	3.0	3.0

For plain evaporation the spherical quadrupole traps ($\delta = 3$) are clearly favourable. Even for a truncation parameter as low as 2.25 the density grows monotonously for these traps. Also the large efficiency parameter $\tilde{\alpha}$, indicating that the temperature falls with the fifth power of the number of atoms remaining in the trap, is spectacular. This means that the spherical quadrupole trap is best suited to load a trap to the onset of evaporative cooling.

Forced evaporative cooling at constant η. Turning to forced evaporative cooling we consider two special cases: (a) forced evaporative cooling at constant n_0 (*i.e.* $\eta = \eta_0$) and (b) forced evaporative cooling at the onset of run-away evaporation (*i.e.* $\eta = \eta_c$). In both cases we consider power-law traps, so we may use the simplifying relation $\tilde{\gamma} + \tilde{\xi} = \delta$ (see Section 3.4.3). The efficiency parameter ($\tilde{\alpha}$) may then be expressed as

$$\tilde{\alpha} = \frac{1}{\delta + \tilde{\beta}}, \tag{147}$$

where $\tilde{\beta}$ has the value 0, for forced evaporative cooling at constant n_0, or $-1/2$, at the onset of run-away evaporation.

(a) For forced evaporative cooling at constant n_0 the evaporation process starts, as in the case of plain evaporation, with $\eta = \eta_0$. Lowering ε_t proportional to T, thus keeping η constant, causes the density n_0 to remain constant as can be seen from Equation (143) where T does not appear explicitly. The difference from plain evaporative cooling shows up in the value for the efficiency parameter ($\tilde{\alpha}$) which is readily calculated with Equation (147) by setting $\tilde{\beta}$ equal to zero.

δ	3	2.5	2	1.5	1	0.5
η_0	2.25	2.39	2.59	2.91	3.50	5.05
$\tilde{\alpha}$	1/3	2/5	1/2	2/3	1	2

Not surprisingly, the particle loss due to the ramp-down procedure is most pronounced for traps with the largest density of states near the truncation edge. This shows up in the efficiency parameter, $\tilde{\alpha}$. Similar behaviour can be found for the spilling parameter, $\tilde{\xi}$.

(b) Choosing η constant and equal to η_c the cooling not only **starts** at constant $n_0\bar{v}$ but also *continues* at constant $n_0\bar{v}$. This is the onset condition for run-away evaporation. As in the previous example the expression (147) holds, but now, in accordance with Equation (144), we require the scaling parameter $\tilde{\beta}$ to be less than $-1/2$. For three dimensional harmonic traps it follows with Equation (145) that $\eta_c = 4.59$ and $E/NkT = C_\eta/Nk = 2.41$. The volume ratios are given by $X_{ev}/V_{ev} = 0.42$ and $V_{ev}/V_e = 1.37$. For the scaling parameters we calculate $\tilde{\gamma} = 0.912$, $\tilde{\xi} = 0.588$ and $\tilde{\alpha} = 1$. The cooling rate may be expressed as

$$-\frac{\dot{T}}{T} = \frac{V_{ev}}{V_e(\tilde{\gamma} - 1/2)}\, n_0 \sigma \bar{v}\, e^{-\eta_c} \approx 0.034\, n_0 \sigma \bar{v}, \tag{148}$$

which is a factor 3 slower than the initial cooling rate given in expression (146) but has the advantage that the exponential throttling with falling quasi-temperature has been eliminated. Note that $\tau_t^{-1} = 1.4\,\tau_{ev}^{-1}$. This means that we are at the edge of the quasi-static ramping regime and that more than half of the atoms are lost by spilling.

Comparing the various power-law traps for evaporation forced at run-away conditions we find the results compiled in the table below:

δ	3	2.5	2	1.5	1	0.5
η_c	3.17	3.43	3.84	4.59	6.59	∞
$\tilde{\alpha}$	2/5	1/2	2/3	1	2	-

Both examples of forced evaporative cooling show that high-δ traps require the smallest truncation parameter to move at the desired slope through the $n_0 - T$ phase diagram. This is a nice feature because it enables fast cooling. However, by comparing the $\tilde{\alpha}$ parameters we see that the low-δ traps are more efficient. This raises the question how the various traps compare (under forced evaporative cooling at constant η) for equal values of the truncation parameter. This comparison was made for $\eta = 4.59$, the critical truncation value for run-away evaporation in harmonic traps. The results are presented below:

δ	3	2.5	2	1.5	1	0.5
$\tilde{\alpha}$	0.54	0.64	0.79	1	1.31	1.8

Note that also at equal η the harmonic traps are more efficient than spherical quadrupole traps. However, since in this case the high-δ traps are operated well inside 'the run-away regime' ($\eta \approx 1.5\,\eta_c$) the evaporation speeds up (due to a strong increase in n_0 and related to the strong T dependence of the effective volume). This run-away may be an important advantage in cases where short cooling times are required.

Appendix: Incomplete gamma functions

The incomplete gamma function is defined by

$$P(a,\eta) \equiv \frac{1}{\Gamma(a)} \int_0^\eta dt\; t^{a-1} e^{-t}, \tag{149}$$

where $\Gamma(a)$ is the Euler gamma function. $P(a,\eta)$ grows monotonically from zero to one for increasing $\eta \geq 0$. $P(a,\eta)$ can be expressed as a series expansion

$$P(a,\eta) = e^{-\eta}\eta^a \sum_{m=0}^{\infty} \frac{\eta^m}{\Gamma(m+a+1)} \tag{150}$$

which reduces for integer $a = 1, 2, 3, \cdots$ to

$$P(a,\eta) = 1 - e^{-\eta} \sum_{m=0}^{a-1} \frac{\eta^m}{m!}. \tag{151}$$

Expansion in terms of $1/\eta$:

$$P(a,\eta) = 1 - e^{-\eta}\frac{\eta^{a-1}}{\Gamma(a)}\left[1 + \frac{(a-1)}{\eta} + \frac{(a-1)(a-2)}{\eta^2} + \cdots\right]. \tag{152}$$

The derivative of $P(a,\eta)$ with respect to η is given by

$$\frac{\partial P(a,\eta)}{\partial \eta} = \frac{a}{\eta}[P(a,\eta) - P(a+1,\eta)] = \frac{e^{-\eta}\eta^{a-1}}{\Gamma(a)} \tag{153}$$

and hence

$$R(a,\eta) \equiv \frac{P(a+1,\eta)}{P(a,\eta)} = 1 - e^{-\eta}\frac{\eta^a}{\Gamma(a+1)\,P(a,\eta)}. \tag{154}$$

A useful recursion relation for $R(a,\eta)$ functions is

$$\frac{1}{R(a,\eta)} = 1 + \frac{a+1}{\eta}[1 - R(a+1,\eta)]. \tag{155}$$

The derivative of $R(a,\eta)$ with respect to η is given by

$$\frac{\partial R(a,\eta)}{\partial \eta} = \frac{R(a,\eta)}{\eta}[1 + aR(a,\eta) - (a+1)R(a+1,\eta)]. \tag{156}$$

Acknowledgements

Many discussions with the members and guests of the atomic hydrogen group at the University of Amsterdam have been invaluable for preparing this paper. These are warmly acknowledged. In particular I would like to thank Meritt Reynolds for detailed comments and for double checking some numerical values presented in the evaporative cooling section of the paper. The author wishes to thank his colleagues at the ENS in Paris for their exceptional hospitality and stimulating discussions during the preparation of the manuscript. Further I would like to thank the organisers and editors of SUSSP44 for creating an interesting summer school and for patience with the author when the manuscript was delayed undesirably long.

The hydrogen research in Amsterdam is supported by the Stichting voor Fundamenteel Onderzoek der Materie (FOM), by the Nederlandse Organisatie voor Wetenschappelijk Onderzoek (NWO-PIONIER).

References

[1] *Fundamental Systems in Quantum Optics*, Les Houches Summer School LIII, 1990, edited by J Dalibard, J-M Raimond and J Zinn-Justin (Elsevier Science, Amsterdam 1992) course 9

[2] T J Greytak and D Kleppner, in *New Trends in Atomic Physics*, Vol II, edited by G Grynberg and R Stora (Elsevier Science, Amsterdam 1984) page 1125

[3] I F Silvera and J T M Walraven in *Prog Low Temp Phys*, Vol X, edited by D F Brewer (Elsevier Science, Amsterdam 1986) page 139

[4] *Bose-Einstein Condensation*, edited by A Griffin, D W Snoke and S Stringari (Cambridge University Press, Cambridge 1993)

[5] M H Anderson, J R Ensher, M R Matthews, C E Wieman, and E A Cornell, *Science*, **269**, 198 (1995)

[6] G Breit and I I Rabi, *Phys Rev*, **38**, 2082 (1931)

[7] I F Silvera and J T M Walraven, *Phys Rev Lett*, **44**, 164 (1980)

[8] W Kołos and L Wolniewicz, *J Chem Phys*, **43**, 2429 (1965); *Chem Phys Lett*, **24**, 457 (1974); *J Mol Spectr*, **54**, 303 (1975)

[9] W Kołos, K Szalewicz, and H J Monkhorst, *J Chem Phys*, **84**, 3278 (1986)

[10] I F Silvera, *Rev Mod Phys*, **52**, 393 (1980)

[11] H T C Stoof, J M V A Koelman, and B J Verhaar, *Phys Rev B*, **38**, 4688 (1988)

[12] Yu Kagan, I A Vartanyantz, and G V Shlyapnikov, *Sov Phys JETP*, **54**, 590 (1981)

[13] A Lagendijk, I F Silvera, and B J Verhaar, *Phys Rev B*, **33**, 626 (1986)

[14] I D Setija, H G C Werij, O J Luiten, M W Reynolds, T W Hijmans, and J T M Walraven, *Phys Rev Lett*, **70**, 2257 (1993)

[15] R Van Roijen, J J Berkhout, S Jaakkola, and J T M Walraven, *Phys Rev Lett*, **61**, 931 (1988)

[16] N Masuhara, J M Doyle, J C Sandberg, D Kleppner, and T J Greytak, *Phys Rev Lett*, **61**, 935 (1988)

[17] B J Verhaar, in *Atomic Physics 14*, edited by C E Wieman, D J Wineland and S J Smith (American Institute of Physics, New York 1995)

[18] J De Boer and R Bird, in *Molecular Theory of Gases and Liquids*, edited by J O Hirschfelder, R F Curtiss and R B Bird (Wiley, New York, 1954) page 424

[19] L H Nosanow, *J de Physique*, **41**, C7-1 (1980)

[20] W L McMillan, *Phys Rev*, **138A**, 442 (1965)

[21] M D Miller and L H Nosanow, *Phys Rev B*, **15**, 4376 (1977)

[22] R D Etters, D V Dugan, and R W Palmer, *J Chem Phys* , **62**, 313 (1975)

[23] L D Landau and E M Lifshitz, *Quantum Mechanics*, (third edition) (Pergamon, New York 1977)

[24] D G Friend and R D Etters, *J Low Temp Phys*, **72**, 1414 (1980)

[25] Y H Uang and W C Stwalley, *J de Physique*, **41**, C7-33 (1980)

[26] M Abramowitz and I A Stegun, *Handbook of Mathematical Functions*, (Dover, New York 1965)

[27] K Huang, *Statistical Mechanics* (Wiley, New York, 1963)

[28] J T M Walraven and I F Silvera, *Phys Rev Lett*, **44**, 168 (1980)

[29] V V Goldman, I F Silvera, and A J Leggett, *Phys Rev B*, **24**, 2870 (1982)

[30] D A Huse and E D Siggia, *J Low Temp Phys* , **60**, 137 (1982)

[31] H F Hess, *Phys Rev B*, **34**, 3476 (1986)

[32] R V E Lovelace, C Mehanian, T J Tommila, and D M Lee, *Nature*, **318**, 30 (1985)

[33] R V E Lovelace and T J Tommila, *Phys Rev A*, **35**, 3597 (1987)

[34] V Bagnato, D E Pritchard, and D Kleppner, *Phys Rev A*, **35,** 4354 (1987)

[35] T W Hijmans, Yu Kagan, G V Shlyapnikov, and J T M Walraven, *Phys Rev B*, **48**, 12886 (1993)

[36] P A Ruprecht, M J Holland, K Burnett, and M Edwards, *Phys Rev A*, **51**, 4704 (1995)

[37] Yu Kagan, G V Shlyapnikov, and J T M Walraven, to be published (1995)

[38] L D Landau and E M Lifshitz, *Statistical Physics*, (third edition, Part I by E M Lifshitz and L P Pitaevskii) (Pergamon, New York, 1980)

[39] V L Ginzburg and L P Pitaevskii, *Sov Phys JETP*, **7**, 858 (1963)

[40] E P Gross, *J Math Phys*, **4**, (1963)

[41] T Tommila, *Europhys Lett*, **2**, 789 (1986)

[42] E L Surkov, J T M Walraven, and G V Shlyapnikov, *Phys Rev A*, **49**, 4778 (1994); *ibidem*, accepted for publication (1995)

[43] J M Doyle, J C Sandberg, I A Yu, C L Cesar, D Kleppner, and T J Greytak, *Phys Rev Lett*, **67**, 603 (1991)

[44] J M Doyle, PhD thesis, Massachusetts Institute of Technology (1991), unpublished

[45] O J Luiten, H G C Werij, I D Setija, M W Reynolds, T W Hijmans, and J T M Walraven, *Phys Rev Lett*, **70**, 544 (1993)

[46] O J Luiten, PhD thesis, University of Amsterdam (1993), unpublished

[47] K B Davis, M-O Mewes, and W Ketterle, *Appl Phys B*, **60**, 155 (1995)

[48] K B Davis, M-O Mewes, M A Joffe, M R Andrews, and W Ketterle *Phys Rev Lett*, **74**, 5202 (1995)

[49] W Petrich, M H Anderson, J R Ensher, and E A Cornell, *Phys Rev Lett*, **74**, 3352 (1995)

[50] H F Hess, G P Kochanski, J M Doyle, N Masuhara, D Kleppner, and T J Greytak, *Phys Rev Lett*, **59**, 672 (1987)

[51] J M Doyle, J C Sandberg, N Masuhara, I A Yu, D Kleppner, and T J Greytak, *J Opt Soc Am B*, **6**, 2244 (1989)

[52] C C Bradley, C A Sackett, J J Tollett, and R G Hulet, *Phys Rev Lett*, **75**, 1687 (1995)

[53] J M Doyle, J C Sandberg, I A Yu, C L Cesar, D Kleppner, and T J Greytak, *Physica B*, **194-196**, 13 (1994)

[54] G P Kochanski, PhD thesis, Massachusetts Institute of Technology (1987), unpublished

[55] K Helmerson, A Martin, and D A Pritchard, *J Opt Soc Am B*, **9**, 1988 (1992)

[56] O J Luiten, M W Reynolds, and J T M Walraven, *Phys Rev A*, accepted for publication (1995)

[57] D W Snoke and J P Wolfe, *Phys Rev B*, **39**, 4030 (1989)

[58] B V Svistunov, *J Moscow Phys Soc*, **1**, 373 (1991)

[59] Yu Kagan, B V Svistunov, and G V Shlyapnikov, *Sov Phys JETP*, **75**, 387 (1992)

Poster Abstracts

- **Spontaneous emission in confined geometries**
  M A Rippin and P L Knight
  We describe the modification of electromagnetic field structures in microlaser geometries, paying particular attention to modified spontaneous emission in cylindrical cavities and DBR resonators.

- **Vacuum fluctuations of radiation in dispersive and lossy media**
  D G Welsch and T Gruner
  The correlation of the radiation-field ground-state fluctuations in a dispersive and lossy dielectric are studied in terms of the symmetrised autocorrelation function of the electric field strength. Effects of dispersion and absorption are discussed in comparison to the vacuum case with special emphasis on the optical frequency domain.

- **Experiments with spherical micro-cavities**
  V Sandoghdar, D S Weiss, V Lefévre, L Collot, J Hare, F Treussaut, J M Raimond and S Haroche
  Optical whispering gallery modes in silica spheres (diameter around 30–200μm) are studied and quality factors up to $2\text{x}10^9$ are measured at $\lambda = 780$nm. Due to a very low mode volume $\approx 10^3\lambda^3$, the dipole force of a single photon inside the cavity leads to a macroscopic angle of deflection. Experiments studying such a phenomenon and also construction of a low threshold micro-laser are discussed.

- **Semiclassical dynamics of a bound system in a high-frequency field**
  N Brenner and S Fishman
  The quantal behaviour of a particle in one-dimensional triangular potential well, driven by a monochromatic electric field, is studied semiclassically. The Floquet evolution operator is calculated in the unperturbed basis and the quasi-energy states are investigated. They are found to be asymptotically extended, with a finite region of exponential decay. The system studied here is suggested as a prototype model for a class of driven one-dimensional bound systems whose main characteristic is an increasing density of states as a function of energy.

- **Non-poissonian pumping of a micromaser using a pre-cavity**
  G Kastelewicz and H Paul
  Experimentally micromasers are pumped by a thermal atomic beams, which is described by a poissonian statistic. In order to get even more non-classical states of the field other pump statistics are considered theoretically. We investigate the probability of pumping a micromaser with atoms which have left another cavity.

- **State scarring by "ghosts" of periodic orbits**

  P Bellomo and T Uzer

  We uncover a novel structure in the eigenfunctions of quantum billiards, namely scarring by families of stable periodic orbits. We call such orbits 'ghosts' because they also scar states residing in a neighbouring billiard where there are no periodic orbits to support the state.

- **The cloud of virtual photons around a free electron**

  G M Salamone and G Compagno

  The spatial structure and the time evolution of the cloud of virtual photons surrounding a free electron are described by means of average values of correlation operators in non-relativistic QED. The connection between the electron's cloud and observable physical quantities is discussed.

- **High resolution spectroscopy of YbF**

  B E Sauer, Jun Wang and E A Hinds

  We are pursuing a program of high-resolution laser spectroscopy of YbF. We have measured the details of the spin-rotation and hyperfine interactions, *i.e.* the information that is needed for our planned experiment to measure the electric dipole moment of the electron.

- **Quantum gates**

  A F Barenco and A K Ekert

  Quantum gates will probably be the basic constituents of quantum computers. The controlled-NOT is the simplest non trivial gate in which one input selectively controls the NOT operation performed on the other input. We outline theoretical applications and present realistic physical models for the realisation of the quantum controlled-NOT operation.

- **Quantum statistics of three mode interaction**

  J Perina Jr

  Nonclassical behaviour of light produced in the optical parametric process is studied in the approximation of small fluctuations around a stationary point. The depletion of the pump mode as well as the influence of reservoir and internal losses are considered.

- **Quantum trajectories of atomic systems**

  W L Power

  Quantum state diffusion and Monte-Carlo wave-function simulations are demonstrated for two-level and three-level atomic systems. These are used to demonstrate quantum jumps and the Zeno effect. A simulation method based on Markov chains is demonstrated which is more computationally efficient than calculating individual trajectories.

- **Non-linear coupling of two laser beams using a trapped-atom medium**

  T Van der Veldt, J-F Roch, Ph Grelu, J-P Poizat and P Grangier

  The coupling of two laser beams in a crossed Kerr regime is able to lead to non-linear and quantum effects such as bistability, squeezing and quantum non demolition measurements. Trapped atoms make a promising medium to achieve the required non-linearities. We are currently investigating non-linear dispersion and absorption effects in a Rubidium magneto-optical trap.

- **Quantum mechanical diffusion of the polarisation of a laser**

  M A van Eijkelenborg, C A Schrama and J P Woerdman

  We have observed the influence of spontaneous emission on the polarisation of a miniature gas laser. The polarisation exhibits a diffusion-like motion with a rate related to that of Schawlow-Townes phase diffusion.

- **Schemes for atomic state teleportation**

  A S Parkins and J I Cirac

  We present two schemes employing cavity QED phenomena to realise the teleportation of quantum states following the principle outlined by Bennett et al. [Phys Rev Lett **70**, 1895, (1993)].

- **Models for cold collisions in laser fields**

  K-A Suominen, M J Holland, K Burnett and P S Julienne

  When laser cooled and trapped atoms collide they form quasimolecules, which can experience laser-induced excitation and subsequent spontaneous decay during the cold collision. Dynamical evolution of the quasimolecule between excitation and decay can make the process inelastic. We have prepared simple models which describe such situations.

- **Achromatic lenses for atoms based on light-induced dipole forces**

  M Drewsen, R J C Spreeuw and J Mlynek

  Light-induced dipole lenses for atoms that are achromatic to first order in the velocity spread is proposed. The lenses rely on the possibility of making the laser detuning and hence the light-atom interaction potential velocity dependent.

- **Semiclassical interpretation of tunnelling times**

  M Kira, I Tittonen and S Stenholm

  We use our earlier developed semiclassical description of tunnelling to obtain interpretations of the physics behind various suggested tunnelling time definitions. Investigations identify particle and wave properties of each tunnelling time. The identification explains tunnelling time behaviour at the low transmission opaque limit and the above barrier transmission limit.

- **Quantum theory of retarding plates**

  D Pegg

  The quantum theory of the action of a retarding plate on a propagating field is investigated in terms of the concept of a 'dressed' field. It is found that the action of the plate is mathematically equivalent to that of a phase-shift operator.

- **Periodic orbits and molecular photo dissociation**

  Oliver Zobay and G Alber

  A method for the semiclassical evaluation of molecular photoabsorption rates in terms of contributions of periodic orbits is developed and applied to a collinear model of CO_2. Effects originating from the finite extension of the Franck-Condon transition region in comparison to the Fresnel zone of the contributing periodic orbits are shown to be important.

- **The quantum dynamics of a simple nonholonomic system**

  A E Pratt and J H Hannay

  The problem of quantising nonholonomic systems is described. The meaning of the term nonholonomic is discussed. Difficulties inherent in obtaining a quantum mechanical wave equation with the correct classical limit are mentioned, and a model system aimed at solving the problem for a special case is presented.

- **Quantum theory of microlasers**

  B J Dalton, E S Guerra, B Garraway and P L Knight

  A simple model of a microlaser involving two level atoms in a Fabry-Perot cavity of dimension comparable to the optical wavelength is treated. Threshold conditions, laser statistics and special width are examined.

- **Bright squeezed light by second-harmonic generation & parametric oscillation**
  S Schiller, S Kohler, R Paschotta and J Mlynek

  Continuous-wave bright squeezed light has been successfully generated by singly-resonant second-harmonic generation. We analyse the requirements for achieving stronger squeezing and discuss extensions of this scheme to three-wave mixers with injected sub-harmonic and harmonic waves, where the relative powers and phases are adjustable.

- **Pattern formation in the transverse section of lasers**
  G Huyet, M C Martinoni, S Rica and J R Tredicce

  We discuss the transverse dynamic of lasers with large Fresnel number using the Maxwell-Bloch equations. We show that the dynamic may be described by two fields; the former is 'turbulent' while the latter represents a periodic modulation in space. Analysis of the data recorded from a CO_2 laser shows that the signal measured in one point of the pattern is chaotic in time while the average mean structure is periodic. These experimental results seem to confirm our theoretical approach.

- **Non-classical photon statistics at multiport beam-splitter**
  K U Mattle, M Michler, H Weinfurter, A Zeilinger and M Zukowski

  Non-classical two particle statistics occurs when a pair of quantum particles (fermions, bosons) interfere at a beam splitter. The generalisation of the standard 4-port beam-splitter to 2N-multiports (N input and N output ports) now gives additional degrees of freedom like internal phase shifts and topological variability. The extension to a higher dimensional Hilbert-space results in new quantum mechanical two particle interference effects. Experimentally we investigated the quantum distributions for 6-ports and 8-ports.

- **An atom interferometer with phase gratings**
  M Oberthaler, G Gostner, E Rasel, J Schmiedmayer and A Zeilinger

  Phase gratings for atoms can be realised by standing light waves with three such gratings one can build a Mach-Zehnder type interferometer for atoms. We present our latest experimental results.

- **Improved trapping in a vapour cell magneto-optical trap using multiple laser frequencies**
  A G Sinclair, E Riis and M J Snadden

  A novel method has been demonstrated for increasing the number of ^{85}Rb atoms in a vapour cell MOT. A second set of trapping laser beams is overlapped with the original beams and detuned further from resonance. This increases the filling rate while the blocking the centres of the new beams together with a reduction in intensity of the original beams, reduces the loss rate. Overall, we can obtain a fivefold increase in the number of trapped atoms.

- **Temperature measurements of laser-cooled samples by coherent optical transients**
  A V Durrant, E Usadi, K E Hill and S A Hopkins

  A technique is proposed for measuring velocity distributions in laser-cooled samples by creating and probing a ground-state population grating using a sequence of short laser pulses.

- **Quantum theory of simple optical instruments**
  U Leonhardt

  The action of simple optical instruments is described in terms of simple geometrical transformations in phase space. The formalism is applied to measurements of phase-space distribution functions.

• Detachment energy of H⁻ in superintense laser fields

E van Duijn and H G Muller

We calculate the eigenvalues of the Hamiltonian describing a H⁻-ion in a linearly polarised laser field. High frequency Floquet theory is used. We observe crossings between different kinds of light induced excited states.

• Dressed states analysis of dispersion lineshapes in driven two-level systems

C Szymanowski, C H Keitel, B J Dalton and P L Knight

We investigate the possibility of tailoring lineshapes and in particular enhancing the dispersion while simultaneously minimising absorption in driven two-level systems. The different influences of coherences and populations of the corresponding dressed states govern the structure of Lorentzian and Rayleigh like curves.

• Resonant interaction of negative ions and strong laser fields

P Kristensen, H Stapelfeldt, P Balling, J D Voldstad, T Andersen, H K Haugen, L D Noordam and U Ljungblad

The negative rubidium ion was used in an experiment to prove that photoionisation of an autoionising state occurs when a strong probe field is used. In a preliminary experiment a light induced state close to the threshold of the negative selenium ion was examined. The coupling to the continuum, which determines the shape of the embedded structure, varies orders of magnitude as the position is changed.

• Observation of dynamical localisation in atomic momentum transfer: a new testing ground for quantum chaos

J C Robinson, C Bharucha, F L Moore and M G Raizen

We report a direct observation of dynamical localisation: a quantum suppression of diffusion that is classically chaotic. Our experiment measures the momentum transferred from a modulated standing-wave of a near-resonant laser to a sample of ultra-cold atoms.

• Polarisation correlation in unpolarised light

H Paul and J Wegmann

Even if a light beam is unpolarised, correlation of the two modes of orthogonal polarisation can be measured using a simple set-up. A classical and a quantum mechanical treatment is used to study the behaviour of different types of unpolarised light.

• Dynamics of a two-mode field coupled to a two-level atom by two-photon processes

A Napoli, G Benivegna and A Messina

We propose an effective Hamiltonian describing the coupling of a two- level atom to a two-mode electromagnetic field. Studying the field population evolution, we show that the field dynamics manifests sensitivity to the initial field statistics.

• Optical Ramsey interferences in the presence of magnetic and electric fields

V Rieger, F E Dingler, K Sengstock, U Sterr and W Ertmer

Atom interferometry in the Ramsey-Borde set-up was performed. A magnetic field was used to create different atomic polarisations. The crossover resonance between different sublevels of the excited state eliminates the low-frequency recoil component. With an additional electric field the DC-Stark shift of the Mg intercombination transition was measured.

- **Phase operators on an infinite-dimensional Hilbert space spanned by non-negative photon number states**
  J A Vaccaro

  A simple formalism in introduced for describing the phase of a single mode field on a particular infinite-dimensional Hilbert space H_{sym} whose basis includes states of infinite energy. The Pegg-Barnett unitary phase operators are shown to converge strongly on H_{sym}. We compare H_{sym} with Hilbert space used by Newton.

- **Laser manipulation of atoms with finite closed families of states**
  P M Visser and G Nienhuis

  Atom-field configurations with a Hamiltonian having closed finite sub-spaces of states allow a fully quantum mechanical description of manipulation processes. Physical quantities as force and momentum fluctuations are completely determined by spontaneous decay.

- **Third-man quantum cryptography**
  B Erckert, H Weinfurter and A Zeilinger

  Secret information transfers have an unavoidable weakness: keeping the key used to encode crypto-texts secret. However, a number of systems have been designed using fundamental laws of quantum mechanics to find out whether the key generation process was eavesdropped or not. Here an extended system consisting of three instead of two communication partners and its practical realisation is proposed.

- **Two-photon spectroscopy of the 6S1/2 $\rightarrow$ 6D5/2 transition in trapped atomic Cesium**
  N Ph Georgiades, E S Polzik and H J Kimble

  Two-photon spectroscopy of atomic Cesium confined and cooled in a magneto-optical trap is reported. Hyperfine-structure constants of the 6D5/2 state are determined with 1% accuracy. New capabilities for studying ac Stark shifts and kinetic transport for cold atoms are suggested.

- **Time evolution of multi-component superposition states in quantum amplifiers**
  K S Lee, M S Kim and V Buzek

  We study the time evolutions of multicomponent superposition states in a sub-poissonian amplifier and a phase-sensitive amplifier. Some non-classical properties can appear or survive for a longer period as we choose an appropriate amplifier. The non-classical properties of the field in quantum amplifier is compared with that in the classical amplifier.

- **Interferometer as an optical network**
  P Törmä and S Stenholm

  Recently the analysis of optical network (built for instance out of beam-splitters and phase shifters) attracted considerable interest. For passive networks an effective linear theory can be used. We analyse, using such a theory, interferometers and point out how it can be used to improve certain measurement schemes.

- **Operational entropies in phase space**
  V Buzek, C H Keitel and P L Knight

  We define entropies in phase space via a mapping of the measured state with the set of detector states. Particular emphasis is put on the Wehrl entropy as an appropriate measure of information in quantum optics which we relate to uncertainty in phase space rather than to purity.

- **Quantum motion of a trapped atom interacting with a quantised cavity mode**

V Buzek, G Drobny and M S Kim

We propose a simple model for a two-level atom confined in a trapping potential and interacting with a quantised light field mode. The vibrational motion of the atom within the trap is treated quantum mechanically. When the vibrational mode is prepared in a coherent state and the cavity mode is in a Fock state we recover results of Blockley et al. [Europhys Lett **17**, 509 (1992)]. If the cavity field is prepared in a coherent state new purely quantum-mechanical effects can be observed.

- **Classical scaling in H-atom ionisation and other models**

A Rabinovitch and M Zaslavsky

The validity of the classical scaling in a quantum-mechanical approach to the problems of H-atom ionisation, X9-potentials and Fermi accelerator model is checked. The result may be used to justify the semiclassical and adiabatic approximations in some interval of small frequencies.

- **Squeezing in the self-pulsing domain of lasers with a two-photon saturable absorber**

A Sanchez-Diaz and P Garcia-Fernandez

The best conditions for quantum noise reduction arise when an instability threshold is approached, beyond the phase-instability threshold, and the output intensity breaks into spontaneous oscillations. Squeezing has been defined in the presence of a limit cycle. We will consider a laser with a two-photon saturable absorber when the laser medium and the absorber are placed in the same cavity.

- **Wave-packet dynamics in molecules**

B M Garraway, W K Lai, S Stenholm and K A Suominen

We show how simple models of molecules subjected to laser pulses lead to the generation and diversion of wave-packets. Dynamical effects of spontaneous emission are also presented.

- **Compensation of Berry's phase by another geometric phase**

S Klein and W Dultz

Berry's phase is defined as a phase proportional to an area enclosed by a curve on which a state is transported around. An easy realisation of such a cycle for the state of polarisation of light is given by a $\lambda/2$ plate inserted in a Michelson interferometer. We show theoretically and experimentally that if the closed curve on the Poincarè sphere, describing the states of polarisation, consists only of non-great circles, Berry's phase is compensated by a second geometric phase.

- **Non-dissipative optical lattices**

B P Anderson, T L Gustavson and M A Kasevich

Lithium atoms have been spatially confined and adiabatically cooled in a far-detuned 3-D optical dipole-force lattice, formed by the intersection of four laser beams with either red or blue detunings.

- **Squeezed bath and macroscopic quantum superpositions**

G M D'Ariano, M Fortunato and T Tombesi

We show that a squeezed bath is more efficient than a thermal one in preserving the generation of 'quasi-superposition' states. Furthermore, we suggest a new kind of isotropic squeezing, much more efficient in preserving quantum coherence of a Schrödinger-cat input state than a thermal bath and even than an ordinary squeezed one. .

- **Doubly excited $^1S^e$ states of the of the Helium atom**

  A Bürgers and D Wintgen

  We calculate highly precise ab initio parameters of both energies and widths of doubly excited resonances in the nonrelativistic helium atom using the complex scaling technique. A quantum defect analysis shows a great number of overlapping and interacting Rydberg series.

- **Molecular excitation by chirped pulses**

  A Paloviita, K-A Suominen and S Stenholm

  We study numerically the excitation process of a molecular wave packet when the laser pulse frequency undergoes time-dependent change (CHIRP). We show that the wave dynamical effects during the excitation process can strongly affect the validity of the standard area theorem.

- **Distance between density operators: quantum optical applications**

  A Orlowski and L Knoell

  A measure for the distance between arbitrary density operators, based on the Hilbert-Schmidt norm, is introduced and applied to the Jaynes-Cummings model.

- **Atom interferometry using adiabatic transfer**

  B C Young, M Weitz and S Chu

  We present results of adiabatic population transfer between the ground-state hyperfine levels of laser cooled cesium atoms. We have observed transfer efficiencies of more than 95% for Doppler-sensitive transitions, and more than 98.6% for Doppler-free transitions. We also report a working atom interferometer using adiabatic beam splitters.

- **Non-causal effects in quantum optics**

  O Steuernagel and H Paul

  Starting from the Weisskopf-Wigner treatment of a spontaneously radiating atom the integration over the proper physical field modes, leading to a partially non-causal signal, and the standard summation over negative frequencies too are compared.

- **Optical fibre sensing using the maximum entropy method**

  L Stergioulas, A Vourdas and G R Jones

  The spatial distribution of the intensity of light at the exit of an optical fibre can be useful in optical fibre sensing applications. In this work, we use the maximum entropy method to obtain the most likely spatial distribution consistent with the noisy data (speckle pattern). The preservation of the first N moments of the distribution provides the constraints for the application.

- **Wave-packet-dynamics in the Coulomb potential**

  R Dehnen and V Engel

  We study wave-packet-dynamics in the singular Coulomb potential solving the time-dependent Schrödinger equation with fast-Fourier techniques. We concentrate on the simple radial motion of hydrogen in a strong laser pulse and show how above threshold ionisation can be interpreted in terms of momentum space wave functions.

- **Degree of nonclassical behaviour**

  N Lütkenhaus and S M Barnett

  Negative values of the phase-space quasiprobability distributions are evidence of nonclassical behaviour. All pure states except the squeezed and coherent states exhibit this feature for distributions narrower than the Q-function.

- **Bell's inequality and quantum entanglement**

  A Chefles, M Brownlie and S M Barnett

  We investigate the link between the violation of Bell's inequality and quantum entanglement, and show that the well-known absolute bound on this violation follows from the Heisenberg Uncertainty Principle. We also show that entangled states can be produced without direct interaction, and present some simple models which illustrate this.

- **Emission and absorption spectra of trapped two-and three-level atoms**

  P E MacKay and G-L Oppo

  The resonance fluorescence emission and absorption spectra of harmonically trapped two-and three-level atoms are investigated. Additional sharp features appear at integral steps of the trap frequency from line centre because of the low confinement.

- **Quantised transport in chaotic billiards**

  P N Walker

  We consider a two-dimensional electron gas pierced by two flux tubes. Varying one flux induces an emf and hence electrons are transported. It is known that if one flux varies slowly over one flux quantum, the electrons are transported adiabatically around the second flux such that after averaging each electron completes an integer number of circuits N_n. For generic quantum billiards, the dynamics is chaotic and hence N_n is unpredictable, its statistics however, are universal.

- **Order, chaos and nuclear dynamics**

  J Blocki, C Jarzynski, J Skalski and W J Swaitecki

  The transition from ordered to chaotic nucleonic motions in a nucleus idealised as a gas of independent particles in a slowly deforming mean-field potential well, is paralleled by a transition from an elastic to a dissipative response to deformations.

Participants

•Oded Agam
Department of Physics
Technion Israel Institute of Technology
Haifa 32000
Israel

•Richard M Amos
Thin Films & Interfaces
Deptartment of Physics
University of Exeter
Stocker Road
Exeter EX4 4QL
England, UK

•Brian Anderson
Department of Physics
Stanford University
Stanford, CA 94305
USA

•Vitali Averbukh
Deptartment of Physics
Technion Israel Institute of Technology
Haifa 32000
Israel

•Adriano F Barenco
Claredon Laboratories
Oxford University
Oxford OX1 3PA
England, UK

•Paolo Bellomo
School of Physics
Georgia Institute of Technology
Atlanta Georgia 30332-0430
USA

•J Bestle
Abteilng Quantenphysik
University of Ulm
89069 Ulm
Germany

•Naama Brenner
Deptartment of Physics
Technion Israel Institute of Technology
Haifa 32000
Israel

•Andre Burgers
Fakultät für Physik
Universität Freiburg
Hermann Herder str. 3
79104 Freiburg
Germany

•Lowell Carson
4001 Edgely Drive
Philadelphia PA 19131
USA

•Anthony Chefles
Department of Physics & App Physics
University of Strathclyde
107 Rottenrow
Glasgow, G4 0NG
Scotland, UK

•Jonathan Crombie
Department of Physics & Astronomy
University of Wales
College of Cardiff
P.O. Box 913
Cardiff CF2 3QD
Wales, UK

•Bryan Dalton
Optics Section, Blackett Laboratories
Imperial College
London SW7 2BZ
England, UK

•Iain Dawson
Physics Department
University of Sheffield
Hicks Building, Hounsfield Road
Sheffield, S3 7RU
England, UK

•Richard Dehnan
Fakultät für Physik
Universität Freiburg
Hermann Herder str. 3
79104 Freiburg
Germany

•Ralf Deutschmann
Institut für Angewandte Physik
Universität Bonn
Wegeler str. 8
D-53115 Bonn
Germany

•Michael Drewsen
Faculty of Physics
University of Konstanz
Postfach 5560, M696
D-78434 Konstanz
Germany

•Gabriel Drobny
Institute of Physics SAS
Dubravska cesta 9
842 28 Bratislava
Slovak Republic

•Beatrix Erckert
University of Innsbruck
Technikerstrasse 25
A-6020 Innsbruck
Austria

•Mauro Fortunato
Scuola del Dottorato di Ricerca
Dipartimento di Fisica
Universitá di Roma 'La Sapienza'
Piazzale A Moro
00185 Roma
Italy

•Barry Garraway
Optics Section, Blackett Laboratories
Imperial College
London SW7 2BZ
England, UK

•Alessandra Gatti
Dipartimento di Fisica
Universitá degli studi di Milano
Via Celoria 16
20133 Milano
Italy

•Nikos Georgiades
California Institute for Technolgy
Department of Physics 103-33
Pasadena CA 91125
USA

•Philippe Grelu
Institut d' Optique
Université Paris Sud
Bât. 503 BP 147
91403 Orsay Cedex
France

•Toralf Gruner
Theoretical Physics Institute
'Friedrich Schiller'
University of Jena
Max Wien Platz 1
D-07743 Jena
Germany

•Troy Hammond
Massachusetts Institute of Technology
77 Massachusetts Avenue
Cambridge MA 02139
USA

•Stephen A Hopkins
Physics Department
Open University, Walton Hall
Milton Keynes, MK7 6AA
England, UK

•Matthias Hug
Abteilng Quantenphysik
University of Ulm
89069 Ulm
Germany

•Guillaume Huyet
Institut Non Lineaire de Nice
1361 Route des Lucioles
06560 Valbonne
France

•Alexander Iomin
Deptartment of Physics
Technion Israel Institute of Technology
32000 Haifa
Israel

•John Jeffers
Department of Physics & App. Physics
University of Strathclyde
107 Rottenrow
Glasgow, G4 0NG
Scotland, UK

•Thomas G Jorgenson
Chemical Physics DTH-301
Chemistry Department B
Technical University of Denmark
DK-2800 Lyngby
Denmark

•Georg Kastelewicz
AG 'Nichtklassische Strahlung'
Max-Planck-Gesellschaft
Universität zu Berlin
Rudower Chaussee 5
12484 Berlin
Germany

•Ritchie Kay
Department of Physics & App. Physics
University of Strathclyde
107 Rottenrow
Glasgow, G4 0NG
Scotland, UK

•Christoph H Keitel
LASP, Physics Deptartment
Imperial College
London SW7 2BZ
England, UK

•Mackillo Kira
Institute for Theoretical Physics
University of Helsinki
PO Box 9
00014 Helsinki
Finland

•Susanne Klein
Physikalishes Institut
Johann Wolfgang Göthe Univeristät
Robert Mayer str. 2-4
60054 Frankfurt am Main
Germany

•Poul Kristensen
Institute of Physics & Astronomy
Aarhus University
NY Munkegade
DK-8000 Aarhus C
Denmark

•Kang-soo Lee
Physics Deptartment
Sogang University
CPO Box 1142
Seoul 100-611
South Korea

•Vassilis Lembessis
Physics Department
University of Essex
Wivenhoe Park
Colchester, CO4 3SQ
England, UK

•Ulf Leonhardt
AG 'Nichtklassische Strahlung'
Max-Planck-Gesellschaft
Universität zu Berlin
Rudower Chaussee 5
12484 Berlin
Germany

•Norbert Lütkenhaus
Department of Physics & App. Physics
University of Strathclyde
107 Rottenrow
Glasgow, G4 0NG
Scotland, UK

•Graham M MacFarlane
Department of Physics & App. Physics
University of Strathclyde
107 Rottenrow
Glasgow, G4 0NG
Scotland, UK

•Peter E MacKay
Department of Physics & App. Physics
University of Strathclyde
107 Rottenrow
Glasgow, G4 0NG
Scotland, UK

•Alois Mair
Institute for Experimental Physics
University of Innsbruk
6020 Innsbruck
Austria

•Ullrich Martini
Physics Section
University of Munich
Teresienstr. 37
80333 Munchen
Germany

•Klaus U Mattle
University of Innsbruck
Technikerstrasse 25
A-6020 Innsbruck
Austria

•Alain Michaud
College de France Chaire de Physique
24 rue Lhomond
75231 Paris Cedex 05
France

•Klaus B Moller
Chemical Physics DTH-301
Chemistry Department B
Technical University of Denmark
DK-2800 Lyngby
Denmark

•Allard P Mosk
Van der Waals-Zeeman Laboratories
Valckenierstrat 65-67
1018 Xe Amsterdam
The Netherlands

•Anna Napoli
Istituto di Fisica
Universitá di Palermo
Via Archirafi 36
90123 Palermo
Italy

•Markus K Oberthaler
University of Innsbruck
Technikerstrasse 25
A-6020 Innsbruck
Austria

•Arkadiusz J Orlowski
AG 'Nichtklassische Strahlung'
Max-Planck-Gesellschaft
Universität zu Berlin
Rudower Chaussee 5
12484 Berlin
Germany

•Asta M Paloviita
Institute for Theoretical Physics
University of Helsinki
PO Box 9
00014 Helsinki
Finland

•Andrew S Parkins
Fakultät für Physik
Universität Konstanz
D-78434 Konstanz
Germany

•David Pegg
Faculty of Science & Technolgy
Griffith University
Nathan Brisbane 4111
Australia

•Jan Perina
Kmochov 3
779 00 Olomouc
Czech Republic

•Pepyn Pinkse
Van der Waals-Zeeman Laboratories
Valckenierstrat 65-67
1018 Xe Amsterdam
The Netherlands

•William L Power
Optics Section, Blackett Laboratories
Imperial College
London SW7 2BZ
England, UK

•Alan E Pratt
H.H. Wills Physics Laboratory
University of Bristol
Royal Fort, Tyndall Avenue
Bristol BS8 1TL
England, UK

•Marcos Protopapas
Laser Optics Group
Blackett Laboratories
Imperial College
London SW7 2BZ
England, UK

•Volker Rieger
Institut für Angewandte Physik
Universität Bonn
Wegeler str. 8
D-53115 Bonn
Germany

•Michael A Rippen
LASP, Physics Deptartment
Imperial College
London SW7 2BZ
England, UK

•Roberts Matthew
National Phisics Laboratories
Bld. 3 Queens Road
Teddington, Middlesex TW11 0LW
England, UK

•John C Robinson
Department of Physics
University of Texas at Austin
Austin Texas 78712-0883
USA

•Frank Ruschewitz
Institut für Angewandte Physik
Universität Bonn
Wegeler str. 8
D-53115 Bonn
Germany

•Giuseppa M. Salamone
Mathematical & Physical Sciences
University of Sussex
Falmer, Brighton BN1 QH
England, UK

•Angel Sanchez Diaz
Instituto de Estructra de la Materia
CSIC Nuclear Fisica & Estadistica
C / Serrano 123
28006 Madrid
Spain

•Vahis Sandoghdar
Ecole Normale Supèrieure
Laboratoire Kastle - Brossel
24 rue Lhomond
75231 Paris Cedex 05
France

•Benjamin Sauer
Physics Department
Yale University
PO Box 208120
New Haven CT 06520-8120
USA

•Stephen Schiller
Fakultät für Physik
Universität Konstanz
D-78434 Konstanz
Germany

•David Shelton
Department of Theoretical Physics
University of Oxford
1 Keble Road
Oxford OX1 3NP
England, UK

•Alastair G. Sinclair
Department of Physics & App. Physics
University of Strathclyde
107 Rottenrow
Glasgow, G4 0NG
Scotland, UK

•Euan C Sinclair
H.H. Wills Physics Laboratories
University of Bristol
Royal Fort, Tyndall Avenue
Bristol BS8 1TL
England, UK

•Michael J Snadden
Department of Physics & App. Physics
University of Strathclyde
107 Rottenrow
Glasgow, G4 0NG
Scotland, UK

•Ole Steuernagel
AG 'Nichtklassische Strahlung'
Max-Planck-Gesellschaft
Universität zu Berlin
Rudower Scheissee 5
12484 Berlin
Germany

•Lampros Stergioulas
Deptartment of Electrical
Engineering & Electronics
University of Liverpool
Liverpool L69 3BX
England, UK

•Kalle-Antti Suominen
Department of Theoretical Physics
University of Oxford
1 Keble Road
Oxford OX1 3NP
England, UK

•Wladslaw Swiatecki
Lawerence Berkeley Laboratories
Berkeley CA
USA

•Carten Szymanowski
Optics Section, Blackett Laboratories
Imperial College
London SW7 2BZ
England, UK

•Päivi E Törmä
Institute for Theoretical Physics
University of Helsinki
PO Box 9
0014 Helsinki
Finland

•John Vaccaro
AG 'Nichtklassische Strahlung'
Max-Planck-Gesellschaft
Universität zu Berlin
Rudower Chaussee 5
12484 Berlin
Germany

•Ernst van Duijn
FOM-Institute AMOLF
Kruislaan 407
Amsterdam
The Netherlands

•M van Eijkelenborg
Huygens Laboratories
University of Leiden
Afdeling MFQO
2333 CA Leiden
The Netherlands

•Harald van Kampen
Huygens Laboratories
University of Leiden
Postbus 9504
2300 RA Leiden
The Netherlands

•L Vejby-Christensen
Danish Institute of Physics &
Astronomy
Aarhus University
NY Munkegade
DK-8000 Aarhus C
Denmark

•Tonny van de Veldt
Institut d' Optique
Université Paris Sud
Bât. 503 BP 147
91403 Orsay Cedex
France

•Paul Visser
Huygens Laboratories
University of Leiden
Postbus 9504
2300 RA Leiden
The Netherlands

•Paul Walker
Department of Physics & App. Physics
University of Strathclyde
107 Rottenrow
Glasgow, G4 0NG
Scotland, UK

•Jana Wegmann
AG 'Nichtklassische Strahlung'
Max-Planck-Gesellschaft
Universität zu Berlin
Rudower Chaussee 5
12484 Berlin
Germany

•Martin Wilkens
Fakultät für Physik
Universität Konstanz
D-78434 Konstanz
Germany

●Gordon Yeoman
Department of Physics & App. Physics
University of Strathclyde
107 Rottenrow
Glasgow, G4 0NG
Scotland, UK

●Brenton Young
PO Box 9254
Stanford CA 94309-9254
USA

●Michael Zaslavsky
Department of Physics
Ben Gurion University of the Negev
PO Box 653
Beer Sheva, 84105
Israel

●Oliver Zobay
Fakultät für Physik
Universität Freiburg
Hermann Herder str. 3
79104 Freiburg
Germany

Index